*Probability and Mathematical Statistics (continued)*

ROHATGI · An Introduction to Probability Theory and Mathematical Statistics
SCHEFFE · The Analysis of Variance
SEBER · Linear Regression Analysis
WILKS · Mathematical Statistics
ZACKS · The Theory of Statistical Inference

*Applied Probability and Statistics*

BAILEY · The Elements of Stochastic Processes with Applications to the Natural Sciences
BAILEY · Mathematics, Statistics, and Systems for Health
BARNETT and LEWIS · Outliers in Statistical Data
BARTHOLOMEW and FORBES · Statistical Techniques for Manpower Planning
BARTHOLOMEW · Stochastic Models for Social Processes, *Second Edition*
BECK and ARNOLD · Parameter Estimation in Engineering and Science
BENNETT and FRANKLIN · Statistical Analysis in Chemistry and the Chemical Industry
BHAT · Elements of Applied Stochastic Processes
BLOOMFIELD · Fourier Analysis of Time Series: An Introduction
BOX · R. A. Fisher, The Life of a Scientist
BOX and DRAPER · Evolutionary Operation: A Statistical Method for Process Improvement
BOX, HUNTER, and HUNTER · Statistics for Experimenters: An Introduction to Design, Data Analysis, and Model Building
BROWN and HOLLANDER · Statistics: A Biomedical Introduction
BROWNLEE · Statistical Theory and Methodology in Science and Engineering, *Second Edition*
BURY · Statistical Models in Applied Science
CHAMBERS · Computational Methods for Data Analysis
CHATTERJEE and PRICE · Regression Analysis by Example
CHERNOFF and MOSES · Elementary Decision Theory
CHOW · Analysis and Control of Dynamic Economic Systems
CLELLAND, deCANI, BROWN, BURSK, and MURRAY · Basic Statistics with Business Applications, *Second Edition*
COCHRAN · Sampling Techniques, *Third Edition*
COCHRAN and COX · Experimental Designs, *Second Edition*
COX · Planning of Experiments
COX and MILLER · The Theory of Stochastic Processes, *Second Edition*
DANIEL · Biostatistics: A Foundation for Analysis in the Health Sciences, *Second Edition*
DANIEL · Application of Statistics to Industrial Experimentation
DANIEL and WOOD · Fitting Equations to Data
DAVID · Order Statistics
DEMING · Sample Design in Business Research
DODGE and ROMIG · Sampling Inspection Tables, *Second Edition*
DRAPER and SMITH · Applied Regression Analysis
DUNN · Basic Statistics: A Primer for the Biomedical Sciences, *Second Edition*

*continued on back*

# Statistical Techniques for Manpower Planning

# Statistical Techniques for Manpower Planning

DAVID J. BARTHOLOMEW

*London School of Economics and Political Science*

*and*

ANDREW F. FORBES

*Institute of Manpower Studies, University of Sussex*

JOHN WILEY & SONS

Chichester · New York · Brisbane · Toronto

***Library of Congress Cataloging in Publication Data*:**

Bartholomew, David J.
Statistical techniques for manpower planning.

(Wiley series in probability and mathematical statistics)
Bibliography: p. 273
Includes index.
1. Manpower planning—Statistical methods.
2. Manpower planning—Mathematical models.
I. Forbes, Andrew F., joint author. II. Title.
HF5549.5.M3B37 331.1′1′0182 78–8604
ISBN 0 471 9967 0 X

Printed in Great Britain by Page Bros (Norwich) Ltd,
Mile Cross Lane, Norwich, Norfolk

# Preface

Manpower planning in the narrow sense is concerned with matching the supply of people with the jobs available. It is a multi-disciplinary activity and in so far as it involves numbers and uncertainty there is a clear role for the statistician. This role was being increasingly recognized ten years ago when we were at the University of Kent developing statistical methodology for manpower planning. As a result we decided to offer a one-week course on Statistical Techniques of Manpower Planning at the University. This proved successful and has been repeated once or twice a year since—latterly at the London School of Economics under the auspices of the Institute of Manpower Studies. The object of the course, then as now, was to equip people with knowledge of the statistical techniques becoming available. Most people attending the courses were working in operational research, management services, and personnel management and they usually had or were about to acquire manpower planning responsibilities. Many came from industry and the public service in the United Kingdom, but some were from Canada, Germany and South America.

One of the difficulties we encountered was that the literature of the subject was widely scattered and often disguised in unfamiliar notation and concerned with applications far removed from manpower planning. From the beginning, therefore, lecture notes were provided to accompany the course and later a special edition of the *Statistician* (1971) appeared containing articles written by the course team. Over the years these notes have been revised and enlarged in the light of experience and advances in the subject. We have frequently been asked if copies could be purchased and this led us to think of publishing the notes in book form. The present book is based on the lecture notes but has been enlarged and somewhat re-organized in order to make it suitable for self-study.

Since the early days, the literature of the subject has grown rapidly and there are now several books devoted to manpower planning. However, although several of these deal with methodology, none provides a systematic account of the technical aspects such as we have aimed to provide here.

The book is primarily concerned with elementary methods which have proved their worth in practice. It is designed to provide enough information for those who wish to use the techniques and, for this reason, a good proportion of most chapters is taken up with examples, exercises and their solutions. Where convenient these are based on real data but as their purpose is pedagogical we have often constructed artificial (though realistic) data to make our point. Listings are provided of computer programs for evaluating some of the models. Such programs are intended to encourage the reader to undertake the exercises and generally to experiment with the models in order to master fully the techniques and their applications.

One of the challenges of applied work is that the practical situation rarely corresponds exactly with the methods of the text book. We make no pretence that this book contains techniques for all occasions. Most of the methods rest on assumptions which will not always be satisfied. Where appropriate we have emphasized the limitations of the techniques and indicated how they might be overcome. What we would claim is that the person who has mastered the methods presented here will, as a result, be better equipped to cope with the unexpected and non-standard situations of real life.

Emphasis throughout has been laid on simple robust methods with the use, for example, of graphical methods in place of more sophisticated analyses. There is no exaggerated concern with fully efficient estimators and such like. In our experience of applying the techniques it is rarely necessary to use highly sophisticated theory on rather crude data. It is much more important to have a reliable, simple, and easily understood methodology which does not lean too heavily on distributional assumptions which are more appropriate to controlled experimentation than observation of social phenomena.

As remarked above, we have tried to make the book suitable for self-study. We hope also that it will find a place as a text in universities, polytechnics, and management trianing centres where formal teaching on manpower planning is becoming increasingly common.

As an aid to the reader who wishes to become acquainted with the wider background or to take his studies further, we have provided a 'Complements' section to most chapters, and a comprehensive bibliography will be found at the end of the book.

Although we shall consider the application of the techniques solely in the field of manpower planning, many of them have uses in other fields. Indeed, most have been developed from methods used in such fields as demography, actuarial science, and life testing. The Markov-chain models which play a central role in this book have been widely applied in many parts of the social sciences including education planning, hospital planning, and the study of buying behaviour and occupational mobility.

In order to obtain full benefit from the book an elementary knowledge of mathematics and statistics is required. However, graduates in these subjects will find the theory easy and the interest for them will be in the applied aspects of the work. Anyone who has taken a first-year university course in statistics should be able to work through the book acquiring such further knowledge as is needed in the process. Those, including many managers, who do not have the basic technical training or need to master the techniques should nevertheless find the book useful in showing what kinds of problem are amenable to statistical analysis and what kinds of answer can be expected.

*London* D. J. BARTHOLOMEW
*Brighton, Sussex* A. F. FORBES

**Acknowledgements**

Our greatest debt is to the few hundred people who have attended our courses serving as unwitting guinea-pigs for this undertaking. Without their criticism and encouragement this book would have been the poorer.

Likewise our thanks are due to Malcomn Marshall and Pauline Sales, who were closely involved in the organization and conduct of the courses for several years. Clive Purkiss regularly contributed a session on demand forecasting and our present Chapter 8 owes a great deal to the foundation which he laid.

While we have been writing the book, numerous colleagues have read drafts of chapters and offered suggestions for improvement. Of these we would mention especially Gordon Keenay, Roger Morgan, and John Webb, but to all we would tender our warmest thanks.

Finally, the brunt of the secretarial work involved in producing the final version has been borne by Helen Chapman of the Institute of Manpower Studies and Susan Hayden of the London School of Economics. We owe them a particular debt which we gladly acknowledge.

# Notation

The familiar problem of devising a simple and consistent notation for use throughout a book is compounded in the present case by the diverse origins of much of the material. The conflicting desires to have a notation agreeing with current usage and one which is internally consistent cannot be reconciled and we have therefore had to compromise. For certain quantities which occur throughout the book the same symbols have been used. Where possible these are the initial letters of the terms which they represent. Thus $n$ is always a number of people, $p$ a probability, and $R$ a number of recruits. Subscripts and arguments are added to identify classes and times to which these quantities relate. Bold type is used for vectors (written as rows) and matrices. For quantities which occur only locally within the text, and where there is little risk of confusion, the same symbol may be used in different parts of the book with different meanings. This is especially true of the Complements sections where, in referring to the work of others, we have felt it best to follow their notation. Variables of integration and indices of summation are defined by their context. A particular difficulty arises from our use of computer printouts where we are restricted to upper case roman letters. This means that symbols appearing on the printouts cannot always match those in the text. Wherever possible we have used the same letters in the printouts even though the distinctions represented by different type faces and cases are lost.

Logarithms to the base e are denoted by 'ln'. Where 'log' is used with no base shown it means that the base is immaterial. The symbol ◀ is used to mark the end of examples. In the solutions to exercises which do not use the computer programs, we have often carried more decimal places than the data justify. This is to help readers who wish to follow through the steps in the calculation and is not intended as an example of good practice in presenting results in real life.

# Contents

CHAPTER 1

# Statistics and Manpower Planning

## 1.1 INTRODUCTION

Manpower planning is often defined as the attempt to match the supply of people with the jobs available for them. This problem may be posed at the national or regional level, in which case it is likely to be an aspect of planning undertaken by government. Equally, the problem arises in the management of individual firms or occupational groups. It is with the latter that we shall be primarily concerned in this book although some of the methodology is useful at both levels. Indeed, many other planning and forecasting problems outside the manpower field also have the same basic structure.

There are two features of most manpower planning problems which render them suitable for statistical treatment. The first is the concern with aggregates. Manpower planning, unlike individual career planning, is concerned with numbers, that is, with having the right numbers in the right places at the right time. The individual and aggregate aspects are, of course, intimately related and, in practice, cannot be separated but statistical methods are of most direct relevance for handling the aggregate side. It cannot be too strongly emphasized that our present concern with the statistical approach is in no sense a denial of the importance of other dimensions of human and organizational behaviour. Much harm has been done in the past by setting up a false antithesis between statistical manpower planning as a soulless 'numbers game' and the personnel manager's concern with individual welfare. Neither approach can succeed without the other and the statistical approach of the present book is expounded in full recognition of the fact that although it is necessary it can never be sufficient. The second feature of manpower planning which calls for statistical expertise is the fact of uncertainty. This arises both from the uncertainty inherent in the social and economic environment in which the firm operates and from the unpredictability of human behaviour. Any attempt to construct a theoretical base for manpower planning must therefore reckon with the element of uncertainty by introducing probability ideas.

In some respects the statistical aspects of manpower planning are no different from those of other fields of applied statistics. In so far as this is the case our concern will be to show how established statistical methods can be brought to bear in the present context. In other respects, however, there are special features arising in manpower work which have brought into being a variety of methods and models which are peculiar (or almost so) to manpower planning. The aim of this book is to give an account of the theory, as it now exists, in a form which makes it accessible to practical manpower planners.

As already noted, the emphasis throughout will be on manpower planning in the firm, where the word 'firm' is being used in a broad sense to cover any system under a common management or working with a common purpose. Thus we use it to include large multinational companies, the civil service, or parts thereof, a single factory, or a specialist profession like medicine or university teaching. Because of our concern with aggregate properties, the methodology is mainly applicable to fairly large organizations. How large depends on the amount of subdivision required for planning purposes and the degree of precision necessary. To give a rough guide, an organization numbered in tens is likely to be too small whereas one running to hundreds, or better still, thousands offers much more scope.

It is useful at the outset to distinguish four purposes which statistical methods serve in manpower planning, as follows.

### Description

The first step in any investigation will usually be to describe the system in numerical terms and to summarize the results in an easily understood manner. Statisticians are sometimes tempted to neglect this aspect in their desire to pass on to the apparently more challenging intricacies of modelling. However, experience shows that a great deal can be learnt from a careful examination of the system as it is. For example, an inspection of the current age distribution will often draw attention to incipient problems like promotion bottlenecks.

### Forecasting

This is often seen as the main activity of the statistician in planning though its purpose is often misunderstood. Forecasts should never be interpreted as what *will* happen but as what *would* happen if the assumed trends continue. They therefore provide a guide to the management action required to achieve a desired objective. The means by which such objectives might be achieved belongs to the province of:

### Control

Some parts of a manpower system are subject to control by management action; for example, the numbers and points of entry of recruits. The object of a theory of control is to devise strategies for ensuring that change takes place in the desired direction.

**Design**

When a new organization is being established or an old one re-organized it may be possible, within limits, to design its structure and mode of operation. A theory of manpower systems provides a tool for evaluating different policies and choosing between them.

## 1.2 CONCEPTS, TERMINOLOGY, AND NOTATION

The central idea which underlies all our analyses is that of regarding an organization as a dynamic system of *stocks* and *flows*. At any time the members of a system can be classified into groups on the basis of whatever attributes are relevant for the exercise in hand. The numbers in such categories will be called the *stocks* at that time. If we denote the number of categories by $k$, the stock in category $i$ at time $T$ will be written $n_i(T)$ and the set of stocks as a row vector thus

$$\mathbf{n}(T) = (n_1(T), n_2(T), \ldots, n_k(T))$$

The stock vector provides a 'snapshot' of the system but tells us nothing directly about change over time. The stock picture must therefore be supplemented by the numbers that have moved between categories over an interval of time. We consider an interval of unit length from $T - 1$ to $T$ and denote the number of individuals moving between categories $i$ and $j$ in this period by $n_{ij}(T - 1)$. It is important to observe that flows relate to an interval and not a point as is the case with stocks. This interval may be at choice but is more often determined by the manpower accounting practice of the organization. A more explicit notation would be $n_{ij}(T - 1, T)$ showing the end-points of the interval but we have adopted the simpler form in the interests of brevity. With $k$ categories there are $k(k - 1)$ possible flows but it is useful to adopt the convention that those who stay in the same category are also counted with the flows. The flows can then be set out in a square matrix as follows:

$$\begin{bmatrix} n_{11}(T-1) & n_{12}(T-1) \ldots n_{1k}(T-1) \\ n_{21}(T-1) & n_{22}(T-1) \ldots n_{2k}(T-1) \\ \cdot & \cdot \qquad\quad \cdot \\ \cdot & \cdot \qquad\quad \cdot \\ \cdot & \cdot \qquad\quad \cdot \\ n_{k1}(T-1) & n_{k2}(T-1) \ldots n_{kk}(T-1) \end{bmatrix} = \mathbf{N}(T-1), \text{ say}$$

In practice many of the cells will have zero entries because, for example, it is obviously impossible to move from one age group to a lower one. The flows listed in the matrix above are internal to the system and will be described as *transfers* or, where applicable, as promotions or demotions. In addition there is a two-way flow between the system and the outside world. The wastage flow from category $i$ in the interval $(T - 1, T)$ will be denoted by $n_{i,k+1}(T - 1)$

and the vector of wastage flows by $\mathbf{n}_{k+1}(T-1)$. The recruitment flow into category $i$ in the same interval is $n_{0i}(T)$ with corresponding vector $\mathbf{n}_0(T)$. In the latter case we have departed from the convention of identifying the time interval by its starting-point and have used the end-point instead. This is because we shall wish to think of recruitment as taking place after wastage and transfer, and it is helpful to emphasize this ordering of events in the notation.

The stock and flow framework is extremely general and flexible since the categories can be formed in any way whatsoever to suit the purpose in hand.

Sometimes it is more informative to express the flows as proportions. To see how this might be done we first set out all the quantities defined above in a table of manpower accounts as in Table 1.1. It should be clear from the definition of the flows that the row and column totals give the stocks at the two end-points of the interval.

***Table 1.1 Manpower accounts for the interval*** $(T-1, T)$

| | $n_{01}(T)$ | $n_{02}(T)$ | $\ldots n_{0k}(T)$ | — | *Row totals* |
|---|---|---|---|---|---|
| | $n_{11}(T-1)$ | $n_{12}(T-1)$ | $\ldots n_{1k}(T-1)$ | $n_{1,k+1}(T-1)$ | $n_1(T-1)$ |
| | $n_{21}(T-1)$ | $n_{22}(T-1)$ | $\ldots n_{2k}(T-1)$ | $n_{2,k+1}(T-1)$ | $n_2(T-1)$ |
| | $\vdots$ | $\vdots$ | $\vdots$ | $\vdots$ | $\vdots$ |
| | $n_{k1}(T-1)$ | $n_{k2}(T-1)$ | $\ldots n_{kk}(T-1)$ | $n_{k,k+1}(T-1)$ | $n_k(T-1)$ |
| *Column totals* | $n_1(T)$ | $n_2(T)$ | $\ldots n_k(T)$ | — | — |

If now we express the table elements as proportions of their row total we obtain the *flow rates* out of each category. Thus, for example, $n_{i,k+1}(T-1)/n_i(T-1)$ is the wastage rate for category $i$. Such rates are a useful way of expressing the pattern of flows and of identifying problem areas. They will play a particularly important part in the class of Markov models introduced in Chapter 4.

It is also useful, though less common, to express the elements in the table as proportions of the column totals. This would be appropriate if we were interested in seeing the pattern of flows into a given category rather than the pattern of flows out.

The accounting relationships set out in Table 1.1 can also be expressed algebraically in a form which will be necessary for subsequent developments. Thus

$$n_j(T) = \sum_{i=1}^{k} n_{ij}(T-1) + n_{0j}(T) \qquad (j = 1, 2, \ldots, k) \tag{1.1a}$$

or

$$\mathbf{n}(T) = \mathbf{1N}(T-1) + \mathbf{n}_0(T) \tag{1.1b}$$

where $\mathbf{1} = (1, 1, \ldots, 1)$. Notice that the wastage flows do not appear in these equations because individuals who have left cannot contribute to future stocks.

*Example* 1.1 Stock and flow information can be set out in a network diagram as shown in Figure 1.1 for a simple three-grade hierarchy.

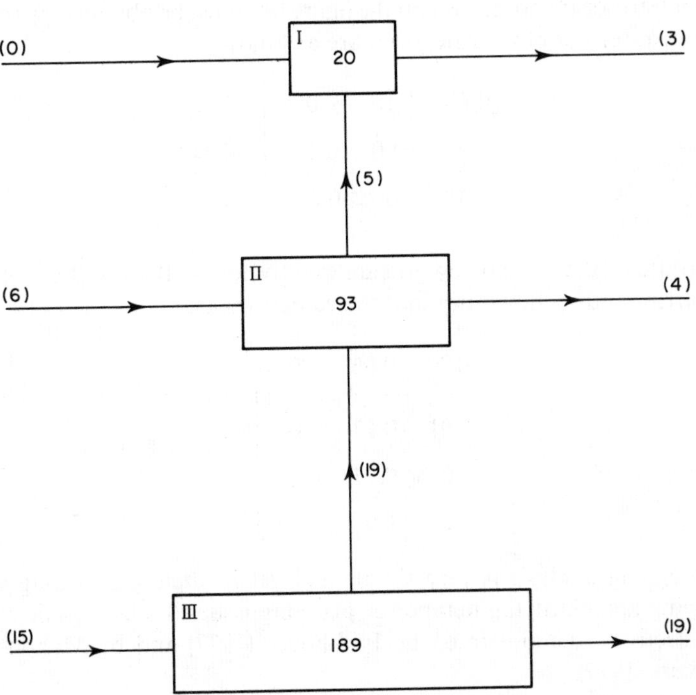


Figure 1.1 Stocks and flows for a three-grade hierarchy

The rectangles represent three grades and the numbers inside them are the stocks at the beginning of the interval. The arrows represent the paths along which flows take place, the numbers in brackets being the numbers involved. Table 1.2 shows this information is set out in a flow table in the format of Table 1.1.

***Table 1.2 Manpower accounts for the system shown in Figure 1.1***

| | 15 | 6 | 0 | *Wastage* | *Initial stocks* |
|---|---|---|---|---|---|
| | 151 | 19 | 0 | 19 | 189 |
| | 0 | 84 | 5 | 4 | 93 |
| | 0 | 0 | 17 | 3 | 20 |
| *Final stocks* | 166 | 109 | 22 | | |

The final stocks are not shown on the figure but must be obtained by addition. The flow rates out of the categories are as follows:

$$\begin{array}{ccc:c} 0.80 & 0.10 & 0 & 0.10 \\ 0 & 0.90 & 0.05 & 0.04 \\ 0 & 0 & 0.85 & 0.15 \end{array}$$

The failure of the second row to sum exactly to 1 is the result of rounding. The corresponding flow rates into the categories are:

$$\begin{array}{ccc:c} 0.09 & 0.06 & 0 & \\ \hdashline 0.91 & 0.17 & 0 & \\ 0 & 0.77 & 0.23 & \\ 0 & 0 & 0.77 & \end{array}$$

The foregoing analysis is very simple and yet it often goes a long way to identifying and clarifying manpower planning issues. Its use in this way has been effectively demonstrated by Parkhouse (1977) and by Mahoney and coworkers (1977).

In passing, we remark that the stock-and-flow representation of a manpower system serves to specify two basic requirements for a personnel record system. It must be capable of providing information, firstly, about stocks over a period of time and, secondly, about flows. It is in the second aspect that records are most often found to be deficient.

## 1.3 MODELS OF MANPOWER SYSTEMS

A manpower model is a mathematical description of how change takes place in the system. First of all this requires the specification of any constraints under which the system operates. Very often, for example, as with the renewal models discussed in Chapter 5, the grade sizes are fixed. This implies that the left-hand side of the basic accounting equation is fixed and hence that flows are constrained by the necessity of satisfying these equations. In more complicated cases the constraints may take the form of inequalities to be satisfied by the stocks and flows. Secondly, a model must specify the mechanism which generates flows. Some flows, such as promotion or demotion, are under the direct control of management and in such cases the assumptions to be made will be about the decisions which managers will make. If there are well-defined operating rules or a corporate plan which is to be implemented, this may be relatively straightforward. In other cases one may have to rely on managerial opinions and the evidence of past practice. Other flows, such as voluntary wastage, are not under direct management control and assumptions about their future levels is likely to be based on a blend of historical data and management judgement.

Assumptions about flows can be classified in various ways. Two such classifications which are central to the treatment given in this book are based on the following dichotomies.

### Stochastic/deterministic

If we were to assume that exactly 10 per cent of those in a particular category would leave in a given year then we would have made a deterministic assumption about that flow. Given the stock in question this means that there is no uncertainty about how many people will actually leave. If, on the other hand, we were to suppose that each individual in the category had a 10 per cent probability of leaving in the year then we would have made a stochastic assumption. If we were given the stock in this case we could not predict precisely how many would leave but only the probability distribution. In particular, if the individuals behave independently the actual number has a binomial distribution whose average is equal to 10 per cent of the stock. In that sense, therefore, we may say that the deterministic and stochastic assumptions are equivalent *in an average sense*. Voluntary wastage is almost always a stochastic flow since it results from a number of more or less independent individual decisions. Promotion may or may not be stochastic. If as a matter of policy exactly 5 per cent of a grade are promoted, that would require a deterministic assumption. If, on the other hand, the proportion varies from year to year it would be more realistic to treat it as stochastic. In practice we shall often deliberately suppress the distinction between stochastic and deterministic flows because our main interest will be in averages. For the sake of greater generality and realism we shall usually define our models in stochastic terms even though our analysis of them will be largely confined to the determination of expected values.

**Push/pull**

Flows can also be classified according to whether the impetus for a move lies at its starting-point or at its destination. Thus if an individual moves into a higher grade *because* it was necessary to fill a vacancy arising at that level we can think of the person as being pulled into the higher grade. If, on the other hand, the move to the higher grade is automatic as a result of acquiring a new qualification the move takes place because of an event occurring at the point of origin. Such flows are called *push* flows.

The distinction between push and pull flows is often not as clear-cut as this account might suggest. There can, for example, be both a push and a pull element involved in a move as when a vacancy arises which can only be filled by a suitably qualified person. Nevertheless, we have found the distinction a useful one in practice and it lies behind the division of system models into 'Markov' models in Chapter 4 and 'renewal' models in Chapter 5.

Wastage will always be treated as a push flow on the ground that its statistical characteristics can be adequately described in terms of factors concerning the present state of an individual like sex and length of service (see Chapters 2 and 3). Promotion, transfer, and recruitment can be treated in either way according to the type of system being modelled.

In the stochastic case, push flows will be modelled by specifying the probability of the event in question. In the case of a pull flow a move is determined by the creation of a vacancy together with the choice of an individual to fill it. There may be a chance element at either or both of these stages which can be expressed in probability terms. In a deterministic model the probabilities become proportions.

There is an interesting duality between the movement of people in a system and the associated movement of vacancies, which we shall meet again in Chapter 5. A move can be described either in terms of a person going from $A$ to $B$ or of the vacancy going from $B$ to $A$. A pull flow for a person can thus also be viewed as a push flow for the corresponding vacancy since the start and finish of the move are then interchanged. This makes it possible, if we wish, to work almost entirely in terms of models embodying push flows.

## 1.4 STRUCTURE OF THE BOOK VIEWED IN A HISTORICAL PERSPECTIVE

The statistical treatment of manpower systems must be as old as the planning of the military and building exploits of the ancient world. However, the origins of a systematic approach using modelling techniques can be traced to the work of Seal (1945), Vajda (1947 and 1948), and others during the Second World War. Even then the potential of this new approach could not be fully exploited until cheap computing facilities became widely available in the 1960s. This is not to say that nothing worth while can be achieved without computers. Indeed, several pioneering studies like that of Lane and Andrew (1955) did not depend on large-scale computations for their success and many of the methods described later, especially in Chapters 2 and 3, can be used with very modest facilities. However, the speed and ease with which a large range of policy options can be explored using computer-based models make it possible to incorporate the statistical approach in day-to-day management practice. A computerized personnel record system also removes one of the main obstacles which beset many of the early attempts at manpower planning, namely lack of readily available data. For these reasons the exposition of the techniques in this book has been intimately linked to the use of the computer. We believe that the techniques are best learnt by using them in a practical way from the very beginning. For this reason, also, examples, exercises and their solutions occupy a large proportion of the text.

Wastage plays a key role in any manpower system and it is not surprising to find that much of the earliest statistical work was directed to this topic. The pioneering studies of Rice and coworkers (1950), Silcock (1954), and Lane and Andrew (1955) are still worth reading as they laid the foundation of much current practice. We have followed the historical chronology by treating wastage in some depth at the start of the book in Chapters 2 and 3.

Following the early work on wastage, attention moved on to the construction of transition models for systems of stocks and flows. Early papers which foreshadowed much of what was to follow were those of Young and Almond (1961) and Gani (1963). Initially the emphasis was on predicting stocks and flows—usually under the assumption of constant flow rates and using the theory of Markov chains. We have followed this pattern in Chapter 4 before going on to deal with later developments. The next main class of models to be discussed was designed for systems in which the grade sizes were fixed and for these renewal theory provided the mathematical foundation. The application of renewal theory to the study of manpower systems was foreshadowed in Bartholomew (1963a) and developed in some depth in the first edition (1967) of Bartholomew (1973a). Since then the emphasis has been on the construction of algorithms for computing flow forecasts in discrete time as in the Civil Service's KENT model described in Hopes (1973) and Smith (1976). Chapter 5 of this book contains an attempt to integrate these two approaches into a single and relatively simple treatment in which the close parallel with Markov models is emphasized. During the same period Morgan and others developed techniques for studying career prospects in graded systems; see for example, Morgan (1971), and Morgan and coworkers (1974). Some of these methods, together with others deriving from elementary probability arguments and Markov-chain theory, are covered in Chapter 6.

Much of the work on career patterns is based on assumptions of stationarity. If one wishes to determine desirable long-term manpower policies it is sensible to concentrate on the stationary behaviour of the system. This approach is an old one with its origins in demography and actuarial science. We prefer to regard it as an analysis of the limiting states approached by both Markov and renewal models. The theory of stationarity is closely linked with that of control and optimization, which is one of the main areas of current research. Much of this work lies outside the scope of the present book but an introductory account containing some of the most useful results is included in Chapter 7.

The demand side of manpower forecasting calls for a rather different approach since it is concerned with jobs rather than with people. This difference is reflected in the style of presentation which we have adopted in Chapter 8. Because of the very varied circumstances which obtain in practice it is not possible to define a collection of basic demand forecasting techniques. Instead we must cast around for statistical methods which may be helpful in particular cases. Our approach is, therefore, to review some of the techniques which have proved useful in the past and to provide a commentary on the pitfalls and problems which may arise in their application. It would be impracticable as well as unnecessary to give a detailed exposition of the techniques mentioned as they are treated in standard texts to which reference is made as the need arises. For similar reasons there are no worked examples in Chapter 8. We do not regard it as necessary for the would-be manpower planner to master all the techniques, many of which may never be necessary in his own work. Rather, we wish him to know what is available and where to turn if the need does arise.

It is not easy to draw the boundaries of the subject of statistics and the same is equally true of its application to manpower planning. The boundaries marked out by this book are clearly arbitrary and reflect our own interests. We have excluded, for example, applications of linear programming and of goal programming. The latter have been very fully developed by others, notably Charnes, Cooper, and Niehaus whose book (1972) provides a useful starting-point.

We have also omitted any discussion of individual simulation models. In such a model the behaviour of each member of the system is simulated on the computer using random numbers to determine the events which occur. Simulation models are particularly useful in studying small systems where random fluctuations are likely to be more serious. They also have the important advantage that it is much easier to incorporate realistic operating rules and so avoid much of the oversimplification which is inevitable with mathematical models. Their main disadvantages lie in the amount of computer time and space which they require for moderate or large systems and the difficulty of gaining the degree of insight into the basic dynamics of a complex system which a simpler mathematical model affords. Our omission is deliberate and for two reasons. Firstly the basic methodology of simulation is well known and is widely used in operational research and statistics. Just as we have excluded an exposition of the relevant statistical theory in Chapter 8 on the grounds that it is adequately covered elsewhere, so here we have taken the same view about simulation. In the second place, the essential concept of a manpower system as a dynamic and stochastic system of stocks and flows is equally relevant to both simulation and mathematical models. We therefore think that the best preparation for a would-be simulation model builder is to first master the methodology set out here and then to combine the knowledge gained with the necessary computing expertise. One of the earliest published examples of a simulation model will be found in Price (1971) and a more recent one is described in Blom and Knights (1976). An account of the Civil Service's MANSIM model is given by Wishart, in Smith (1976).

Newcomers to the field have often been handicapped by the scattered nature of the literature on the subject. In recent years this deficiency has been largely remedied by the appearance of several books and bibliographies. Of these we may mention Bartholomew (1977a) and (1977d), and Grinold and Marshall (1977). The Complements sections which appear at the end of most chapters will, among other things, lead the reader to some of the literature lying beyond the bounds of this book and the Bibliography contains not only all of the references occurring in the text but also most of the other relevant work of which we are aware.

CHAPTER 2

# Analysis of Wastage I: Rates and Life-table Methods

## 2.1 BACKGROUND

Of all the flows in a manpower system, wastage is the most fundamental for manpower planning. To a large extent it is the result of individual decisions to leave, and is hence outside the direct control of management. Furthermore, in organizations where the number of jobs is controlled it is wastage which creates vacancies and so provides opportunities for promotion and recruitment. Successful manpower planning therefore depends in large measure on being able to describe and predict patterns of wastage. Measures of wastage are often used as indicators of organizational health along with such things as absenteeism and industrial disputes. The correct interpretation of such indices requires an understanding of the underlying processes and this, in turn, depends on statistical analysis. It is therefore appropriate to begin our study of manpower systems by dealing with the statistical aspects of wastage and describing the methods which are available for its analysis. In the present chapter we shall concentrate on an empirical approach to the description and measurement of the wastage process. In the following chapter we turn to questions of summary measures and control, and for this it will be necessary to consider probability models for the wastage process.

We shall use the term 'wastage' to refer to the total loss of individuals from a system for whatever reason. Other terms are often used in the same sense, and 'attrition' and 'turnover' are among the most common. Turnover is sometimes distinguished from the others by saying that it refers to the combined process of recruitment and loss but this distinction is rarely made in practice and we shall treat all these terms as synonyms. The notion of stability is sometimes used as an alternative to wastage but the two ideas are essentially equivalent: high wastage implies low stability, and conversely.

Within the total wastage there are important subdivisions which must be carefully distinguished in practice. The main division is between voluntary and involuntary wastage. Involuntary wastage covers all losses for reasons beyond the control of the individual, such as death, ill-health, redundancy, and retirement. To a large extent it is predictable, at least in aggregate terms, and presents few difficulties for the manpower planner. Voluntary or natural wastage on the other hand arises whenever an individual leaves of his own choice—usually to take another job. It is a highly variable phenomenon and, except when total wastage is very low, is usually the dominant component. For this reason we shall be concerned almost exclusively with voluntary wastage in this and the following chapter. This does not mean that involuntary wastage can be ignored. If it is a substantial part of total wastage it should be dealt with as a separate exercise. Retirements, for example, can be dealt with by subtracting the individual's age from the normal retirement age and making an actuarial allowance for death. Where involuntary wastage is a minor component there will usually be little error in combining it with voluntary wastage. If the records do not distinguish reasons for leaving there will, of course, be no alternative but to combine them.

Any attempt to describe wastage patterns must reckon with the fact that an individual's propensity to leave a job depends on a great many factors, both personal and environmental. Later we shall formalize the notion of propensity to leave as a function of such factors. As early as 1954 Silcock summarized the state of statistical knowledge on the matter by giving a list of factors affecting propensity to leave, including length of service, age, sex, marital status, and the level of unemployment. Work since that date has corroborated Silcock's conclusions (see, for example, Hedberg (1961)) and identified further factors, some specific to particular industries. Some factors like age and length of service are partially confounded in that a person who is young cannot have a long length of service whereas one who is old may. However, the separate effects of age and length of service have been demonstrated by Young (1971) among others. In general, propensity to leave decreases with age and length of service, is higher for women than men, and decreases with increasing salary, status, or skill. These are, of course, generalizations and subject to qualifications and exceptions.

It is clear from these considerations that any theory of wastage must take account of these relationships and this may be approached in two ways. One, which we shall follow here, is to base the analysis on *homogeneous groups* of individuals. By this we mean groups of people who are similar with respect to all known factors which affect propensity to leave. This is a counsel of perfection which can never be attained in practice but the important thing is to ensure that there is no major source of heterogeneity. Thus, for example, we would not consider mixed groups of men and women unless one sex was very much more numerous than the other, and similarly we would try to avoid large disparities in age. In a prediction exercise we would make separate forecasts for each group and conflate the results at the end if need be. Since we shall often wish to test whether groups are sufficiently alike to be amalgamated, some statistical tests for this purpose are given in Section 2.4.

A second approach to the problem of heterogeneity is to express propensity to leave as a mathematical function of those variables on which it is believed to depend. The statistical problem is then to estimate the functional form of the relationship rather in the manner of classical regression analysis. The end-product of such an analysis would be an equation enabling us to predict what level of wastage would result from any particular combination of factors. Such methods are still in their infancy as far as wastage analysis is concerned but we expect them to play an increasing role. The topic will be taken up again in the Complements section of the present chapter (Section 2.5), where references will be found by the reader who wishes to explore this new territory.

The traditional way of approaching wastage analysis is via rates. Wastage rates may be defined in slightly different ways as we shall see in Section 2.2, but in essence a 'crude' rate is obtained by dividing the number of leavers from a group in some interval of time by the number of those at risk of leaving. Rates are often expressed as percentages. They have the advantages that they are easy to calculate and widely known. However, they are often misused and misunderstood in that they are frequently calculated for groups which are not homogeneous, especially with respect to length of service. Such difficulties can be avoided by following the demographers' example of computing standardized rates, but this idea does not seem to have commended itself to manpower planners. Whatever the reason for this, we shall not pursue the idea because there is another approach based directly on length of service which is more satisfactory statistically.

We noted above that propensity to leave depends on length of service and in practice this seems to be the most important factor of all. It also differs from the other variables mentioned in that it is intrinsic to the wastage process itself, in the sense that the length of time that individuals spend in their jobs is the factor that determines wastage rates. In particular, if most people in an organization have short lengths of service then other things being equal, the overall wastage will be high, and vice versa. This is illustrated in Example 2.1.

*Example 2.1.* Consider the two firms A and B both having the same wastage pattern as shown in Table 2.1. (We shall generally use the term wastage pattern to refer to the way in which propensity to leave is related to length of service or age.) This pattern, which has high wastage rates at short service (15 per cent per annum at 0–3 years) and low rates at long service (3 per cent for 3 + years), is typical of those observed in practice. In terms of length of service, A is the older organization with relatively more people in the 3 + class. Table 2.1 shows that despite having the same wastage pattern, the crude or overall wastage rate for A is considerably lower at 5.4 per cent than for B at 12.6 per cent. This difference is due only to the different length-of-service structures of the two firms. Organizations such as A are typical of older well-established or contracting systems, while younger organizations such as B are usually associated with new or expanding enterprises. ◀

***Table 2.1 Showing how two organizations with the same wastage pattern can have quite different overall wastage***

| *Length-of-service class* | *Per annum wastage rates for both A and B* | *Organization A ('old')* | | *Organization B ('young')* | |
|---|---|---|---|---|---|
| | | *number in service* | *expected leavers* | *number in service* | *expected leavers* |
| 0–3 | 0.15 | 200 | 30 | 800 | 120 |
| 3 + | 0.03 | 800 | 24 | 200 | 6 |
| *Totals* | | 1000 | 54 | 1000 | 126 |
| *Overall wastage* | | | 24<br>0.054 | | 0.126 |

† These are transition rates—see Section 2.2.

Example 2.2 provides a further illustration of how the length of service structure needs to be taken into account when analysing wastage.

*Example 2.2* Company C had a factory which they wished to re-tool with more modern machinery over 5 years. At the start of the period they employed 1000 operators although when the re-tooling was completed they would require only 600. Previously, both their total losses and total recruitment each year had been constant at 100. Since these losses represented a wastage of 10 per cent they calculated that with natural wastage and no recruitment they could just reach their target of 600 in 5 years. This expected reduction, which implied no redundancies, is shown in Table 2.2. In fact the planned reduction was not achieved because they did not allow for the change in the length of service structure and its consequent effect on the overall wastage rate. As the work force was run down the structure aged, and thus the overall

***Table 2.2 Illustrating how an expected contraction was not achieved***

| *Time* | 0 | 1 | 2 | 3 | 4 | 5 |
|---|---|---|---|---|---|---|
| Expected reduction at 10% per annum | 1000 | 900 | 810 | 729 | 656 | 590 |
| Actual reduction | 1000 | 925 | 873 | 828 | 788 | 751 |
| Actual reduction rate (compare this with 10%) | | 7.5% | 5.6% | 5.2% | 4.8% | 4.7% |

wastage rate decreased as implied in Example 2.1. The actual rates of reduction and the numbers achieved are also shown in Table 2.2. This result could have been anticipated given the structure of the length of service of the organization and its wastage pattern, as Exercise 2.4 at the end of the chapter shows. ◀

As these two examples indicate, length of service is a more fundamental quantity than the wastage rate, and for this reason our discussion will be based mainly on the search for patterns in the distribution of completed lengths of service. Nevertheless, rates do play a basic role as input to and outputs from manpower planning models and, when used with care, can be useful as indicators of organizational well-being.

## 2.2 COHORT ANALYSIS

Perhaps the most natural way of collecting data to investigate the pattern of wastage is to observe a homogeneous group of entrants and note how long each remains in the organization before leaving. Such a group, joining at about the same time, is known as a cohort. In practice the data will often come in grouped form as, for example, if we were given for each leaver the number of completed years of service. Even where we have the exact length of service most of our methods involve grouping the data. From the outset, therefore, we shall suppose that the completed lengths of service are grouped. Suppose the scale on which we measure length of service is divided at the points $x_1, x_2, \ldots, x_k$ and that we only know that $L_i$ members of our cohort left in the interval $(x_i, x_{i+1})$ $(i = 0, 1, 2, \ldots, k; x_0 = 0$, and $x_{k+1}$ is the maximum length of service which is often unknown). The last interval may span a very long range and may, in practice, contain a large proportion of the cohort. In statistical terminology our sample would then be said to be censored at the point $x_k$. The following table gives some cohort data for 800 entrants to Company X in 1971; the distribution is censored at the end of 1977 so that $x_k = x_6 = 6$.

***Table 2.3 Distribution of completed length of service for Company X, 1971 cohort***

| Completed length of service | 0– | 1– | 2– | 3– | 4– | 5– | 6– | *Total* |
|---|---|---|---|---|---|---|---|---|
| Number of leavers ($L_i$) | 364 | 162 | 78 | 60 | 31 | 21 | 84 | 800 |

For these data $x_1 = 1$, $x_2 = 2$, $x_3 = 3$, etc., and in writing '3–' for example we mean the interval from 3 to include 3 but exclude 4. It is common for the $x$'s to be equally spaced as in this case but this is by no means necessary. An alternative but equivalent way of presenting the same data is in terms of the number surviving to each of the times $x_i$. We shall denote this number by $Z_i$ which is related to the $L_i$'s by the formula

$$Z_i = \sum_{j=i}^{k} L_j$$

Thus for the data given above $Z_0 = 800, Z_1 = (162 + 78 + \ldots + 84) = 436$, and so on.

The data in Table 2.3 are an example of an empirical CLS (completed length of service) distribution and we shall first consider a number of probability functions of interest in manpower planning which can be estimated from such data.

**The probability density function**

We can obviously and easily estimate the probability that an entrant with this wastage pattern will leave with length of service in the interval $(x_i, x_{i+1})$. Thus for the interval (2–) the estimate for Company X would be 78/800. However, if the intervals $(x_i, x_{i+1})$ are of different lengths such probabilities depend on the length of the interval and it is often more useful, especially when making comparisons, to define them as referring to an interval of standard length. We shall take as this standard length the unit of measurement of the $x$'s, choosing it so that all intervals are multiples of this unit in length. Assuming that individual CLS's are uniformly distributed in each interval, the expected proportion of the cohort who will leave in any interval of unit length in $(x_i, x_{i+1})$ is $L_i/c_iZ_0$ where $c_i = (x_{i+1} - x_i)$. This may be thought of as an estimate of the probability

$$f_i = \Pr\{\text{entrant leaves during a unit interval within } (x_i, x_{i+1})\} \quad (i = 0, 1, 2, \ldots, k) \tag{2.1}$$

and we refer to the set of such probabilities as the discrete probability density function of CLS. The standard error of our estimate $\hat{f}_i$ of $f_i$ can be found from the fact that, if the group are homogeneous and behave independently, $L_i$ is a binomial random variable. Thus $L_i$ has variance $Z_0 f_i c_i(1 - f_i c_i)$ and hence

$$\text{var}(\hat{f}_i) = f_i(1 - f_i c_i)/c_i Z_0 \tag{2.2}$$

The standard error of the estimator may thus be estimated by replacing $f_i$ by $\hat{f}_i$ giving

$$\hat{\text{se}}(\hat{f}) = \left[L_i\left(1 - \frac{L_i}{Z_0}\right)\right]^{\frac{1}{2}} \tag{2.3}$$

There is a problem about calculating this quantity for the last group when $i = k$ since we do not usually know the upper limit of service and in any case the assumption of a uniform distribution of leavers throughout a long interval is unrealistic. This group is best dealt with by the survivor function method given below.

### The survivor function

Another useful quantity is the probability that a recruit will survive for a length of time $x$. This probability can only be estimated directly for values of $x$ coinciding with the $x_i$'s. Let us therefore define

$$G_i = \Pr\{\text{an entrant survives to } x_i\} \qquad (i = 1, 2, \ldots, k) \tag{2.4}$$

We may estimate $G_i$ by $\hat{G}_i = Z_i/Z_0$ and since $Z_i$ is binomial the estimated standard error will be

$$\hat{\text{se}}(\hat{G}_i) = \{\hat{G}_i(1 - \hat{G}_i)/Z_0\}^{\frac{1}{2}} \qquad (i = 1, 2, \ldots, k) \tag{2.5}$$

These standard errors must be interpreted with care since pairs of $\hat{G}_i$'s will be correlated because they have some $Z_i$'s in common.

### Conditional probability of leaving at length of service $x$

We shall sometimes wish to consider the leaving prospects of individuals who have already attained a certain length of service. Let us therefore consider the probability

$$q_i = \Pr\{\text{individual with length of service } x_i \text{ leaves before } x_i + 1\}$$
$$(i = 0, 1, 2, \ldots, k) \tag{2.6}$$

As when defining $f_i$, and for the same reasons, it is more convenient to define the probability for an interval of unit length but this time it is not any such interval but the one beginning at $x_i$. On the uniform leaving assumption made before, the expected number of leavers in $(x_i, x_i + 1)$ is $L_i/c_i$ and so we adopt the estimator

$$\hat{q}_i = L_i/c_iZ_i \qquad (i = 0, 1, 2, \ldots, k) \tag{2.7}$$

Since $L_i$ and $Z_i$ are both random variables the calculation of the standard error is more difficult, but it seems more relevant here to treat $Z_i$ as given since the probability is only of real interest when the point $x_i$ is reached and $Z_i$ is known. Under these circumstances the binomial argument applies and

$$\hat{\text{se}}(\hat{q}_i) = [c_i\hat{q}_i(1 - c_i\hat{q}_i)/Z_i]^{\frac{1}{2}}/c_i \qquad (i = 0, 1, \ldots, k) \tag{2.8}$$

but care must be taken to remember the conditional basis of the argument when applying the result.

*Example 2.3* Estimates of all the foregoing functions have been made using

***Table 2.4* Estimates of the various probability functions for Company X and their standard errors (given in brackets)**

| $x_i$ | 0 | 1 | 2 | 3 | 4 | 5 |
|---|---|---|---|---|---|---|
| $\hat{f}_i$ | 0.455 (0.018) | 0.203 (0.014) | 0.098 (0.010) | 0.075 (0.009) | 0.039 (0.007) | 0.026 (0.006) |
| $\hat{G}_{i+1}$ | 0.545 (0.018) | 0.342 (0.017) | 0.245 (0.015) | 0.170 (0.013) | 0.131 (0.012) | 0.105 (0.011) |
| $\hat{q}_i$ | 0.455 (0.018) | 0.372 (0.023) | 0.285 (0.027) | 0.306 (0.033) | 0.228 (0.036) | 0.200 (0.039) |

the data for Company X and they are brought together in Table 2.4. The standard errors are given in brackets below the corresponding estimate. The frequency function decreases rapidly, with short service times being much more common than long. The estimated standard errors show that there is considerable uncertainty about the true values but the general pattern is clear enough. It is the values of $\hat{q}_i$ which are, perhaps, the most interesting. These are estimates of the probabilities of leaving in the next year given survival up to each point. In general this is a decreasing function indicating that propensity to leave falls away with increasing service and this is what is usually found. Sometimes there is an initial rise before the decline sets in, suggesting the need for time on the part of individuals to decide whether they wish to stay. In this particular example there is an apparent rise in the 3- to 4-year interval though the size of the standard errors suggest that this should not be taken too seriously. However, it may mean that three to four years' service was seen by some as about the right length of experience before seeking advancement elsewhere. ◀

### A continuous-time representation of the probability functions

In the earlier part of this section we adopted an empirical approach in which the motivating idea was to find what probabilities of interest to the manpower planner could be estimated from a grouped cohort distribution. We must now set these in a theoretical framework both to achieve a clearer perspective of what we have done and to prepare the ground for the census analyses of the following section.

Leaving is a process which can occur at virtually any time in a person's career and so it is reasonable to treat completed length of service as a continuous variable. We may thus introduce the *survivor function* $G(x)$ as the probability that an individual survives in the job for time $x$, defined for all values of $x$ up to the maximum possible length of service. This function is the complement of the more familiar distribution function $F(x)$ so that

$G(x) = 1 - F(x)$. The probability density function of $x$ is then

$$f(x) = \frac{\mathrm{d}F(x)}{\mathrm{d}x} = -\frac{\mathrm{d}G(x)}{\mathrm{d}x} \tag{2.9}$$

and, conversely,

$$G(x) = \int_x^\infty f(u)\,\mathrm{d}u \tag{2.10}$$

The probability $f_i$ defined in equation (2.1) may be expressed as

$$f_i = \int_{x_i}^{x_{i+1}} f(u)\,\mathrm{d}u/(x_{i+1} - x_i)$$

from which we can see that it is the average value of the density over the interval.

The third probability considered in the discrete case was the conditional probability of leaving in the next unit interval. Here we shall consider a small interval $(x, x + \delta x)$ and define an instantaneous rate of leaving, or force of separation, $m(x)$ by

$$m(x)\delta x = \Pr\{\text{individual with length of service } x \text{ leaves in } (x, x + \delta x)\}$$

This probability may be thought of as a limiting version of $q_i$ as the unit interval is made very small. Notice that $m(x)$ itself is not a probability but a rate or intensity. The form of the function $m(x)$ tells us how propensity to leave changes with length of service. Values of $x$ for which $m(x)$ is relatively large will indicate periods at which the risk of leaving is high and such information is obviously useful to the manpower planner. A simple probability argument gives

$$m(x)\delta x = f(x)\delta x/G(x) \quad \text{or} \quad m(x) = f(x)/G(x) \tag{2.11}$$

A version of (2.11) which is useful for estimation is

$$m(x) = -\frac{\mathrm{d}\ln G(x)}{\mathrm{d}x} \tag{2.12}$$

Furthermore, integrating both sides of (2.11)

$$G(x) = \exp\left[-\int_0^x m(u)\,\mathrm{d}u\right] \tag{2.13}$$

The foregoing results show that $f(x)$, $G(x)$, and $m(x)$ are equivalent in the sense that any one can be obtained from any of the others. Which one we use will therefore be partly a matter of convenience and will partly depend on our objectives.

We have introduced the rate function $m(x)$ rather than a continuous analogue of the probability $q_i$ because it is more useful and, in a sense, more fundamental. In fact if we define $q(x)$ to be the probability that someone with length of service $x$ leaves before $x+1$ this can be simply expressed in terms of $m(x)$ by an extension of the idea behind (2.13). Thus $1-q(x)$ is the probability of surviving to time $x+1$ starting at $x$ and so

$$q(x) = 1 - \exp\left[-\int_x^{x+1} m(u)\,du\right] \tag{2.14}$$

**Estimation of $m(x)$**

We can estimate $m(x)$ as a step function by treating it as constant within each of the intervals $(x_i, x_{i+1})$. Thus let $m_i$ denote the constant approximating value of $m(x)$ in this interval; then using (2.12) we have

$$\begin{aligned} m_i(x_{i+1} - x_i) &= \int_{x_i}^{x_{i+1}} m(x)\,dx = -\int_{x_i}^{x_{i+1}} \frac{d \ln G(x)}{dx}\,dx \\ &= \ln G_i - \ln G_{i+1} \end{aligned} \tag{2.15}$$

and $m_i$ may then be estimated by

$$\hat{m}_i = (\ln \hat{G}_i - \ln \hat{G}_{i+1})/c_i = (\ln Z_i - \ln Z_{i+1})/c_i \quad (i = 0, 1, 2, \ldots, k-1) \tag{2.16}$$

An alternative derivation is by the method of maximum likelihood. If each individual is subject to a constant propensity to leave, $m_i$, during the interval $(x_i, x_{i+1})$ then he will survive with probability $\exp(-m_i c_i)$ and leave with probability $\{1 - \exp(-m_i c_i)\}$. Given $Z_i$ people at the beginning of the interval and $L_i$ leavers during it, the likelihood function is

$$\{\exp(-m_i c_i)\}^{Z_i - L_i}\{1 - \exp(-m_i c_i)\}^{L_i}$$

from which it follows that (2.16) is the maximum likelihood estimator of $m_i$. The approximate standard error can be obtained from the second derivative of the log-likelihood function in the usual way. Its estimate is

$$\begin{aligned} \hat{\text{se}}(\hat{m}_i) &= [(\hat{G}_i - \hat{G}_{i+1})/\hat{G}_i \hat{G}_{i+1} Z_0]^{\frac{1}{2}}/c_i \\ &= (L_i/Z_i Z_{i+1})^{\frac{1}{2}}/c_i \qquad (i = 0, 1, 2 \ldots, k-1) \end{aligned} \tag{2.17}$$

A second method of estimating $m_i$, used by actuaries, may also be derived by maximum likelihood methods. If we knew the times at which each of the $L_i$ leavers left in $(x_i, x_{i+1})$ the likelihood function would be

$$m_i^{L_i} \exp\{-m_i T - m_i c_i (Z_i - L_i)\}$$

where $T$ is the total service time in $(x_i, x_{i+1})$ of those who leave in that interval.

The maximum likelihood estimator in this case is

$$\hat{m}_i = L_i/\{c_i Z_{i+1} + T\} \tag{2.18}$$

If the data are grouped we shall not know $T$ but it can be estimated by assuming that leavers survive, on average, for a proportion $a_i$ of the interval. Unless there is some reason for supposing otherwise it is usual to take $a_i = \frac{1}{2}$, giving the estimator

$$\hat{m}_i = L_i/\{c_i Z_{i+1} + \tfrac{1}{2}c_i L_i\} \tag{2.19}$$

The standard methods applicable to maximum likelihood estimators give for the estimated standard error of $m_i$

$$\hat{\text{se}}(\hat{m}_i) = \hat{m}_i/\sqrt{L_i} \tag{2.20}$$

Although (2.16) and (2.19) do not look alike they are, in fact, equal to a first approximation, and in applications they are usually very close, as are the expressions for the standard errors.

*Example 2.4* For Company X the estimates of $m_i$ and its standard error by both methods are as set out in Table 2.5.

***Table 2.5 Estimates of $m_i$ with standard errors for Company X, 1971 cohort***

| | $i$ | 0 | 1 | 2 | 3 | 4 | 5 |
|---|---|---|---|---|---|---|---|
| $\hat{m}_i$ | (eq. (2.16)) | 0.607 | 0.465 | 0.335 | 0.365 | 0.259 | 0.223 |
| $\hat{m}_i$ | (eq. (2.19)) | 0.589 | 0.456 | 0.332 | 0.361 | 0.257 | 0.222 |
| $\hat{\text{se}}$ | (eq. (2.17)) | 0.032 | 0.037 | 0.038 | 0.047 | 0.047 | 0.049 |
| $\hat{\text{se}}$ | (eq. (2.20)) | 0.031 | 0.036 | 0.038 | 0.047 | 0.046 | 0.048 |

The estimates are very close, especially in the case of the standard errors, and ease of calculation can be the determining factor in choosing a method. We have a slight preference for (2.16) and (2.17). The pattern here is very similar to that of the $q_i$'s for the same data given in Table 2.4 and the comments made there apply here also. ◀

Both $m_i$ and $q_i$ show how propensity to leave changes with length of service and each has its advantages and disadvantages. The main advantage of $q_i$ is that, being a probability, it can be interpreted in a direct way as an expected proportion. The meaning of $m_i$, as a rate, is possibly a little more difficult for the layman to grasp, but its real value will become apparent in the next section.

**The expected further duration**

The three functions $G(x)$, $f(x)$, and $m(x)$ are not the only ways of expressing the dependence of propensity to leave on length of service. Another that is particularly useful is the expected further duration. This is the expected remaining service of a person who has reached service $x$ and we shall denote it by $D(x)$. With grouped data we shall estimate this function at the points $x_i$ and abbreviate $D(x_i)$ to $D_i$. The expected remaining service for an individual with length of service $x$ is

$$D(x) = \int_x^\infty uf(u)\,\mathrm{d}u/G(x)$$

Integration by parts can be used to show that this may also be expressed as

$$D(x) = \int_x^\infty G(u)\,\mathrm{d}u/G(x) \tag{2.21}$$

We now require to estimate this function when $x = x_i$, which resolves itself into estimating

$$\int_{x_i}^{x_{i+1}} G(x)\,\mathrm{d}x$$

for each interval. Assuming the step function approximation $m_i$ then $G(x)$ will have an exponential decay curve in each of those intervals such that

$$G(x) = G_i\exp[-m_i(x - x_i)], \qquad x_i < x < x_{i+1}$$

Thus, integrating

$$\int_{x_i}^{x_{i+1}} \hat{G}(x)\,\mathrm{d}x = \frac{\hat{G}}{\hat{m}_i}\{1 - \exp(-\hat{m}_i c_i)\}$$
$$= (\hat{G}_i - \hat{G}_{i+1})/\hat{m}_i \quad \text{(using 2.16)} \tag{2.22}$$

Beyond $x_k$ we are in difficulties since we usually have very little empirical information about the distribution beyond that point. If a substantial part of the distribution lies beyond $x_k$ we would not proceed with this method but base the calculations on a fitted curve as described in Chapter 3. If, as with Company X, the amount is not too large and if we can make a reasonable guess at the maximum possible length of service ($x_{k+1}$), some progress can be made. Since $G(x_{k+1}) = 0$ we cannot assume $G(x)$ is exponential over this final $(x_k, x_{k+1})$ so we shall treat it as linear. The area under $G(x)$ is then a triangle with area $\frac{1}{2}G_k(x_{k+1} - x_k)$.

The final estimator which emerges is

$$\hat{D}_i = \hat{G}_i^{-1}\left\{\sum_{j=i}^{k-1} (\hat{G}_j - \hat{G}_{j+1})/\hat{m}_j + \tfrac{1}{2}c_k\hat{G}_k\right\} \qquad (i = 0, 1, \ldots, k-1) \tag{2.23}$$

In practice this expression is usually very dependent on the value chosen for $x_{k+1}$. Since this value is often quite arbitrary, $\hat{D}_i$ should be used with care.

*Example 2.4* (*continued*) If, for Company X, we suppose that entrants are, on average, aged 30 on joining and that they retire at 65 we may take $x_{k+1} = x_7 = 35$ which gives the following estimates:

| $x_i$ | 0 | 1 | 2 | 3 | 4 | 5 | 6 | 35 |
|---|---|---|---|---|---|---|---|---|
| $D_i$ | 3.5 | 5.0 | 6.7 | 8.1 | 10.5 | 12.5 | 14.5 | 0 |

This pattern is also fairly typical in that it shows that, in the early stages of service at least, the longer a person remains in a job the longer they are likely to stay. Eventually, of course, as retirement approaches $D_i$ must begin to decrease to zero. ◀

## 2.3 CENSUS ANALYSIS

A complete historical picture of the wastage process can be built up if data are available on all cohorts recruited in the past. By assuming that patterns in the recent past will persist into the future, forecasts can be made in the manner described in the last section. In practice, however, it is often only possible to obtain data on stocks and flows in the current accounting period. This means that the only information we have on each cohort represented in the present stock concerns its wastage experience over a short interval of time. Nevertheless, because the present stock comprises cohorts of different ages it is possible to reconstruct a composite picture of a survivor function from this information, and it is the purpose of this section to provide methods for doing so. We shall describe methods appropriate to two different forms of the initial data. Both these forms consist of a record of the stocks and leavers in an interval of time. We shall speak of this interval as a year, as it often is in practice, but any other period can be substituted without affecting the essentials of the argument. For the methods to provide accurate and useful estimators it is necessary for the interval to be reasonably 'short' compared with the typical length of completed service. Census data may also be described as 'current' or 'cross-sectional'.

### Census data: transition and central rates

Wastage rates play a fundamental role in census analysis and in manpower work generally and so we shall begin by discussing the two main kinds. The distinction between them is important and they are easily confused.

Suppose that we have data on the leaving experience of a homogeneous group of employees for a calendar year; then we may define the following annual rates.

(a) The transition wastage rate:

$$\hat{w} = \frac{\text{Number of leavers during the year from amongst those in the class at the start of the year}}{\text{Number in this class at the start of the year}}$$

$$= L'/S', \text{ say.} \tag{2.24}$$

(b) The central wastage rate:

$$\hat{m} = \frac{\text{Number of leavers during the year who were in this class when they left}}{\text{Average number in this class during the year}}$$

$$= L^*/S^*, \text{ say} \tag{2.25}$$

The notation $w$ is used in anticipation of Chapter 4 where this quantity will appear as an estimate of a wastage probability. Likewise the use of $m$ is in line with Section 2.2, since when the classes are defined with respect to length of service it can be thought of as estimating the constant rate of leaving or force of separation during the year. In practice both rates are often expressed as percentages.

Standard errors of the rates can be estimated if we make some assumptions about the wastage process. In the case of $w$, which is similar to the $q_i$'s discussed in the last section, we assume that each individual in the initial stock has the same probability $w$ of leaving during the year. If, also, individuals behave independently then the binomial model applies and the estimated standard error is

$$\hat{se}(\hat{w}) = [\hat{w}(1 - \hat{w})/S']^{\frac{1}{2}} \tag{2.26}$$

The calculation of the standard error of $m$ involves an approximation but otherwise follows the method used earlier. If the number in the class throughout the year were constant and equal to $S^*$ we could think of the class as consisting of $S^*$ 'jobs' in each of which the occupants were subject to a probability $m\delta x$ of loss in each interval $(x, x + \delta x)$. Under these assumptions the number of losses would have a Poisson distribution with mean $S^*m$. The estimated standard error of $\hat{m}$ would therefore be

$$\hat{se}(\hat{m}) = [\hat{m}/S^*]^{\frac{1}{2}} = \hat{m}/\sqrt{L^*} \tag{2.27}$$

In practice the number in the class will fluctuate during the year and so we approximate by treating the group as though the class size were fixed at its average level throughout the year. Ideally we should take $S^*$ to be the time average over the year but we shall usually have to be content to estimate it by taking the average of the number at the beginning and end of the year.

*Example 2.5* Suppose that there are 1000 people in a class at the beginning of the year. The total number of leavers during the year is 350, made up of 250 from those there at the beginning of the year and 100 from those who entered during the course of the year. The total number of entrants during the year is 450. From these data we can calculate both rates and their standard errors. Thus

$$L' = 250, \qquad S' = 1000 \quad \text{so} \quad \hat{w} = \frac{250}{1000} = 0.250$$

and

$$\hat{\text{se}}(\hat{w}) = \{0.250\,(1 - 0.250)/1000\}^{\frac{1}{2}} = 0.014$$

For $m$, $L^* = 350$, the number in the class at the beginning of the year is 1000, and at the end it is $1000 - 250 + (450 - 100) = 1100$, so that $S^* = \frac{1}{2}(1000 + 1100)$, and

$$\hat{m} = \frac{350}{1050} = 0.333, \qquad \hat{\text{se}}(\hat{m}) = [0.333/1050]^{\frac{1}{2}} = 0.018$$ ◀

It is useful to divide entrants to the class into what may be termed the gross and net inflows. The gross inflow refers to all entering during the year (450 in Example 2.5) whereas the net inflow is that part of the gross inflow who remain at the end of the year ($450 - 100 = 350$ in the example). Those counted in the gross but not the net inflow are those who join and leave in the year. These people do not appear in the numerator or denominator of $\hat{w}$, but in $\hat{m}$ they contribute to both. Note also that those who start the year in the class, move to another class, and then leave, all within the year, should be included in the numerator of $\hat{w}$ but not of $\hat{m}$.

The advantage of the transition rate is that it can be used directly for forecasting. Thus if it is applied to a stock at the start of a year it gives a prediction of those who will leave during the year. After adding the net inflow we have a forecast of the stock at the beginning of the next year. This idea is the basis of the transition models discussed in Chapter 4. Although the central rate is not so easy to interpret in this way it does give a better representation of propensity to leave for members of the class as it includes a contribution from the experience of those entering during the year.

In practice, the form of rate used may well depend on the data available. Thus if leavers are classified by their status at the beginning of the year, the transition rate will be indicated. If they are classified by their status on exit, the central rate must be used.

If the assumptions underlying the standard error calculations hold, it is possible to determine a relationship between the two rates by essentially the same argument as used in deriving (2.14). The probability, $w$, that an individual subject to a force of separation $m$ leaves in a unit interval of time is $1 - \exp(-m)$, hence we have

$$w = 1 - \exp(-m) \tag{2.28}$$

In Example 2.5, where $\hat{m} = 0.333$, we would expect $w$ to be $1 - \exp(-0.333) = 0.282$, which is reasonably close to the actual value of 0.250.

In the following sections we shall derive methods of estimating the survivor function using both kinds of rates. It may appear at first sight that if we could rely on the assumptions on which (2.28) is based only one method would be necessary, since either rate could be converted into the other. However, as explained later, length-of-service specific rates do not convert into values applying to exactly the same classes.

## Estimation of the wastage pattern using central rates

In this approach the rates are calculated for classes defined by length of service. Let $S_i$ be the average number in service during the year in length-of-service class $(x_i, x_{i+1})$, and let $L_i$ be the number of leavers during the year who were in length-of-service class $(x_i, x_{i+1})$ when they left. Note that the $x$'s have a slightly different meaning here from that given to them in cohort analysis in that they represent the grouping adopted for the current length-of-service distribution. Using these data we may estimate the function $m(x)$ by making the step function approximation $m(x) = m_i$ for $x_i \leqslant x < x_{i+1}$.

***Table 2.6 Company X administration group 1977: current (central) data and estimates for the length of service specific rates, $m_i$***

| | *Current (central) data* | | | |
|---|---|---|---|---|
| *Length-of-service class* $x_i - x_{i+1}$ | *Leavers by class on exit* $L_i$ | *Average nos. in service* $S_i$ | *Estimated central rates* $\hat{m}_i$ | *Estimated standard error* $\hat{se}(\hat{m}_i)$ |
| 0– | 87 | 124.5 | 0.699 | 0.075 |
| 1– | 23 | 83.0 | 0.277 | 0.058 |
| 2– | 11 | 63.0 | 0.175 | 0.053 |
| 3– | 15 | 50.0 | 0.300 | 0.077 |
| 4– | 8 | 40.0 | 0.200 | 0.071 |
| 5–8 | 17 | 109.5 | 0.155 | 0.038 |
| 8–10 | 5 | 50.5 | 0.099 | 0.044 |
| 10–15 | 12 | 107.0 | 0.112 | 0.032 |
| 15–20 | 5 | 79.5 | 0.063 | 0.028 |

The central rate for the group is an estimate of $m_i$ and so

$$\hat{m}_i = L_i/S_i, \qquad \hat{\text{se}}(\hat{m}_i) = \{m_i/S_i\}^{\frac{1}{2}}. \tag{2.29}$$

*Example 2.6* Census (central) data for Company X administration group are given in Table 2.6 together with the estimates calculated using (2.29). The results show a generally decreasing trend in propensity to leave with a possible rise in the 3- to 4-year service group. This may be another manifestation of the tendency to change jobs at this interval of time as noted earlier. The standard errors would have been smaller if the groups had been broader but this would make the step function approximation cruder and we might then have missed the hump around 3 and 4 years. ◀

The estimate of $m(x)$ can be converted into either of $G(x)$ or $f(x)$ using the relationships of equations (2.9) and (2.13). The *stability curve method* of Lane and Andrew (1955) replaces $m(x)$ by its step-function estimator in (2.13) to obtain

$$\hat{G}(x) = \exp\left[-\sum_{j=0}^{i-1} c_j\hat{m}_j - (x - x_i)\hat{m}_i\right], \qquad x_i \leqslant x < x_{i+1}$$
$$(i = 1, 2, \ldots, k) \tag{2.30}$$

The approximate standard error of the estimator can be found by the standard method for functions of random variables using the known results for the $\hat{m}_i$'s. It will usually be sufficient to calculate them at the points $x = x_i$ ($i = 1, 2, \ldots, k$) for which Forbes (1971b) shows that

$$\hat{\text{se}}(\hat{G}_i) \approx \hat{G}_i\left\{\sum_{j=0}^{i-1} \hat{m}_j c_j/S_j\right\}^{\frac{1}{2}} \tag{2.31}$$

The other wastage functions can be calculated in the usual way. In particular,

$$\hat{f}_i = (\hat{G}_i - \hat{G}_{i+1})/c_i \quad \text{and} \quad \hat{q}_i = (\hat{G}_i - \hat{G}_{i+1})/c_i\hat{G}_i \tag{2.32}$$

The expected further duration $D_i$ is given by (2.23). If standard errors are required for these estimators they can also be obtained using the standard approximation method.

*Example 2.7* The calculation of $\hat{G}_i$ and its standard error are given in Table 2.7 for the data of Table 2.6 together with estimates of $f_i$, $q_i$, and $D_i$. These functions have the same shapes as the estimates obtained by cohort methods in the last section. ◀

The *actuarial method* of estimating the survivor function from the $m_i$'s is slightly different and it is instructive to consider its derivation. We may

***Table 2.7 Company X administration group 1977 census analysis: stability estimates for $G_i$, $f_i$, $q_i$, and $D_i$***

| *Length of service* $x_i$ | *Survivor function* $\hat{G}_i$ | *Estimated standard error* $\hat{se}(\hat{G}_i)$ | *Density function* $\hat{f}_i$ | *Conditional probability of leaving* $\hat{q}_i$ | *Further expected duration* $\hat{D}_i$ |
|---|---|---|---|---|---|
| 0 | 1.000 | 0.000 | 0.503 | 0.503 | 3.3 |
| 1 | 0.497 | 0.037 | 0.120 | 0.242 | 5.2 |
| 2 | 0.377 | 0.036 | 0.060 | 0.160 | 5.7 |
| 3 | 0.316 | 0.034 | 0.082 | 0.259 | 5.6 |
| 4 | 0.234 | 0.031 | 0.042 | 0.181 | 6.4 |
| 5 | 0.192 | 0.029 | 0.024 | 0.124 | 6.8 |
| 8 | 0.120 | 0.023 | 0.011 | 0.090 | 7.0 |
| 10 | 0.099 | 0.021 | 0.008 | 0.086 | 6.3 |
| 15 | 0.056 | 0.015 | 0.003 | 0.054 | 4.3 |
| 20 | 0.041 | 0.012 | — | — | 0 |

write

$$G_i = \prod_{j=0}^{i-1} G_{j+1}/G_j \tag{2.33}$$

where each of the factors in this expression is the survival probability from one length of service point to the next. Now since the expression for $m_i$ given in (2.19) can be written as

$$\hat{m}_j = (1 - \hat{G}_{j+1}/\hat{G}_j)/c_j\{1 - \tfrac{1}{2}(1 - \hat{G}_{j+1}/\hat{G}_j)\}$$

we may invert this expression to give

$$\hat{G}_{j+1}/\hat{G}_j = (1 - \tfrac{1}{2}c_j\hat{m}_j)/(1 + \tfrac{1}{2}c_j\hat{m}_j)$$

Inserting this in (2.33) we have

$$\hat{G}_i = \prod_{j=0}^{i-1} (1 - \tfrac{1}{2}c_j\hat{m}_j)/(1 + \tfrac{1}{2}c_j\hat{m}_j) \tag{2.34}$$

If we suppose that leavers depart, on average after a proportion $a_i$ of the interval has elapsed the estimator becomes

$$\hat{G}_i = \prod_{j=0}^{i-1} (1 - a_j c_j\hat{m}_j)/(1 + (1 - a_j)c_j\hat{m}_j) \tag{2.35}$$

Since the $m_j$'s are central rates which can be estimated from census data the derivation is complete. The corresponding standard error derived in Forbes (1971b) is

$$\hat{se}(\hat{G}_i) = \hat{G}_i\left\{\sum_{j=0}^{i-1} \frac{c_j^2\hat{m}_j}{S_j(1 - \frac{1}{4}c_j^2\hat{m}_j^2)^2}\right\}^{\frac{1}{2}} \tag{2.36}$$

These estimators give numerical values very close to those of the stability curve method as the following example shows. Note, however, that this method only estimates $G(x)$ at the points $x = x_i$ ($i = 1, 2, \ldots, k$).

*Example 2.8* The results of calculations for the actuarial method are given in Table 2.8 ◀

***Table 2.8 Company X administration group 1977 census analysis: actuarial estimates for $G_i$, $q_i$, $f_i$, and $D_i$***

| Length of service $x_i$ | Average proportion of interval served by leavers $a_i$ | Survivor function $\hat{G}_i$ | Estimated standard error $\hat{se}(\hat{G}_i)$ | Density function $\hat{f}_i$ | Conditional probability of leaving $\hat{q}_i$ | Further expected duration $\hat{D}_i$ |
|---|---|---|---|---|---|---|
| 0 | 0.5 | 1.000 | 0.000 | 0.518 | 0.518 | 3.2 |
| 1 | 0.5 | 0.482 | 0.041 | 0.117 | 0.243 | 5.2 |
| 2 | 0.5 | 0.365 | 0.038 | 0.059 | 0.161 | 5.7 |
| 3 | 0.5 | 0.306 | 0.036 | 0.080 | 0.261 | 5.7 |
| 4 | 0.5 | 0.226 | 0.032 | 0.041 | 0.182 | 6.5 |
| 5 | 0.5 | 0.185 | 0.029 | 0.023 | 0.126 | 6.8 |
| 8 | 0.5 | 0.115 | 0.023 | 0.010 | 0.090 | 7.0 |
| 10 | 0.5 | 0.094 | 0.021 | 0.008 | 0.088 | 6.3 |
| 15 | 0.5 | 0.053 | 0.015 | 0.003 | 0.054 | 4.3 |
| 20 | 0.5 | 0.039 | 0.012 | — | — | 0 |

## Estimation of the survivor function using transition rates

For transition data we redefine $S_i$ as the number in service at the start of the year in length of service class $(x_i, x_{i+1})$, and $L_i$ as the number of these $S_i$ who leave before the end of the year. Note first the difference between these data and the central $S_i$ and $L_i$ defined previously, and secondly that they refer (like the central data) to the current length-of-service distribution. From (2.24) and (2.26) the transition wastage rate for the $i$th class and its standard error is

$$\hat{w}_i = L_i/S_i, \; \hat{se}(\hat{w}_i) = [\hat{w}_i(1 - \hat{w}_i)/S_i]^{\frac{1}{2}} \tag{2.37}$$

*Example 2.9* Census (transition) data for a sample of computer programmers during 1975 are shown in Table 2.9, together with the wastage transition rates and their errors calculated from (2.37). The results show that, for these people, wastage during the first 4 years of service is surprisingly constant, although at longer lengths of service there is a decreasing chance as we might expect. ◀

***Table 2.9 Computer programmers 1975: transition data and estimated wastage rates***

| $x_i$–$x_{i+1}$ | 0–1 | 1–2 | 2–4 | 4–8 | 8–16 | 16–35 |
|---|---|---|---|---|---|---|
| $L_i$ | 32 | 27 | 23 | 25 | 11 | 1 |
| $S_i$ | 251 | 204 | 184 | 534 | 335 | 222 |
| $\hat{w}_i$ | 0.127 | 0.132 | 0.125 | 0.047 | 0.033 | 0.005 |
| $\hat{se}(\hat{w}_i)$ | 0.021 | 0.024 | 0.024 | 0.009 | 0.010 | 0.004 |

To estimate the survivor function we might consider, as suggested previously, converting these transition rates into central rates by means of (2.28) and then using the census central methods considered above. However, suppose $w_i$ applies to the unit interval $(x_i, x_{i+1} = x_i + 1)$ so that it is the wastage transition rate over the year for people with an average length of service $(x_i + \frac{1}{2})$ at the start. Using (2.28) the central rate $m_i = -\ln(1 - w_i)$ will therefore be appropriate to the interval $(x_i + \frac{1}{2}, x_i + \frac{3}{2})$, which is not convenient if we wish to estimate the wastage functions at the points $\{x_i\}$. The following method is more convenient in this sense and uses the transition rates directly.

Assume that for classes $(x_i, x_{i+1})$ which are not of unit length, the same rate applies to each of the subclasses $(x_i, x_i + 1), (x_i + 1, x_i + 2), \ldots, (x_{i+1} - 1, x_{i+1})$. If people in these subclasses are treated as having length of service equal to the mid-point of the group, $(1 - w_i)$ can then be regarded as the probability of surviving from $x_i + \frac{1}{2}$ to $x_i + 1\frac{1}{2}$, from $x_i + 1\frac{1}{2}$ to $x_i + 2\frac{1}{2}$, $\ldots$, and from $x_{i+1} - \frac{1}{2}$ to $x_{i+1} + \frac{1}{2}$. The probability of surviving from $x_i + \frac{1}{2}$ to $x_{i+1} - \frac{1}{2}$ is thus $(1 - w_i)^{c_i - 1}$. The probability of surviving through the half-interval $(x_{i+1} - \frac{1}{2}, x_i + 1)$ is $(1 - w_i)^{\frac{1}{2}}$ (since the square of this probability must give $1 - w_i$ as the probability for the whole interval). Similarly, the probability for the half-interval $(x_i, x_i + \frac{1}{2})$ is $(1 - w_{i-1})^{\frac{1}{2}}$. Putting these together we obtain for the $i$th interval that

$$\Pr\{\text{survival from } x_i \text{ to } x_{i+1}\} = (1 - w_{i-1})^{\frac{1}{2}}(1 - w_i)^{c_i - \frac{1}{2}}$$

where $c_i = (x_{i+1} - x_i)$ is the interval length. This estimator does not apply to the first interval $(x_0, x_1)$ since we have no value for $(1 - w_{-1})^{\frac{1}{2}}$ which corresponds essentially to the proportion of the gross inflow who survive to give the net inflow. In the absence of other information the best we can do is to estimate its value by $(1 - w_0)^{\frac{1}{2}}$. The survivor function can then be obtained by multiplying together these survival probabilities for all intervals. Thus

$$\begin{aligned} G_0 &= 1 \\ \hat{G}_1 &= (1 - \hat{w}_0)^{c_0} \\ \hat{G}_i &= \hat{G}_{i-1}(1 - \hat{w}_{i-2})^{\frac{1}{2}}(1 - \hat{w}_{i-1})^{c_{i-1} - \frac{1}{2}} \quad (i = 2, \ldots, k) \end{aligned} \tag{2.38}$$

Since the general form for the survivor function is

$$\hat{G}_i = (1 - \hat{w}_0)^{b_0}(1 - \hat{w}_1)^{b_1} \dots (1 - \hat{w}_{i-1})^{b_{i-1}}$$

the standard error can be obtained using the approximate formula for the variance of a function of random variables used before, which gives

$$\hat{\mathrm{se}}(\hat{G}_i) = \hat{G}_i \left\{ \frac{b_0^2 \hat{w}_0}{(1 - \hat{w}_0)S_0} + \dots + \frac{b_{i-1}^2 \hat{w}_{i-1}}{(1 - \hat{w}_{i-1})S_{i-1}} \right\}^{\frac{1}{2}} \quad (i = 1, 2, \dots, k) \tag{2.39}$$

Once the survivor function has been calculated, the other wastage functions can be obtained from their relationship with $G_i$ using (2.32) and (2.23).

*Example 2.10* Table 2.10 contains estimates of the survivor function and of the other wastage probability functions from the transition data on computer programmers given in Table 2.9. ◀

***Table 2.10 Estimated wastage functions for computer programmers (1975) based on the transition data of Table 2.9***

| $x_i$ | $\hat{G}_i$ | $\hat{\mathrm{se}}(\hat{G}_i)$ | $\hat{f}_i$ | $\hat{q}_i$ | $\hat{m}_i$ | $\hat{D}_i$ |
|---|---|---|---|---|---|---|
| 0 | 1.000 | — | 0.127 | 0.127 | 0.136 | 14.6 |
| 1 | 0.873 | 0.021 | 0.113 | 0.130 | 0.139 | 15.7 |
| 2 | 0.759 | 0.029 | 0.090 | 0.119 | 0.136 | 17.0 |
| 4 | 0.579 | 0.036 | 0.030 | 0.052 | 0.059 | 20.0 |
| 8 | 0.458 | 0.036 | 0.014 | 0.030 | 0.034 | 20.7 |
| 16 | 0.348 | 0.039 | 0.002 | 0.005 | 0.005 | 18.1 |
| 35 | 0.315 | 0.045 | — | — | — | 0 |

It is shown in Exercise 2.3 how the wastage functions can be estimated from stock data only. The method, which requires that stocks at the start and end of the year are categorized into single-year length of service classes, involves estimating the survival transition rates $(1 - w_i)$ and then using (2.38).

## 2.4 COMPARISONS

Having estimated the wastage pattern for several groups we may wish to test whether they differ significantly from one another. This may arise when comparing two groups with a view to amalgamating them or when testing whether the pattern is changing with time. A related problem arises when we wish to compare an estimated pattern with some theoretical model. This is the classical goodness-of-fit problem and it is discussed briefly in Chapter 3. In this section we shall describe some simple ways of testing whether two or more estimated wastage functions differ significantly. First we suggest some informal methods based on graphical representations of the data and then move on to formal statistical tests. In the case of the latter we shall consider separately each kind of data—cohort, census (central), and census (transition)—but the graphical methods are the same in all cases.

### Graphical methods

The most convenient functions to compare graphically are the estimates of the propensity to leave, $m(x)$, and the survivor function, $G(x)$. In the case of census data the estimators $m_i$ are statistically independent and this makes it easier to make comparisons by eye than with the $G_i$'s, which are positively correlated with one another. We find it useful to plot not only the estimates themselves but some indication of their errors. This may be done (as in Figure 2.1) by indicating an interval equal to some suitable multiple of the standard error. If the numbers are large enough for the sampling distribution to be approximately normal, limits set at one standard error will include the true value with probability about 70 per cent and those at two standard errors with probability 95 per cent. With small numbers or when near the ends of the range of $G(x)$ a symmetrical interval will be inaccurate and more precise methods should then be used. Details of such methods and further information on graphical analysis are contained in the documentation on the WASP package described in the Complements of this chapter (Section 2.5).

*Example 2.11* Figure 2.1 shows estimates of the wastage rates $\{\hat{m}_i\}$ and the survivor function $\{\hat{G}_i\}$ for the administration and technical groups of Company X. Census data for the administration group have already been analysed in Examples 2.6, 2.7, and 2.8 and that for the technical group will be found in Exercise 2.2.

The survivor functions appear to be well separated in that few of the intervals set at one standard error overlap, but the high correlation between successive $\hat{G}_i$'s must make us cautious. The position is much clearer for the $\hat{m}_i$'s. This suggests that the two functions are much the same except in the region of one to three years' service where the administration group appears to have a lower propensity to leave. ◀

The interpretation of plots like this is necessarily subjective but the differences are often so clear as to make further analysis superfluous. In any case, diagrams are a very useful way of presenting the results to the layman so the effect of constructing them as a first stage in the analysis is rarely wasted.

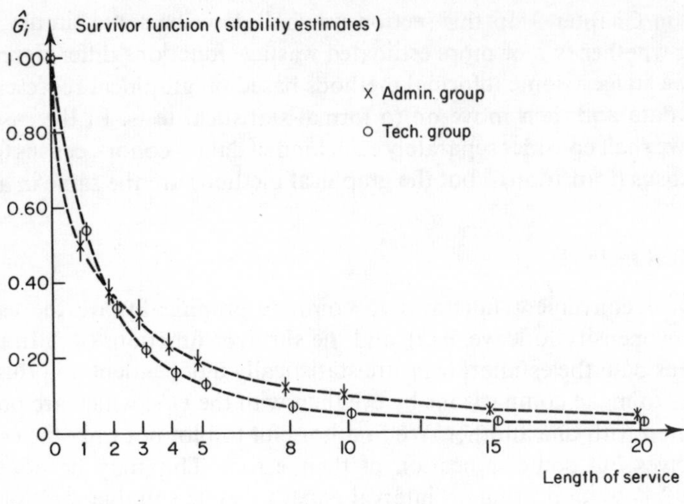


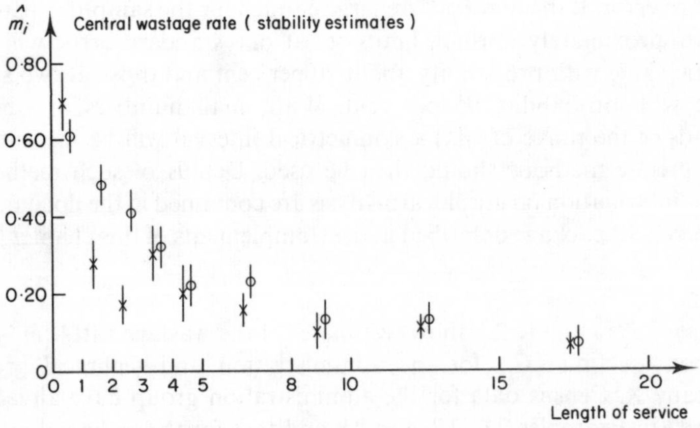


Figure 2.1 Graphical comparison for the administration and technical groups of Company X

Similar comparisons can be made with cohort data but in that case, as will be clear from (2.16), the successive $\hat{m}_i$'s are correlated.

## Formal methods

When comparing the leaving experience of two cohorts we can obtain a test in terms of the frequency distributions. We arrange the data in a two-way table as in Table 2.11.

**Table 2.11**

| *CLS* | $x_0-$ | $x_1-$ | ... | $x_k-$ | *Total* |
|---|---|---|---|---|---|
| Cohort 1 | $L_0^{(1)}$ | $L_1^{(2)}$ | ... | $L_k^{(1)}$ | $Z_0^{(1)}$ |
| Cohort 2 | $L_0^{(2)}$ | $L_0^{(1)}$ | ... | $L_k^{(2)}$ | $Z_0^{(2)}$ |
| *Total* | $L_0$ | $L_1$ | ... | $L_k$ | $Z_0$ |

A test of the hypothesis that the two cohorts are samples from the same underlying distribution can be made using the $\chi^2$-test for a $2 \times (k+1)$ contingency table. The length-of-service groupings must be the same for each distribution and if any cell frequencies are very small (say, less than 5) adjacent groups should be amalgamated as necessary. If there are more than two cohorts to be compared there will be additional rows to the table. The calculations are illustrated in the following example.

*Example 2.12* For Company X the 1970 and 1971 cohorts had the CLS distributions shown in Table 2.12. The expected frequencies are given in brackets.

**Table 2.12**

| *CLS* | 0– | 1– | 2– | 3– | 4– | 5– | 6– | *Total* |
|---|---|---|---|---|---|---|---|---|
| 1970 cohort | 282 (301.5) | 155 (147.9) | 96 (81.2) | 47 (49.9) | 33 (29.9) | 19 (18.7) | 68 (70.9) | 700 |
| 1971 cohort | 364 (344.5) | 162 (169.1) | 78 (92.8) | 60 (57.1) | 31 (34.1) | 21 (21.3) | 84 (81.1) | 800 |
| *Total* | 646 | 317 | 174 | 107 | 64 | 40 | 152 | 1500 |
| *Contribution to* $\chi^2$ *from column* | 2.36 | 0.64 | 5.06 | 0.32 | 0.60 | 0.01 | 0.22 | 9.21 |

The expected value in the top left-hand cell is $700 \times 646/1500$ and similarly for the other cells. With 6 degrees of freedom $\chi^2$ would have to be 10.64 to achieve significance at even the 10 per cent level so we cannot conclude that the wastage pattern has changed on this evidence. Looking at the contributions to $\chi^2$ in the bottom line, most of the total arises from two columns, which might be indicative of some local difference which is not large enough to lead to rejection of the hypothesis with a sample of this size. ◀

With census data, formal tests can be based on the numbers of leavers but the contingency table format is no longer appropriate. The methods we recommend depend on whether we have central or transition data. We shall deal with central data first and then note the modifications necessary to adapt the methods to transition data.

The method involves comparing the estimates of $m_i$ for each group. We have from (2.29)

$$\hat{m}_i = L_i/S_i$$

where $L_i$ has a distribution which is approximately Poisson and if $L_i$ is reasonably large this, in turn, will be nearly normal. Suppose that we wish to compare two estimates of $m_i$ for a given value of $i$. Then using superscripts to index the two groups we may use the fact that

$$\hat{m}_i^{(1)} - \hat{m}_i^{(2)} \mathrel{\dot\sim} N[0, m_i(S_i^{(1)} + S_i^{(2)})/S_i^{(1)}S_i^{(2)}]$$

where $m_i$ is the common, unknown, value of $m_i$ which may be estimated by

$$\hat{m}_i = (L_i^{(1)} + L_i^{(2)})/(S_i^{(1)} + S_i^{(2)})$$

The test for a significant difference may thus be based on the standard normal deviate

$$z_i = (\hat{m}_i^{(1)} - \hat{m}_i^{(2)})/[(L_i^{(1)} + L_i^{(2)})/S_i^{(1)}S_i^{(2)}]^{\frac{1}{2}} \tag{2.40}$$

The $z_i$'s are independent so an overall test of equality may be formed by taking

$$\chi^2 = \sum_{i=0}^{k} z_i^2$$

which will be distributed like $\chi^2$ with $(k + 1)$ degrees of freedom. The $L_i$'s must not be too small and we recommend that if any are 9 or less, adjacent classes should be combined until they are large enough.

*Example 2.13* We shall use the above method on the data used for the graphical comparison of Figure 2.1 in Example 2.11. The values of $z_i$ are as shown in Table 2.13.

***Table 2.13***

| *Length-of-service group* | 0– | 1– | 2– | 3– | 4– | 5– | 8– | 10– | 15– |
|---|---|---|---|---|---|---|---|---|---|
| $z_i$ | 1.9 | −2.42 | −2.72 | −0.23 | −0.25 | −1.40 | −0.51 | −0.34 | −0.27 |

Viewed individually, only two of these reach significance but 8 out of the 9 are negative. The probability that 8 or more independent normal variables out of 9 are negative is 0.02, which strongly suggests that the administration group has a generally lower propensity to leave. The overall value of $\chi^2$ is 17.18 which with 9 degrees of freedom is significant at the 5 per cent level and this tends to confirm our conclusion about the diference between the two groups. ◀

The only difference which arises with transition data is that the $L_i$'s have binomial rather than Poisson distributions. Using the same superscript notation as before

$$\hat{w}_i^{(1)} - \hat{w}_i^{(2)} \mathrel{\dot\sim} N[0, w_i(1 - w_i)(S_i^{(1)} + S_i^{(2)})/S_i^{(1)}S_i^{(2)}]$$

and, therefore, the standard normal deviate is

$$z_i = (w_i^{(1)} - w_i^{(2)})/[\hat{w}(1 - \hat{w})(S_i^{(1)} + S_i^{(2)})/S_i^{(1)}S_i^{(2)}]^{\frac{1}{2}} \tag{2.41}$$

where $\hat{w} = (L_i^{(1)} + L_i^{(2)})/(S_i^{(1)} + S_i^{(2)})$. The sum of squares of the $z_i$'s will thus provide an overall test statistic distributed like $\chi^2$ with $(k + 1)$ degrees of freedom.

## 2.5 COMPLEMENTS

As the title of this chapter suggests, most of its contents lie on the borders between statistics, demography, and actuarial science. Unlike the traditional treatments of demographers and actuaries we have attempted to give standard errors of quantities we have estimated. One of the first major attempts to introduce a stochastic treatment of such problems was the book by Chiang (1968) to which our exposition is heavily indebted. Chiang was concerned with biostatistics, and especially with competing risks but his methods are readily transferable to wastage analysis. This was done in Forbes (1971b) (reproduced in Bartholomew (1977d)) and much of the material has been included in the present chapter.

There is also much in common between wastage analysis and industrial life-testing. Length of service in our case corresponds to the life of a piece of equipment. However, there are important practical differences which make the extensive literature on life-testing of limited relevance in wastage applications. In the industrial context it may be possible to control the experimental conditions. For example, it is sometimes possible to accelerate the wearing-out process and so obtain results more rapidly. Even if the data are censored or truncated they are likely to be recorded in more detail than is commonly the case in our applications. The methods we have described for both cohort and census data presuppose grouped data. If individual CLS's are available it is possible, in principle, to obtain more precise estimates. With large samples this will be hardly worth while but with small samples, for which our methods would require a very broad grouping, a more refined analysis may be desirable. One such distribution-free estimator known as the 'product-limit' estimator is well known in life-testing and could be used for estimating the survivor function in appropriate circumstances. Much of life-testing theory is based on the assumption that the lifetime distribution is exponential or Weibull in form and, as we shall see in Chapter 3, this is unlikely to be the case in wastage analysis.

The use of the survivor function as the key element in wastage analysis goes back at least as far as Rice and coworkers (1950). However, crude wastage rates continue to be widely used in spite of their shortcomings and it is not always explained whether the central or transition rate is intended.

The methods described in this chapter have been made the subject of an interactive computer package called WASP (WAStage Package) prepared by one of the authors (AFF) and available through the Institute of Manpower Studies. In several respects this goes further than the present chapter, especially in the direction of the comparison of wastage patterns between different groups of individuals and of testing goodness of fit. It includes, for example, methods for comparing survivor functions using Kolmogorov–Smirnov statistics.

One of the main reasons for wishing to make such comparisons is to identify factors which influence propensity to leave and to estimate their effects. To do full justice to this problem requires something in the nature of a regression analysis so that the effects of variables such as sex, skill level, salary, etc., can be investigated simultaneously. There have, in fact, been a number of unpublished exercises involving the use of regression analysis. Thus, suppose that we observe a cohort for a fixed length of time and score 1 for any individual who leaves in that period and 0 otherwise. These scores can then be treated as the dependent variable in a multiple regression analysis. The usual distributional assumptions do not hold, of course, and the results have to be interpreted with care as a consequence. An improvement, if sufficient data are available, is to work with homogeneous groups instead of individuals and to use the proportion, $p$, in the group who survive through the interval as the dependent variable. A linear regression is not satisfactory because it may predict values for the dependent variable outside the range (0, 1). A better approach is based on the 'logit' transformation in which the dependent variable is $\log\{p/(1-p)\}$ rather than $p$. For further details the reader should consult Cox (1970), Coleman (1973), Lindsey (1973), or Payne (1977).

Another approach to the same problem was proposed by Cox (1972). His model supposes that the independent variables exert their effect through a multiplicative effect on the function $m(x)$. In particular, his model takes the form

$$m(x) = m_0(x) \exp\left[\sum_{i=1}^{k} \beta_i z_i\right]$$

where the $z_i$'s are the independent variables and the $\beta_i$'s are regression coefficients. The model does not appear to have been applied to wastage analysis and reasons were given by Bartholomew (1972) for believing that the form of the model is not consistent with the empirical evidence about how independent variables affect propensity to leave. Nevertheless, the method merits investigation as a tool for wastage analysis.

## 2.6 EXERCISES AND SOLUTIONS

### Exercise 2.1

The CLS distribution of leavers for the 1970 cohort of Company X is given in Example 2.12. Estimate values for the wastage functions $f_i$, $G_i$, $q_i$, $m_i$, and $D_i$ together with their standard errors.

Plot the survivor functions and the wastage rates $\hat{m}_i$ for both the 1970 and 1971 cohorts with appropriate error intervals and comment on these with respect to the formal comparison of these data in Example 2.12.

### Exercise 2.2

Census-central data for the technical group in Company X are given in Table 2.14.

***Table 2.14***

| *LS class* | 0– | 1– | 2– | 3– | 4– | 5– | 8– | 10– | 15–20 |
|---|---|---|---|---|---|---|---|---|---|
| Leavers | 230 | 107 | 61 | 34 | 18 | 35 | 7 | 11 | 4 |
| Av. stocks | 377.5 | 220.5 | 147.5 | 105.5 | 81.0 | 149.5 | 52.5 | 85.0 | 53 |

Using the stability curve method estimate values for the wastage functions. (The graphical comparison of the administration and technical groups was considered in Example 2.11 and the associated Figure 2.1, and the formal comparison in Example 2.13.)

### Exercise 2.3

Estimate the wastage functions from the stock data in Table 2.15, given that the gross number of entrants during this year was 150.

***Table 2.15***

| *LS class* | 0– | 1– | 2– | 3– | 4– | 5– | 6– |
|---|---|---|---|---|---|---|---|
| Stocks at start of year | 100 | 80 | 65 | 80 | 50 | 40 | 260 |
| Stocks at end of year | 100 | 75 | 65 | 55 | 70 | 45 | 270 |

### Exercise 2.4

Given the data of Table 2.16, showing wastage transition rates and initial length of service structure for Company C of Example 2.2, show that the actual reduction shown in Table 2.2 could have been anticipated.

***Table 2.16***

| *LS class*, $x_i$– | 0– | 1– | 2– | 3– | 4– | 5– | 6+ |
|---|---|---|---|---|---|---|---|
| Initial stocks | 75 | 52 | 45 | 40 | 37 | 37 | 716 |
| Wastage transition rate, $w_i$ | 0.30 | 0.15 | 0.10 | 0.07 | 0.05 | 0.04 | 0.05 |

### Exercise 2.5

Use (2.28) to convert the transition rates into central rates for the computer programmers' data in Table 2.9. Compare these converted central rates with the $\hat{m}_i$ estimated from the census-transition methods and shown in Table 2.10, in the light of the discussion in the text following Table 2.9 concerning the possibility of using these converted rates with census-central methods.

### Solution to Exercise 2.1

The estimated wastage functions for the 1970 cohort are given in Table 2.17, and the graphs of the survivor functions and central wastage rates for this and the 1971 cohort are shown in Figure 2.2.

The survivor functions for the two cohorts appear very similar although as we have noted it is often difficult to compare the wastage patterns from these values. The wastage rates $\hat{m}_i$ provide a clearer picture and show similar patterns except possibly in the (0–1) and (2–3) classes, although the discrepancies in the rates for these classes are probably not large in relation to the errors. These agree with the formal comparison in Example 2.12 and in particular with the contributions to $\chi^2$.

### Solution to Exercise 2.2

The estimated wastage functions for the Company X technical group are shown in Table 2.18. These are compared graphically with the administration group in Examples 2.11 and Figure 2.1, and formally in Example 2.13.

*Table 2.17 Estimated wastage functions for Company X, 1970 cohort*

| $x_i$ | $L_i$ | $Z_i$ | $\hat{f}_i$ $L_i/c_iZ_0$ | $\hat{se}(\hat{f}_i)$ (2.3) | $\hat{G}_i$ $Z_i/Z_0$ | $\hat{se}(\hat{G}_i)$ (2.5) | $\hat{q}_i$ (2.7) | $\hat{se}(\hat{q}_i)$ (2.8) | $\hat{m}_i$ (2.16) | $\hat{se}(\hat{m}_i)$ (2.20) | $\hat{D}_i$ (2.23) |
|---|---|---|---|---|---|---|---|---|---|---|---|
| 0 | 282 | 700 | 0.403 | 0.019 | 1.000 | 0 | 0.403 | 0.019 | 0.516 | 0.031 | 3.4 |
| 1 | 155 | 418 | 0.221 | 0.016 | 0.597 | 0.019 | 0.371 | 0.024 | 0.463 | 0.037 | 4.4 |
| 2 | 96 | 263 | 0.137 | 0.013 | 0.376 | 0.018 | 0.365 | 0.030 | 0.454 | 0.046 | 5.8 |
| 3 | 47 | 167 | 0.067 | 0.009 | 0.239 | 0.016 | 0.281 | 0.035 | 0.331 | 0.048 | 7.8 |
| 4 | 33 | 120 | 0.047 | 0.008 | 0.171 | 0.014 | 0.275 | 0.041 | 0.322 | 0.056 | 9.7 |
| 5 | 19 | 87 | 0.027 | 0.006 | 0.124 | 0.012 | 0.218 | 0.044 | 0.246 | 0.057 | 12.2 |
| 6 | 68 | 68 | — | — | 0.097 | 0.011 | — | — | — | — | 14.5 |

$\hat{D}_i$ has been calculated with $x_{k+1} = 35$.

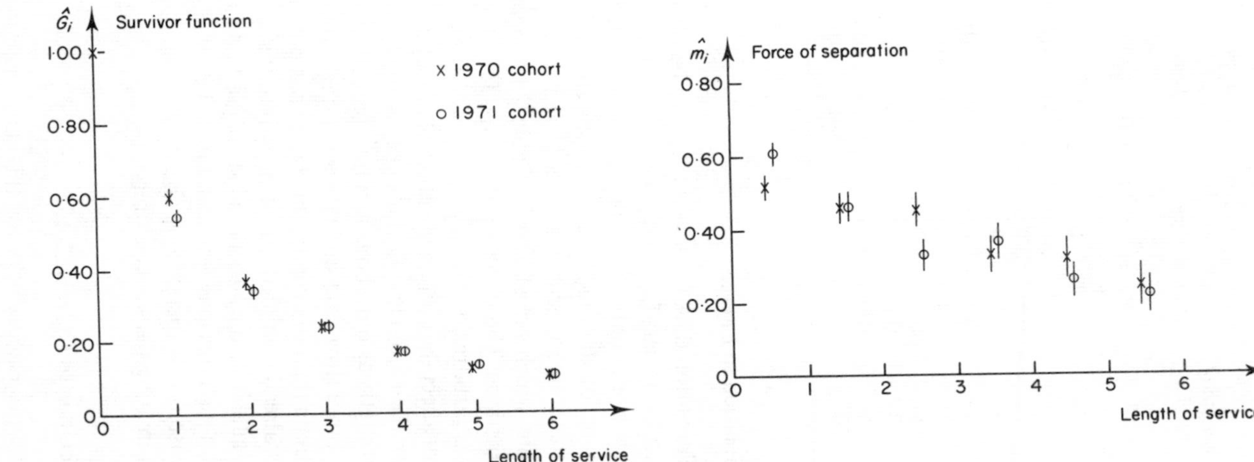


Figure 2.2 Graphical comparison for the 1970 and 1971 cohorts of Company X

***Table 2.18 Company X technical group 1977: estimated wastage functions using the stability curve method***

| Length of service | Central wastage rate and standard error | | Survivor function and standard error | | Density function | Conditional probability of leaving | Further expected duration |
|---|---|---|---|---|---|---|---|
| $x_i^-$ | $\hat{m}_i$ | $\hat{se}(\hat{m}_i)$ | $\hat{G}_i$ | $\hat{se}(\hat{G}_i)$ | $\hat{f}_i$ | $\hat{q}_i$ | $\hat{D}_i$ |
| 0– | 0.609 | 0.040 | 1.000 | 0.000 | 0.456 | 0.456 | 2.5 |
| 1– | 0.485 | 0.047 | 0.544 | 0.022 | 0.209 | 0.384 | 3.1 |
| 2– | 0.414 | 0.053 | 0.335 | 0.021 | 0.113 | 0.339 | 3.8 |
| 3– | 0.322 | 0.055 | 0.221 | 0.018 | 0.061 | 0.276 | 4.5 |
| 4– | 0.222 | 0.052 | 0.160 | 0.016 | 0.032 | 0.199 | 5.1 |
| 5– | 0.234 | 0.040 | 0.128 | 0.014 | 0.022 | 0.168 | 5.2 |
| 8– | 0.133 | 0.050 | 0.064 | 0.010 | 0.007 | 0.117 | 6.2 |
| 10– | 0.129 | 0.039 | 0.049 | 0.009 | 0.005 | 0.095 | 5.9 |
| 15– | 0.075 | 0.038 | 0.026 | 0.007 | 0.002 | 0.063 | 4.2 |
| 20– | — | — | 0.017 | 0.006 | — | — | 0 |

## Solution to Exercise 2.3

Transition data for leavers ($L_i$) can be obtained by subtracting the end year stocks in class ($x_{i+1}$, $x_{i+2}$) from the start year stocks in ($x_i$, $x_{i+1}$). Note that this works only if the classes are in single years. We can avoid this requirement for the final class by assuming that the same transition rate applies after $x = 5$ in this example. Using these methods we obtain the transition data which are shown in Table 2.19 together with the estimated wastage functions. These show no unusual features. The maximum length of service was (arbitrarily) taken as $x_{k+1} = 35$: this value is required for $D_i$ and does not affect the estimated values of any of the other functions. The wastage rate for the (4–5) class is the same as that estimated for the (5 +) class, which strengthens our above assumption concerning the final class.

Since flow data on leavers is often either nonexistent or more difficult to obtain than stock data, this method is potentially extremely useful. However, as already pointed out it does require single-year classes. If these are not available it may be possible to graduate the data into single-year classes, although this may well result in values of $w_i$ greater than 1. For this and other reasons it is therefore probably sensible in practice to smooth the $w_i$ before estimating the wastage functions.

## Solution to Exercise 2.4

The projected stocks for the actual rundown, as shown in Table 2.20, are obtained by applying the wastage rates to the initial stocks and assuming

**Table 2.19** ***Estimated wastage functions from stock only data given in Exercise 2.3***

| $x_i$ | $L_i$ | $S_i$ | $w_i$ | $\hat{se}(w_i)$ | $\hat{G}_i$ | $\hat{se}(\hat{G}_i)$ | $\hat{f}_i$ | $\hat{q}_i$ | $\hat{m}_i$ | $\hat{D}_i$ |
|---|---|---|---|---|---|---|---|---|---|---|
| 0 | 25 | 100 | 0.250 | 0.043 | 1.000 | 0.000 | 0.250 | 0.250 | 0.288 | 8.7 |
| 1 | 15 | 80 | 0.188 | 0.044 | 0.750 | 0.043 | 0.165 | 0.219 | 0.248 | 10.5 |
| 2 | 10 | 65 | 0.154 | 0.045 | 0.585 | 0.053 | 0.100 | 0.171 | 0.187 | 12.3 |
| 3 | 10 | 80 | 0.125 | 0.037 | 0.485 | 0.051 | 0.068 | 0.140 | 0.150 | 13.7 |
| 4 | 5 | 50 | 0.100 | 0.042 | 0.418 | 0.049 | 0.047 | 0.113 | 0.119 | 14.8 |
| 5–35 | 30 | 300 | 0.100 | 0.017 | 0.371 | 0.046 | — | — | — | 15.6 |

*Table 2.20 The manpower rundown for Company C*

| *Time* | | 0 | 1 | 2 | 3 | 4 | 5 |
|---|---|---|---|---|---|---|---|
| | 0– | 75 | — | — | — | — | — |
| Stocks | 1– | 52 | 52 | — | — | — | — |
| in | 2– | 45 | 45 | 45 | — | — | — |
| each | 3– | 40 | 40 | 40 | 40 | — | — |
| LS | 4– | 37 | 37 | 37 | 37 | 37 | — |
| class | 5– | 35 | 35 | 35 | 35 | 35 | 35 |
| | 6 + | 716 | 176 | 716 | 716 | 716 | 716 |
| *Total* | | 1000 | 925 | 873 | 828 | 788 | 751 |

zero recruits. Note that the numbers in each class repeat themselves, showing that the initial stocks are in a steady state.

## Solution to Exercise 2.5

The converted and estimated central rates are shown in Table 2.21. It follows from the discussion in the text following Table 2.9 that $\hat{m}_i^{(1)}$ should apply to length of service classes approximately one half-year later than $\hat{m}_i^{(2)}$. In other words, in Table 2.21 $m_i^{(1)}$ should correspond with $m_i^{(2)}$ when the former has been shifted slightly to the right. This is certainly consistent with these values, providing the peak propensity to leave lies between $\hat{m}_1^{(2)}$ and $\hat{m}_2^{(2)}$, and bearing in mind that the census-transition method assumes that the net inflow rate can be determined from $\hat{w}_0$ so that we would expected $\hat{m}_0^{(1)} = \hat{m}_0^{(2)}$.

*Table 2.21 Central rates from computer programmers' data*

| *LS* | 0– | 1– | 2– | 4– | 8– | 16–35 |
|---|---|---|---|---|---|---|
| Converted $\hat{m}_i^{(1)}$ using (2.28) | 0.136 | 0.142 | 0.134 | 0.048 | 0.033 | 0.005 |
| Estimated $\hat{m}_i^{(2)}$ from census transition | 0.136 | 0.139 | 0.136 | 0.059 | 0.034 | 0.005 |

Note: $\hat{m}_i^{(1)} = \ln(1 - \hat{w}_i)$ (the $\hat{w}_i$ used had more than the 3 decimal places shown in Table 2.9)
$\hat{m}_i^{(2)} = (\ln \hat{G}_i - \ln \hat{G}_{i+1})/c_i$ ($m_i^{(2)}$ was taken directly from Table 2.10).

CHAPTER 3

# Analysis of Wastage II: Modelling, Prediction, and Measurement

## 3.1 THE NEED FOR THEORY

The approach to the analysis of wastage adopted in Chapter 2 was empirical and, in particular, it required no assumption about the mathematical form of the survivor function. For many practical purposes this is entirely adequate as the examples and exercises in the last chapter show. However, there are many instances where we wish to take the analysis further than this purely empirical treatment allows, and for this a firmer theoretical base is necessary. For example, if we have data on the leaving experience of a cohort over a period of, say, three years we know how to estimate the survivor function over that interval of time. But if we wish to project the leaving pattern forward in time beyond three years we need some rational basis for any extrapolation that this involves. It is the purpose of this chapter to provide the theoretical foundation for such exercises and to show how it can be put to practical use. For the most part it will be convenient to conduct the discussion in terms of the frequency function, $f(x)$, which we shall refer to, loosely, as the CLS distribution.

Perhaps the most striking thing about estimated CLS distributions is that they almost always have the same basic shape. An example, based on a cohort of chemical engineers, is shown in Figure 3.1. The distribution is censored with 36 individuals having completed lengths of service in excess of 20 years. It has a high degree of positive skewness reflecting the fact that, while most people have a relatively short stay, a few remain for a very long time. Such distributions usually have a mode near the origin. When this is not apparent from the histogram it may merely mean that the grouping is too coarse to reveal it; an example of this kind is provided by the distributions in Figure 3.2. The time scale may vary a great deal from one set of data to another. For example, unskilled factory workers often stay a matter of weeks rather than years as in the case of the chemical engineers. What does appear to persist across all kinds of occupation and also across national boundaries is the *shape* of the distribution and it is this constant feature on which we shall base our theorizing. We shall start by asking whether there is any simple curve capable of graduating such distributions and, if so, what light it can throw on the underlying social process.

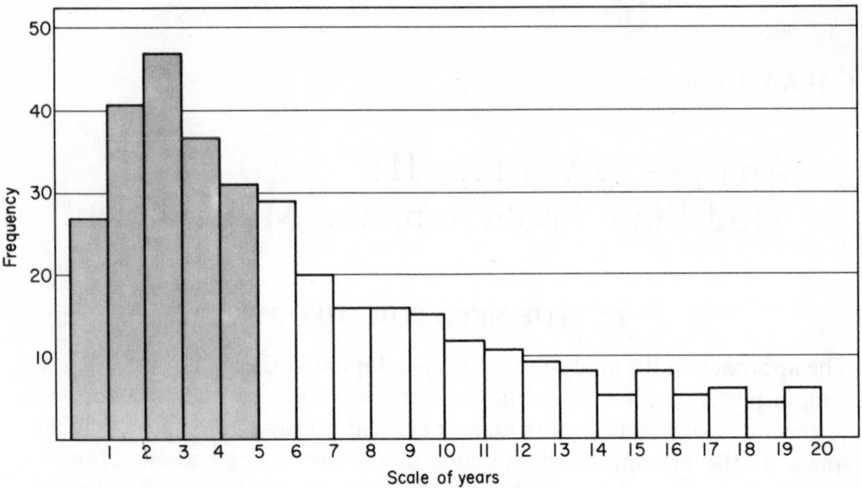


Figure 3.1 A completed length of service (CLS) distribution for a cohort of graduate engineers (censored at 20 years)

Before embarking on this quest we shall motivate the discussion further by listing some of the uses to which such curves could be put.

(a) *Graduation and interpolation*. The methods of estimating survivor (and related) functions in the last chapter provide estimates only at particular points. It would often be useful to interpolate between these points in order to give estimates over the whole range of the function. This can be done, as in Chapter 2, by drawing a freehand curve through the points or by some technique of smoothing. An alternative is to fit a curve and use that as a basis for interpolation. An added advantage of having such a curve is that it facilitates the transition between the three basic functions $f(x)$, $G(x)$, and $m(x)$.

(b) *Prediction*. We have already noted the value of a fitted curve for extrapolation. Suppose the data illustrated in Figure 3.1 had been censored at 5 years so that we only had the shaded part of histogram. The problem of prediction is that of deducing the shape of the unshaded part of the distribution from the shaded part. One way to do this is to fit a curve to the known part and then to extrapolate it. A knowledge of what curves are likely to be successful in graduating CLS distributions will enable us to do this with some degree of confidence.

(c) *Measurement*. Later in the chapter we shall be arguing that indices of labour wastage should be obtained by summarizing the CLS distribution in some relevant fashion. Such measures can be conveniently expressed in terms of the parameters of a fitted distribution.

(d) *Explanation.* Having observed that the shape of CLS distributions is remarkably constant it is natural to ask whether there is any explanation that can be offered for this fact. In statistical language this means searching for a probability model of the leaving process which predicts the shape found in practice. Such a discovery would not merely satisfy our intellectual curiosity but provides a more secure base for the extrapolation involved in any prediction exercise. It might also give sufficient insight into the process to indicate ways in which some degree of control might be exercised over wastage.

(e) *Modelling manpower systems.* Later in the book we shall turn to modelling manpower systems as a whole where wastage is only one component. For this purpose it will be necessary to make assumptions about levels of wastage and it is sometimes convenient to do this by specifying the CLS distribution.

## 3.2 MODELS FOR THE WASTAGE PROCESS

### The exponential (or constant risk) model

The earliest attempts to graduate length-of-service distributions were by Rice and coworkers (1950) and Silcock (1954). Both started from the fact that the distributions known to them had J-shaped survivor functions. To Rice and coworkers this suggested a theoretical survivor function of the form

$$G(x) \propto x^{-a} \qquad (a > 0) \tag{3.1}$$

for $x$ not too close to zero. This suggestion did not give rise to any theoretical development although it is very close to the work of Silcock which followed. He began with the idea of graduating the survivor function by an exponential curve of the form

$$G(x) = e^{-\lambda x} \tag{3.2}$$

Using the relationships given in Section 2.2, it follows immediately that

$$f(x) = \lambda e^{-\lambda x} \quad \text{and} \quad m(x) = \lambda, \tag{3.3}$$

The constancy of $m(x)$ implies that, for this model, propensity to leave does not depend on length of service. This would mean, for example, that knowledge of an individual's current length of service would be of no value in predicting when he would leave. There would therefore be no point in concentrating efforts to reduce wastage at any particular point in an individual's career.

In practice, however, it turns out that the exponential distribution very rarely provides a satisfactory fit (see Figure 3.2 and the associated discussion for an example). Its main interest lies in the scope which it offers for generalization in the direction of greater realism.

## Mixed exponential models

The idea that a mixture of exponential distributions might be more successful goes back to Silcock (1954) but the simplest version of such a model was proposed in Bartholomew (1959). The idea stems from observing the way in which empirical CLS distributions depart from the simple exponential form. Two typical examples are given in Figure 3.2, where the dotted curves are the fitted exponential distributions.

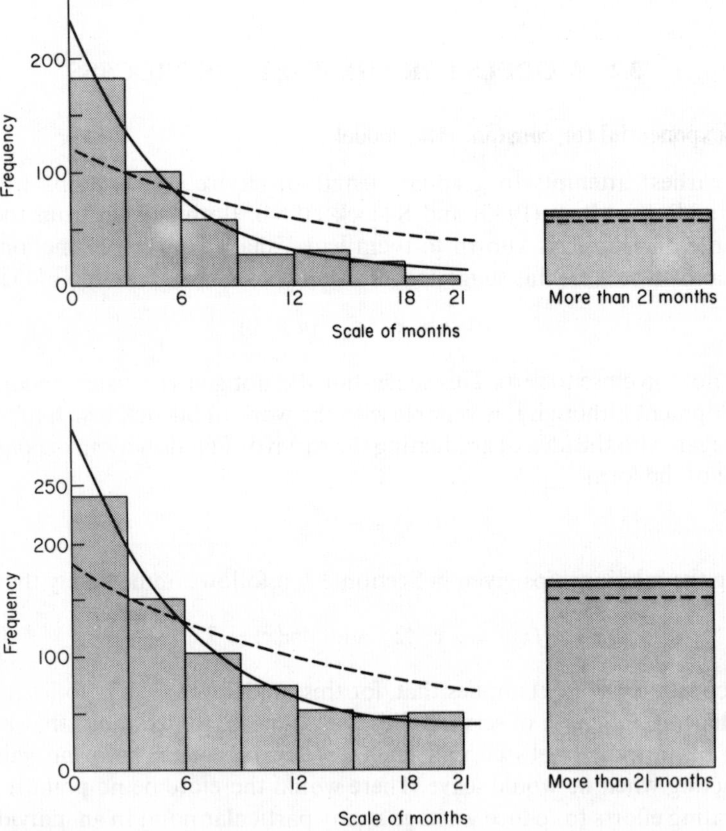


Figure 3.2 Completed length of service distributions for two cohorts censored at 21 weeks with fitted exponential (broken) and Type XI (solid) curves

In both cases the empirical distribution shows an excess of people with very short lengths of service. The detached portion of the diagram representing those with more than 21 months' service also has an excess over expectation and there is a compensating deficiency in the middle of the range. This could mean that we really have a mixture of two exponential distributions, one with a high value of $\lambda$ which dominates at the lower end and one with a low $\lambda$ accounting for the upper tail. This argument may be formalized as follows. Let the proportions of the two types of individual be $\theta$ and $(1-\theta)$ with associated values of $\lambda$ equal to $\lambda_1$ and $\lambda_2$; then a simple probability argument shows that

$$f(x) = \theta\lambda_1 \, e^{-\lambda_1 x} + (1-\theta)\lambda_2 \, e^{-\lambda_2 x} \qquad (x \geqslant 0) \tag{3.4}$$

A model of this kind has some plausibility on general grounds since it would be surprising if there was no variation of propensity to leave among individuals The two-term exponential does, in fact, provide a very much better fit, as reference to Bartholomew (1973a, Table 6.1) shows. Silcock (1954) made a different and, perhaps, more reasonable assumption about the way in which $\lambda$ might vary. He supposed that its variation could be described by a continuous distribution with frequency function $h(\lambda)$, say. Thus if $x$ is exponential for a given $\lambda$ then, clearly,

$$f(x) = \int_0^\infty \lambda \, e^{-\lambda x} h(\lambda) \, d\lambda, \tag{3.5}$$

This distribution cannot be fitted to data without making some assumption about the form of $h(\lambda)$ but certain general properties can be shown to hold for any $h(\lambda)$. In particular it is shown in Bartholomew (1973a, p. 186) that $f(x)$ has the required excess of frequency over the exponential near the origin and in the tails. Silcock (1954) assumed $\lambda$ to have a gamma distribution with density function

$$h(\lambda) = c^q \lambda^{q-1} \exp(-c\lambda)/\Gamma(q) \qquad (q > 0, c > 0, \lambda \geqslant 0) \tag{3.6}$$

This form is capable of representing a wide variety of shapes ranging from extreme positive skewness when $q$ is near zero, to near normality when $q$ is large. The parameter $c$ governs the scale and its value does not affect the shape of the distribution. On substituting (3.6) into (3.5) and integrating we find

$$f(x) = \frac{q}{c}\left(1 + \frac{x}{c}\right)^{-(q+1)}, \qquad G(x) = \left(1 + \frac{x}{c}\right)^{-q}, \qquad m(x) = \frac{q}{c}\left(1 + \frac{x}{c}\right)^{-1} \tag{3.7}$$

The survivor function is very close to the hyperbolic curve suggested by Rice and coworkers (1950) and we note that $m(x)$ is monotonic decreasing. At first sight this last observation might seem at variance with our initial assumption of a constant $\lambda$ for each individual. However, $m(x)$ has to be thought of in an aggregate sense. Early in the life of a cohort the proportion of people with a high value of $\lambda$ will be relatively large whereas, later on, the residue will consist mainly of those with low $\lambda$'s. This change in the mix will create an apparent decline in the propensity to leave. The distribution of (3.7) is sometimes known as a Pearson Type XI or a Pareto Type II curve and it has been successfully fitted to many CLS distributions. Examples are given in Silcock's original paper and in Bartholomew (1973a, Table 6.1). The solid curves of Figure 3.2 are Type XI curves and it is clear that they provide a much better fit to the data shown there than the exponential.

If a two-term exponential distribution had been fitted and shown on the same diagram it would have been hardly distinguishable from the Type XI curve. This suggests that the important thing about mixed exponential models is not the form of the distribution of $\lambda$ but the fact that variation of some kind is introduced.

If a model such as the mixed exponential has any basis in reality there are obvious implications for the control of the wastage process. Since, according to the model, propensity to leave is a 'constant' for the individual the obvious way to influence wastage would be at the selection stage. Attempts would have to be made to distinguish potential long- and short-stayers at the point of entry and to select on that basis. Many selectors do, in fact, attempt to give weight to factors like time spent in previous jobs, sex, age, attitudes, etc., although the statistical basis for the belief that some of these factors affect $m(x)$ is not well documented. Unfortunately, as we shall discover in the next section, there are other models which are equally good at accounting for the occurrence of mixed exponential distributions. This means that we cannot infer from the good fits obtained that the model underlying them corresponds to the realities of the situation.

## Network models

We use this term to describe a class of models which postulate that an individual's stay in an organization can be described in terms of passage through a network of psychological states. Such models have been proposed by Herbst (1963) and Clowes (1972) among others. We shall describe the simplest such model, due to Clowes, by reference to Figure 3.3.

The three rectangles represent the states that an individual may occupy and the arrows indicate the changes of state which can occur. On entry everyone is assumed to be uncommitted to the organization. This stage may be identified with the 'induction' period when individuals are 'finding their feet' and forming a view about their prospects in the job. After the initial state they either become committed (state 2 in Figure 3.3) or they leave

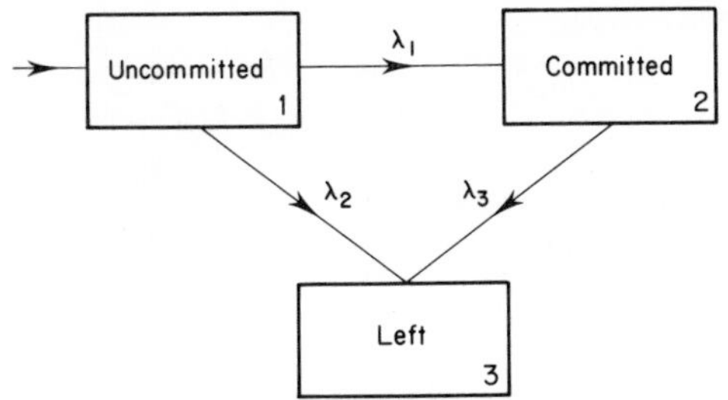


Figure 3.3 Network diagram for Clowes' model

(state 3). Movement to state 2 is, of course, eventually followed by leaving. To complete the specification of the model we must make an assumption about the timing of transitions between states. We assume that the propensities to move between states are denoted by $\lambda_1$, $\lambda_2$, and $\lambda_3$ as shown in Figure 3.3. These mean, for example, that

$$\Pr\{\text{person in state 1 moves to state 2 in } (x, x + \delta x)\} = \lambda_1 \delta x$$

and similarly for the other two paths. As the notation suggests, the $\lambda$'s are essentially the same as the single $\lambda$ of the exponential model which may be thought of as a special case arising when states 2 and 3 are combined.

Clowes' model is, in fact, a continuous-time Markov process and as such it can be analysed by the general methods available for such processes (see, for example, Bartholomew (1973a, Chapter 5)). For our present purpose we shall use a deterministic argument which leads to the same end. Consider a cohort of size $N(0)$ whose members join the system at exactly the same time. Let $N_i(x)$ be the expected number in state $i$ after time $x$; the proportion surviving to $x$ (i.e. the survivor function) is then

$$\{N_1(x) + N_2(x)\}/N(0) = G(x)$$

To find $G(x)$ we set up a system of differential equations for $N_1(x)$, $N_2(x)$, and $N_3(x)$ as follows. Consider first the expected change in $N_1(x)$ between $x$ and $x + \delta x$. The probability of a move to state 2 in that interval is $\lambda_1 \delta x$ and to state 3, $\lambda_2 \delta x$. Hence the expected number of transitions out of state 1 is $N_1(x)(\lambda_1 + \lambda_2)\delta x$ so we have

$$N_1(x) - N_1(x + \delta x) = N_1(x)(\lambda_1 + \lambda_2)\delta x$$

Dividing by $\delta x$ and letting $\delta x \to 0$ gives

$$\frac{\mathrm{d}N_1(x)}{\mathrm{d}x} = -N_1(x)(\lambda_1 + \lambda_2) \tag{3.8a}$$

By a similar argument

$$\frac{dN_2(x)}{dx} = -\lambda_3 N_2(x) + \lambda_1 N_1(x) \tag{3.8b}$$

and

$$\frac{dN_3(x)}{dx} = \lambda_2 N_1(x) + \lambda_3 N_2(x) \tag{3.8c}$$

Notice that

$$\frac{d}{dx}\{N_1(x) + N_2(x) + N_3(x)\} = 0$$

as it must because the total size of the system is fixed.

We now have a set of simultaneous linear differential equations for which methods of solution exist. This particular set can easily be solved from first principles using the initial conditions $N_1(0) = N(0)$, $N_2(0) = N_3(0) = 0$. Equation (3.8a) integrates directly to give

$$N_1(x) = N(0)\,e^{-(\lambda_1+\lambda_2)x}$$

and then from (3.8b)

$$N_2(x) = \frac{N(0)\lambda_1}{(\lambda_1 + \lambda_2 - \lambda_3)}\{e^{-\lambda_3 x)} - e^{-(\lambda_1+\lambda_2)x}\}$$

Hence

$$G(x) = \frac{N_1(x) + N_2(x)}{N(0)} = \frac{\lambda_2 - \lambda_3}{\lambda_1 + \lambda_2 - \lambda_3}e^{-(\lambda_1+\lambda_2)x} + \frac{\lambda_1}{\lambda_1 + \lambda_2 - \lambda_3}e^{-\lambda_3 x} \tag{3.9}$$

Notice that this function has exactly the same form as that of the mixed exponential distribution of (3.4). The coefficients $(\lambda_2 - \lambda_3)/(\lambda_1 + \lambda_2 - \lambda_3)$ and $\lambda_1/(\lambda_1 + \lambda_2 - \lambda_3)$ sum to 1 and provided that they are both positive it will be impossible to distinguish between this model and the mixture model by looking at the empirical survivor function alone. This is an important result and it limits the conclusions which can be drawn from the good fit of mixed exponential distributions in practice. To distinguish between the two models it is necessary, in the case of the mixture, to identify the two (or more) types of individual and in the case of the network, the two states of commitment. In practice this is not an easy task and we do not know of any case where it has been done convincingly.

From the point of view of graduating the survivor function the indistinguishability of the two models is a positive advantage. The knowledge that two quite different models lead to the same mathematical form increases our confidence in using that form for extrapolation. Indeed the more models there are which predict the same curve, the more secure we shall feel.

It can happen that one of the coefficients in (3.9) is negative, and in such a case the distribution cannot be interpreted as a mixture but experience suggests that this circumstance is not likely to occur in wastage studies.

The network approach easily generalizes to more complicated systems. Herbst (1963), for example, considered a five-state system of which Clowes' can be regarded as a special case. As long as the transition rates are constant the expected numbers in the states will always satisfy linear differential equations and the resulting survivor function is always a mixture of exponentials with one fewer component than the number of states. Hence the same problems of interpretation arise as in the two-component case.

If a network model provides an adequate account of the wastage process the policies which are indicated for controlling wastage are quite different to those required in the mixture model. In the network model an individual's propensity to leave depends on which state he is in. There is thus no point in being selective at the point of entry but, rather, the effort must be directed to changing conditions within the system. In Clowes' model this would mean creating conditions designed to increase $\lambda_1$ relative to $\lambda_2$, perhaps by introducing financial incentives to new entrants to stay.

In reality, of course, one would expect heterogeneity to be present in the network model for the same reasons as in the constant-risk model. Thus if a proportion $\theta$ of people in the network system are subject to propensities $\{\lambda_{11}, \lambda_{12}, \lambda_{13}\}$ and the remainder to $\{\lambda_{21}, \lambda_{22}, \lambda_{23}\}$ then the resulting survivor function would be a mixture of two functions like (3.9). This would be a four-component mixed exponential distribution. In general any such mixture of survivor functions will be another mixed exponential distribution but with more terms. The class of survivor functions produced by a network model may thus be described as 'closed' under (finite) mixing. This implies that we can never separate the 'network' and the 'mixing' aspects of such a model by inspection of the survivor function alone.

## The lognormal model

Many empirical CLS distributions have a mode near the origin like that in Figure 3.1 and we have surmised that this may sometimes be the case even when it is not apparent from the histogram. No member of the mixed exponential family is capable of producing this feature as their CLS density functions are all monotonic decreasing. It is possible for a network model to yield a CLS distribution with a mode as can be seen from the solution to Exercise 3.3b, but this is unlikely to be the case in practice. In view of this it is natural to enquire whether there is any other distribution which both has a non-zero mode and successfully graduates observed CLS distributions. The discovery that the lognormal distribution meets these requirements was made by Lane and Andrew (1955) and this has led to the distribution becoming one of the main practical tools of wastage analysis.

To say that CLS is lognormally distributed is the same as saying that the logarithm of CLS has a normal distribution. This is very convenient fact because statistical methods for handling normal variables are very well developed. From the point of view of ease of handling the lognormal distribution is almost unrivalled. The main drawback is on the theoretical side, where it has not been possible to find a model which explains the occurrence of the distribution in an entirely satisfactory way. Two related attempts have been made as follows.

The first is based on a remark of Aitchison (1955) in the discussion of Lane and Andrew's paper. It is, in essence, a translation of a standard derivation of the lognormal distribution into wastage language. Suppose that $x_1$, $x_2$, $x_3, \ldots$ denote the lengths of time that an individual spends in successive jobs. In the model it is supposed that the $x$'s are related according to the equation

$$x_{j+1} = x_j u_{j+1} \qquad (j = 1, 2, \ldots) \tag{3.10}$$

If the $u_i$'s were constants the successive $x$'s would be completely determined by the initial $x$. This seems hardly plausible but if the $u_i$'s are random variables the relationship expressed by (3.10) builds in a degree of correlation between successive lengths of service of the kind that one might expect if one long stay tends to be followed by another or vice versa. If (3.10) is accepted it follows that

$$x_{j+1} = x_1 u_1 u_2 \ldots u_{j+1}$$

or

$$\log x_{j+1} = \log x_1 + \sum_{i=2}^{j+1} \log u_i \tag{3.11}$$

If $j$ is reasonably large we may appeal to the central limit theorem to deduce that, under very general conditions, $\log x_{j+1}$ is approximately normal. This model has two obvious defects. The requirement that $j$ should be 'large' means that the model only explains lognormality of CLS for people who have already had several jobs. Yet lognormality is also found for the case of people like new graduates who have had no previous jobs. The second difficulty is with the practical meaning of (3.10). This may be viewed as an attempt to describe the way in which past experience might be built into future behaviour. Its validity could be tested empirically but appears not to have been. However, (3.10) is not derived from behavioural assumptions as were the network models and hence it provides no indication about how wastage might be influenced. All that can be said is that the length of stay in the first job is critical.

The second model of lognormality is the same mathematically but differs in the meaning given to the variables. Thus $x_1$ becomes the time that an individual expects to stay in the job at the time he joins. This expectation is subsequently modified by a succession of impressions and experiences in the new environment which have the effect of scaling the original expectation up or down according to whether or not they are favourable. These effects are represented by the $u$'s. This version avoids the requirement of having had several jobs but it would be even more difficult to test empirically.

A rather different approach may be used to explain the persistence of the lognormal form by showing that the family is closed under a certain kind of mixing. There is a well-known result in statistical distribution theory which says that if $\ln x$ has a $N[\omega_0, \sigma^2]$ distribution and if $\omega_0$ is itself a random variable with a distribution $N[\omega, \tau^2]$ then $\ln x$ will be distributed like a $N[\omega, \sigma^2 + \tau^2]$ variable. In other words heterogeneity of this kind merely increases the variability of $\ln x$ without changing the form of the distribution. There is some empirical evidence that the variance of $\ln x$ is approximately constant for jobs at a given level of skills. This would suggest that variation between individuals would manifest itself in variation in $\omega$; whether the form of that variation would be normal as the above result requires is more problematical.

A fully satisfactory explanation of the widespread occurrence of lognormal CLS distributions is still awaited. In the meantime the empirical evidence on which it rests is sufficiently strong to justify methods of analysis based on the assumption of lognormality. We shall turn to some of these in the following sections.

## 3.3 FITTING CLS DISTRIBUTIONS

If a parametric form is to be used for any of the purposes listed in Section 3.1 it is first necessary to fit the chosen curve to the empirical distribution. In this section therefore we shall prepare the ground by describing some methods which are widely applicable and sufficiently accurate for most practical purposes. In this connection it must be remembered that the data with which we shall be working are likely to be censored, grouped, or otherwise incomplete. This means that the standard 'best' estimators appropriate for random samples will not be available. Fortunately, high precision is not of the first importance as other, less quantifiable, uncertainties are likely to outweigh those of a statistical character (see Bartholomew and coworkers (1976) for further discussion of this point in a wider context).

### Graphical methods

Perhaps the simplest methods are those based on fitting survivor functions by eye to the values estimated by the methods of Chapter 2. One cannot, of course, easily draw an exponential or lognormal survivor function freehand so, instead, we transform the problem so that the survivor function is a straight line. We demonstrate the method first for the exponential distribution. This is not a practically important case for reasons given earlier but its mathematical simplicity is well suited for explaining the method.

In the exponential case $G(x) = e^{-\lambda x}$ and hence

$$-\ln G(x) = \lambda x \tag{3.12}$$

Thus if we were to plot $-\ln \hat{G}(x)$ against $x$ we would expect to obtain a straight line through the origin with slope $\lambda$. By drawing a line through the points we can obtain an estimate of $\lambda$ from the slope of the line. If we use logarithms to the base 10 instead of base $e$ the slope will be $\lambda \log_{10} e$. By using graph paper with a logarithmic scale we can avoid the need to calculate the logarithms but this is hardly worthwhile especially if a calculator with a logarithmic function is available.

*Example 3.1* In Chapter 2 we estimated a survivor function as follows:

| $x$ | 0 | 1 | 2 | 3 | 4 | 5 | 6 |
|---|---|---|---|---|---|---|---|
| $\hat{G}(x)$ | 1.000 | 0·545 | 0.342 | 0.245 | 0.170 | 0.131 | 0.105 |
| $-\ln\hat{G}(x)$ | 0.000 | 0.607 | 1.073 | 1.407 | 1.772 | 2.033 | 2.254 |

These estimates have been plotted in Figure 3.4. The diagram shows a marked curvature which is typical of CLS distributions. The slope of the line drawn on the figure is 0.424 and this would be our estimate of $\lambda$ if we felt justified in making one with this degree of departure from linearity. ◀

A similar method can be used for the lognormal distribution as follows.

$$G(x) = \Pr\{\text{CLS} \geqslant x\} = \Pr\{\ln \text{CLS} \geqslant \ln x\} = 1 - \Phi\left(\frac{\ln x - \omega}{\sigma}\right)$$

where $\Phi^{-1}(.)$ is the abscissa of the standard normal curve below which Hence,

$$\Phi\left(\frac{\ln x - \omega}{\sigma}\right) = 1 - G(x)$$

and

$$\ln x = \sigma\Phi^{-1}(1 - G(x)) + \omega \tag{3.13}$$

where $\Phi^{-1}(.)$ is the abscissa of the standard normal curve below which the proportion of the distribution specified by the argument lies. If $\ln x$ is plotted against $\Phi^{-1}(1 - \hat{G}(x))$ the points should therefore lie on a straight line if the underlying distribution is lognormal. Special graph paper is available for this purpose with a logarithmic scale on one axis and a normal probability scale on the other. The logarithmic scale is usually to the base 10

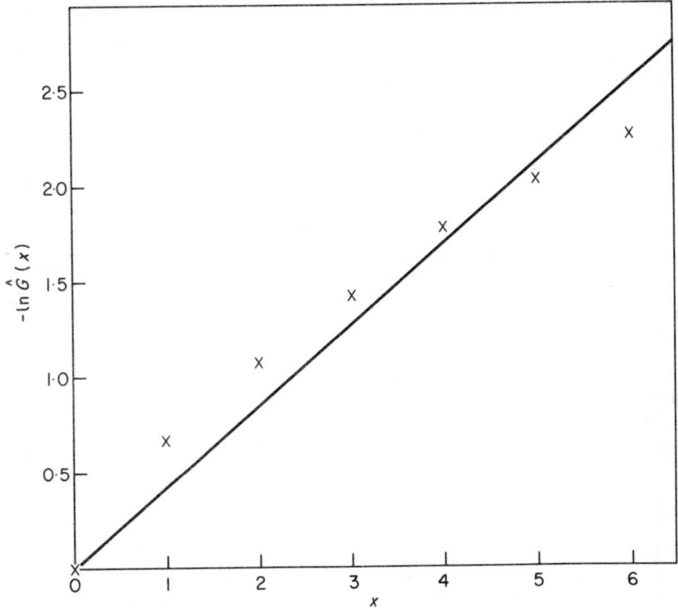


Figure 3.4 Estimated survivor function plotted on logarithmic paper for Example 3.1

but this does not affect the question of linearity. Note that (3.13) involved the complement of $G(x)$ but the graph paper usually has the axis on one side labelled for $G(x)$ and that on the other for $1 - G(x)$.

*Example 3.2* The data used in the previous example have been plotted on lognormal probability paper in Figure 3.5. The points now lie very close to a straight line and we conclude that a lognormal distribution provides a very much better fit to these data than did the exponential. We shall return to the question of estimating $\omega$ and $\sigma$ from such plots later in this section. ◀

The lines in both of the preceding examples were fitted by eye and this introduces a subjective element. If one wished to fit lines using a computer a formal method would be required. A line could be fitted by least squares but the usual requirements of independence and constant variance are lacking here. Marshall (1974) has investigated a number of such methods and most seemed to give similar estimates in spite of the lack of any adequate theoretical justification. For most practical purposes fitting by eye is perfectly satisfactory; if the fit is good there is little room for serious error and if the fit is poor no amount of sophistication in the fitting procedure will compensate for an inadequate model.

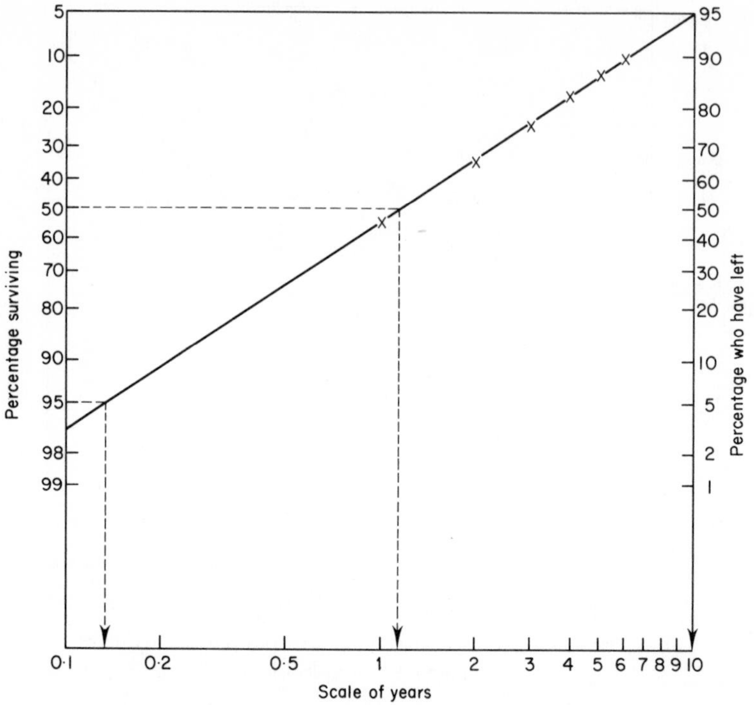


Figure 3.5 Estimated survivor function plotted on lognormal probability paper for Example 3.2

The other main family of CLS distributions which we considered in the previous section were those involving mixtures of exponentials. These cannot be transformed so as to produce a linear relationship with the parameters appearing as coefficients. In the case of Silcock's Type XI distribution, however the fit can be made graphically by a combination of plotting and trial and error. For this distribution

$$G(x) = \left(1 + \frac{x}{q}\right)^{-q}$$

and hence

$$-\log G(x) = q \log\left(1 + \frac{x}{c}\right) \tag{3.14}$$

If $c$ were known $q$ could be estimated from a plot of $\log G(x)$ against $\log(1 + x/c)$ which would be a straight line through the origin. If $c$ is not known, trial values can be taken until the plot obtained is as nearly linear as possible. However, a better approach is to estimate the parameters by the method of percentage points discussed below and then to plot the estimated survivor function and so judge the suitability of the model by the closeness of the observed points to the line.

## The method of percentage points

This involves equating the observed and theoretical values of the survivor function at as many points as there are parameters to be estimated. The resulting equations can then be solved to give unique values for the parameters.

Thus if we wish to fit the exponential distribution by this method we need only one equation because there is only one parameter to be estimated. Let $\hat{G}(x)$ be the estimate of the survivor function at the point $x$. If the distribution is really exponential this will be an estimate of $e^{-\lambda x}$. We choose as our estimate $\hat{\lambda}$ of $\lambda$ that value which makes the two equal; that is

$$\exp(-\hat{\lambda}x) = \hat{G}(x) \quad \text{or} \quad \hat{\lambda} = -\ln \hat{G}(x)/x \tag{3.15}$$

In graphical terms this is equivalent to finding the slope of the line joining the origin to the point $(-\ln \hat{G}(x), x)$. Any value of $x$ could be used and the question naturally arises as to which is the best. The answer depends on how $\hat{G}(x)$ has been estimated but, for a cohort, the maximum precision for $\hat{\lambda}$ is obtained by choosing $x$ so that $\hat{G}(x) \approx 0.20$.

The same approach can be used for the lognormal distribution but in that case there are two parameters, so two equations are required. The relationship between $x$ and $G(x)$ is given by (3.13). If we choose two values $x_1$ and $x_2$ of $x$ the estimating equations will be

$$\left.\begin{aligned} \omega + \sigma\Phi^{-1}\{1 - \hat{G}(x_1)\} &= \ln x_1 \\ \omega + \sigma\Phi^{-1}\{1 - \hat{G}(x_2)\} &= \ln x_2 \end{aligned}\right\} \tag{3.16}$$

Solving these equations we have the estimators

$$\left.\begin{aligned} \hat{\sigma} &= (\ln x_1 - \ln x_2)/[\Phi^{-1}\{1 - \hat{G}(x_1)\} - \Phi^{-1}\{1 - \hat{G}(x_2)\}] \\ \hat{\omega} &= \ln x_1 - \hat{\sigma}\Phi^{-1}\{1 - \hat{G}(x_1)\} \end{aligned}\right\} \tag{3.17}$$

The best choice of $x_1$ and $x_2$ is discussed by Aitchison and Brown (1955). For estimating $\sigma$ they show that the greatest efficiency is obtained by taking the 7th and 93rd percentile and for $\omega$ the 27th and 73rd. When distributions are censored and grouped it will not usually be possible to meet this ideal. In practice we find the most convenient method is to first fit a line to the survivor function plotted on lognormal probability paper and then to determine the parameters of the line. The parameter $\sigma$, which is the slope of the line, can be estimated by taking two conveniently spaced quantiles and $\omega$ can be found from (3.17) by making $x_1$ the median, in which case $G(x_1) = \frac{1}{2}$ and $\hat{\omega} = \ln x_1$.

*Example 3.3* Returning to the distribution of Examples 3.1 and 3.2 we find from Figure 3.5 that the estimated median length of service is 1.15 years. Hence $\hat{\omega} = \ln 1.15 = 0.140$. For $\hat{\sigma}$ we will take $x_1$ and $x_2$ at the 5th and 95th percentiles. Reading off from the graph we find that $x_1 = 0{\cdot}135$ and $x_2 = 10{\cdot}0$ and hence

$$\hat{\sigma} = \frac{\ln 10.0 - \ln 0.135}{1.645 - (-1.645)} = 1.309$$ ◀

The method of percentage points can also be used for the two-term exponential distribution but the equations are not so easy to solve. Since there are three parameters we need three equations which have the form

$$\hat{G}(x_i) = \theta\, e^{-\lambda_1 x_i} + (1 - \theta)\, e^{-\lambda_2 x_i} \qquad (i = 1, 2, 3), \tag{3.18}$$

These may be solved directly by numerical methods but the procedure can be facilitated by choosing $x_2$ and $x_3$ to be integral multiples of $x_1$. In particular let $x_2 = 2x_1$ and $x_3 = ax_1$ where $a$ is a positive integer greater than 2. The equations (3.18) now become

$$\left.\begin{aligned} \hat{G}(x_1) &= \theta(X - Y) + Y \\ \hat{G}(x_2) &= \theta(X^2 - Y^2) + Y^2 \\ \hat{G}(x_3) &= \theta(X^a - Y^a) + Y^a \end{aligned}\right\} \tag{3.19}$$

where $X = e^{-\lambda_1 x_1}$, $Y = e^{-\lambda_2 x_1}$. These equations may be solved as follows. From the first equation

$$\theta = \{G(x_1) - Y\}/(X - Y)$$

Substituting for $\theta$ in the second equation gives

$$V = \hat{G}(x_1)\, U - G(x_2)$$

where $U = X + Y$ and $V = XY$. Substituting the expression for $\theta$ in the third member of (3.19) and eliminating $V$ between the second and third gives

$$\left.\begin{aligned} U &= \{\hat{G}(x_3) - \hat{G}(x_1)\,\hat{G}(x_2)\}/\{\hat{G}(x_2) - \hat{G}^2(x_1)\} \quad \text{if } a = 3 \\ U^2 &= \{\hat{G}(x_3) - \hat{G}^2(x_2)\}/\{\hat{G}(x_2) - \hat{G}^2(x_1)\} \quad \text{if } a = 4 \end{aligned}\right\} \tag{3.20}$$

For higher values of $a$, $U$ is the root of a polynomial equation which will be found for $a = 5$ and 6 in Bartholomew (1959). Once $U$ and $V$ have been determined $X$ and $Y$ and hence $\lambda_1$, $\lambda_2$, and $\theta$ follow. It may happen that one or more estimates lie outside the permissible range which means, of course, that no mixed exponential distribution exists with the required percentage points.

*Example 3.4* Returning again to the estimated survivor function of Example 3.1, let us take $x_1 = 1$, $x_2 = 2$, $x_3 = 3$ and estimate the parameters of the two-term mixed exponential distribution. The estimating equations are

$$0.545 = \theta\, e^{-\lambda_1} + (1 - \theta)\, e^{-\lambda_2}$$

$$0.342 = \theta\, e^{-2\lambda_1} + (1 - \theta)\, e^{-2\lambda_2}$$

$$0.245 = \theta\, e^{-3\lambda_1} + (1 - \theta)\, e^{-3\lambda_2}$$

From (3.20)

$$U = \{0.245 - (0.545)(0.342)\}(\{0.342 - (0.545)^2\}$$

$$= 1.303$$

$$V = (0.545)(1.303) - 0.342 = 0.368$$

Therefore

$$X + Y = 1.303$$

$$XY = 0.368$$

and

$$X - Y = \{(1.303)^2 - 4(0.368)\}^{\frac{1}{2}} = 0.475$$

Hence

$$X = \tfrac{1}{2}(1.303 + 0.475) = 0.889$$

$$Y = \tfrac{1}{2}(1.303 - 0.475) = 0.414$$

$$\hat{\theta} = (0.545 - 0.414)/0.475 = 0.276$$

Finally,

$$\hat{\lambda}_1 = 0.118 \quad \hat{\lambda}_2 = 0.882$$

We know from the method of fitting that this distribution reproduces the empirical distribution exactly at $x = 1, 2$, and 3. We now compute the fitted values at $x = 4, 5$, and 6 and compare them with the actual values. Table 3.1

***Table 3.1***

| $x_i$ | 4 | 5 | 6 |
|---|---|---|---|
| Actual value of survivor function | 0.170 | 0.131 | 0.105 |
| Fitted value | 0.193 | 0.162 | 0.140 |

shows the results obtained by substituting in equation (3.18). These calculations suggest that the overall fit is not particularly good. Some further light will be thrown on this question by Exercise 3.4. ◀

The Pearson Type XI curve can also be fitted by the method of percentage points. With two parameters to be estimated the two equations are

$$\left.\begin{aligned}\hat{G}(x_1) &= \left(1 + \frac{x_1}{c}\right)^{-q} \\ \hat{G}(x_2) &= \left(1 + \frac{x_2}{c}\right)^{-q}\end{aligned}\right\} \tag{3.21}$$

If we take logarithms of both sides of each equation, $q$ can easily be eliminated to give the following equation for $c$:

$$\log(1 + x_1/c) = \kappa \log(1 + x_2/c) \tag{3.22}$$

where $\kappa = \log \hat{G}(x_1)/\log \hat{G}(x_2)$ (the base of the logarithms is immaterial). This equation must be solved by numerical methods and then $q$ can be estimated from

$$\hat{q} = -\log \hat{G}(x_1)/\log(1 + x_1/\hat{c}) \tag{3.23}$$

*Example 3.5* We will illustrate the method using the same data as in the previous examples taking $x_1 = 1, x_2 = 4$ for which $\hat{G}(x_1) = 0{\cdot}545$, $\hat{G}(x_2) = 0.170$.

$\kappa = \log 0.545/\log 0.170 = 0.3425$, hence $\hat{d} = \hat{c}^{-1}$ is the root of the equation

$$\log(1 + d) - 0.3425 \log(1 + 4d) = 0$$

Trial and error shows that there is a root between $d = 0.3$ and $0.4$ and inverse interpolation or a standard iterative technique shows that its value is $\hat{d} = 0.346$, whence $\hat{c} = 2.890$. (There is another root $d = 0$ but this cannot be a solution of our problem.) Equation (3.23) now yields

$$\hat{q} = -\log 0.545/\log(1.346) = 2.043$$

**Other methods**

Graphical and percentage point methods give sufficient accuracy for most practical purposes and are particularly well suited for use with censored cohort and census data. The main alternative methods are maximum likelihood and moments. Both methods are straightforward with complete data from a cohort although the method of moments is liable to be inefficient for distributions as skew as those with which we are concerned. With incomplete data the likelihood function is more complicated and it is doubtful whether the gain in efficiency from using maximum likelihood is worth the effort. Silcock (1954) used this method for fitting the Type XI distribution and reference may be made to that paper for details. He also gave an explicit formula for the maximum likelihood estimator of the exponential parameter for a grouped (and censored) sample. Similar results for the lognormal distribution will be found in Aitchison and Brown (1955).

## Goodness of fit

Having fitted a distribution we face the question of whether the fit is satisfactory. In statistical practice this is answered by a goodness-of-fit test which is concerned with whether the observed deviations from the fitted curve are such as could be accounted for by chance. In our view this is not usually the question of practical importance. We really need to know whether the use of the fitted distribution will be sufficiently accurate for practical purposes and this can only be decided in the context of the problem in hand. It can well happen, for example, that a fit which would be rejected by a formal test of significance leads to predictions of sufficient accuracy for a particular purpose. However, we regard this not as an argument for disregarding goodness of fit altogether but for interpreting the results sensibly, with due regard for the uses to which the analysis is to be put.

If a graphical method has been used, a good deal can be learnt from an inspection of the plot. If the points show a marked departure from linearity as in the exponential fit of Figure 3.4 one would clearly be very wary of using the fitted distribution for any purpose which depended on extrapolating the line. If, on the other hand, the points were linear but subject to a wide scatter about the line we would suspect that the form of the distribution fitted was satisfactory but that its parameters could not be determined very precisely. The extent of the scatter would also give us some indication of the error of estimates of the survivor function made by reading off values from the line. When parameters have been estimated by the method of percentage points it is desirable to compare the fitted and actual values of the survivor function at points other than those used to make the estimates. We did this in Example 3.4 and were thus able to detect what appeared to be a significant deviation in the upper tail.

These methods are often very suggestive and should form a normal part of data analysis. In experienced hands they may be all that is needed but they are subjective and there is much to be said for supplementing them by a formal test of significance. There are several such tests available, of which the well-known $\chi^2$-test is, perhaps, the easiest to use. Once the parameters have been estimated there is no difficulty about computing the expected frequency with which length of service falls in different intervals and the calculation of the criterion is straightforward. There are no problems when dealing with cohort data; the censored portion is treated as a single group and a degree of freedom is subtracted for each parameter fitted and one for the total. (Strictly speaking the distribution theory of the tests depends on the estimates being linear functions of the observed frequencies. This will not be the case if the parameters were estimated by fitting a line by eye but the effect should not be serious.) With census data the position is not quite so straightforward and the reader is referred to Marshall (1974) for details of the method. The fitted distribution is used to obtain 'expected' values for the numbers of leavers in each length-of-service class. If the $\chi^2$-statistic is now calculated in the usual way Marshall (1974) showed that its sampling distribution was approximately that of a $\chi^2$ random variable but with one degree of freedom more than if the calculation had been for a cohort.

## 3.4 PREDICTION OF WASTAGE

The problem of predicting wastage arises in several forms. We begin with the simplest version, basing the discussion on the data illustrated in Figure 3.1 Suppose that data is available for the first five years only so that the shaded part of the distribution is available, and that we wish to predict the future leaving experience of the cohort. The method, already outlined in (b) of Section 3.1 is to fit a suitable distribution to the shaded part of the distribution and extrapolate it into the future. If we plot the survivor function estimated from the first five years' data on lognormal probability paper we find that the points lie very close to a straight line.

*Example 3.6* The solid line on Figure 3.6 is the estimated survivor function for the data of Figure 3.1 on the assumption that the first five years' data only were available. The dotted portion represents an extrapolation over the next

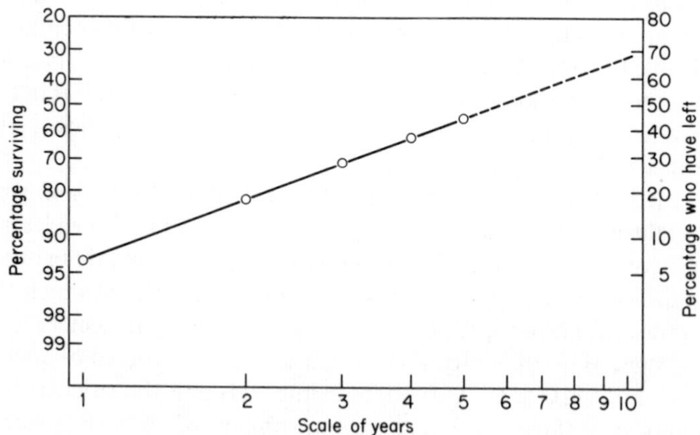


Figure 3.6 Extrapolation of the survivor function fitted to the first five years of the cohort shown on Figure 3.1

five years. Thus we may estimate that by the end of the tenth year almost 70 per cent of the original cohort will have left. The expected numbers leaving within any time interval can easily be found from the difference in the predicted survivor function at the two end-points of the interval. ◀

This type of analysis is very simple but is highly effective in that it can be readily understood by management and can be carried out quickly without computing facilities. It is therefore a very useful technique for those beginning manpower planning. The same method can be used without the help of a diagram using any of the fitted distributions discussed above. Having estimated the parameters of the survivor function we can easily estimate its value at future times as we did in Example 3.4. Since any technique based on the survivor function requires the assumption that the group to which it applies is homogeneous it is important to emphasize again the necessity of ensuring, as far as possible, that this is the case.

The foregoing example was concerned with forecasting the future wastage experience of a cohort for which some data are already available. Sometimes we shall wish to forecast the leaving pattern for cohorts not yet recruited. This can only be done if we are prepared to make an assumption about the form of the CLS distribution. The best guide in these circumstances will often be the experience of recently recruited cohorts or the current leaving experience of all existing cohorts as estimated by a census analysis. As a precaution against faulty assumptions it is advisable to make a range of forecasts by allowing the parameters of the model to vary within the limits of what seems plausible.

The form in which the forecasting problem is most likely to arise is one requiring us to forecast losses arising in some future period from the current stock. If we have complete historical data the existing stock can be divided into cohorts according to date of entry. The earlier leaving pattern of each cohort can be reconstructed from the record and then the future losses from each cohort can be projected exactly as described above. In the absence of historical data this is not possible and we must consider what can be done using current data only. One method of dealing with this problem using a discrete-time Markov-chain model is given in Chapter 4 (Section 4.5) and that will often be the most convenient way of proceeding. However, if the survivor function has already been estimated the following method, which requires very little further calculation, may be preferred.

Our problem will be solved in principle if we can predict what proportion of our present stock will survive for a further length of time $y$, say. We first obtain an expression for this function using a continuous-time formulation and then consider how to approximate it using census data. Let the current length-of-service distribution have density function $S(x)$, meaning that the number with length of service in $(x, x + \delta x)$ is $S(x)\,\delta x$. The probability that an individual now aged $x$ will survive for a further length of time $y$ is $G(x + y)/G(x)$. Hence the expected number of people in the system who will survive for $y$ is

$$\int_0^{x_{k+1}} S(x)\, G(x + y)/G(x)\, \mathrm{d}x \tag{3.24}$$

where, as in Chapter 2, $x_{k+1}$ denotes the greatest possible length of service. It would be possible to fit a curve $S(x)$ to the current length-of-service distribution and to use an estimate of $G(.)$ and then evaluate the integral for each $y$ of interest. In practice it will be easier and sufficiently accurate to approximate the integral by a sum as follows.

The length-of-service distribution is likely to be available in grouped form. If we assume that lengths of service in $(x_i, x_{i+1})$ are uniformly distributed in that interval, this implies that $S(x) = S_i$, say, for $x_i \leqslant x < x_{i+1}$. The expression (3.24) now becomes

$$\sum_{i=0}^{k} S_i \int_{x_i}^{x_{i+1}} G(x + y)/G(x)\,\mathrm{d}x \tag{3.25}$$

Using the trapezoidal rule to approximate the integral we have

$$\tfrac{1}{2}\sum_{i=0}^{k} S_i c_i \left\{\frac{G(x_i + y)}{G(x_i)} + \frac{G(x_{i+1} + y)}{G(x_{i+1})}\right\}$$

$$= \tfrac{1}{2}\sum_{i=0}^{k} (S_i c_i + S_{i-1}c_{i-1})\,G(x_i + y)/G(x_i) \qquad (S_{-1} = S_k = 0) \tag{3.26}$$

*Example 3.7* Suppose that the current length-of-service distribution is as follows:

| $i$ | 0 | 1 | 2 | 3 | 4 | 5 |
|---|---|---|---|---|---|---|
| $x_i$ | 0 | 1 | 2 | 3 | 4 | — |
| $S_i c_i$ | 55 | 40 | 31 | 17 | 6 | — |

There will often be some uncertainty about the maximum length of service in the last, open-ended group, so we have left it unspecified for the moment. Suppose also that we have fitted a Type XI curve to the survivor function estimated from census data and obtained

$$\hat{G}(x) = \left(1 + \frac{x}{3}\right)^{-2}$$

(note that this is very close to the distribution of Example 3.5). We are required to predict the number who will still be in the firm in two years' time. Substituting in (3.26) with $y = 2$ gives the estimated number as

$$\tfrac{1}{2}\Bigg[55(\tfrac{3}{5})^2 + (55 + 40)(\tfrac{4}{6})^2 + (40 + 31)(\tfrac{5}{7})^2 + (31 + 17)(\tfrac{6}{8})^2 + (17 + 6)(\tfrac{7}{9})^2$$

$$+ 6\left(\frac{3 + x_5}{5 + x_5}\right)^2\Bigg] = \tfrac{1}{2}\left[139.16 + 6\left(\frac{3 + x_5}{5 + x_5}\right)^2\right] = 69.58 + 3\left(\frac{3 + x_5}{5 + x_5}\right)^2$$

The smallest reasonable value of $x_5$ would be 5, in which case the last term would be 1.92. At the other extreme as $x_5 \to \infty$, the last term is 3 so the indeterminacy in $x_5$ only makes a difference of 1 in the final forecast which we might reasonably take as 71, or 48 per cent of the current stock. It so happens that in this example the integral in (3.25) can be evaluated exactly without the need for the approximation leading to equation (3.26). This leads to a prediction of 69.78 with $x_5 = 5$ and 72.04 with $x_5 = \infty$. The approximation is clearly quite adequate in this case. ◀

It must be remembered that we have been concerned with predicting wastage from the present stock. These numbers will be augmented in practice by losses from recruits taken in before the expiry of the forecast period. We have already discussed the forecasting of losses from future cohorts and these can be added in if desired. Notice also that we have assumed that all members of the current stock, whatever their length of service, have the same CLS distribution. This is the price which has to be paid for the restriction to current data.

## Forecast errors

It is clearly desirable for wastage forecasts to be accompanied by some measure of the uncertainty to which they are subject. This is not an entirely straightforward matter even if we are prepared to assume that the leaving pattern will be unchanged during the forecast period. If we knew the survivor function, instead of having to work with an estimate, there would be no difficulty.

Thus if we have a group of individuals with current length of service $x$ and if we assume that they behave independently with the same survivor function $G(x)$, then the probability that any individual leaves in the interval $(T_1, T_2)$ (measured from the present) is

$$\{G(x + T_1) - G(x + T_2)\}/G(x) = p, \text{ say}$$

The number of leavers will therefore have a binomial distribution and the standard error of the forecast will be $\sqrt{np(1 - p)}$ where $n$ is the size of the group. This result enables us to deal with a cohort of any age and, by extension, with a projection based on current data only. We would handle this by dealing with each cohort separately and using the result that the total loss is the sum of the losses from each cohort. Assuming independence, the variance of the forecast would be the sum of the cohort variances. The complication arises from the fact that the survivor function is not known but has been estimated. This means that the simple binomial calculation will underestimate the forecast error though the effect of this should not be serious if the estimate is based on a large sample. It is possible to allow for estimation error but the theory required for this would take us too far afield. Even if no formal calculations are made as, for example, when the forecast is made by fitting a line by eye on lognormal probability paper, it will be possible to form some idea of forecast error by reference to the degree of scatter about the line for that part of the distribution where data are available. This must be done with care remembering that the probability scale is not an arithmetical one.

## 3.5 MEASUREMENT OF WASTAGE

There is an extensive literature on the measurement of wastage but from the statistical point of view much of it is unsatisfactory. For this reason we shall begin by trying to clarify the basic ideas and objectives and then continue with a critical discussion of measures in common use.

Measures or indices of wastage are used to compare the propensities of different groups of individuals to leave at the same time or of the same group to leave at different times. Such comparisons are commonly based on the belief that the level of wastage in an organization is an indicator of institutional health. In these terms a high level of wastage is regarded as 'bad' and a low level as 'good'. This, of course, is an oversimplification but it provides a useful starting-point for the discussion. Some writers, for example Lane and Andrew (1955), pose the problem as one of measuring stability rather than wastage, and Bowey (1974) argues that the two concepts are distinct. In our view, as expressed at the beginning of Chapter 2, this is an unnecessary and potentially confusing distinction and we shall treat the two concepts of wastage and stability as equivalent—high wastage meaning low stability and vice versa. (The distinction that Bowey seeks to draw is really a manifestation of heterogeneity which, we have argued, should first be removed.)

We start the discussion from the fact that for a homogeneous group the propensity to leave as a function of length of service is completely described by any one of the three equivalent functions $m(x)$, $f(x)$, and $G(x)$. To compare the wastage of two groups it is thus sufficient to consider estimates of these functions. For example, suppose that we plot $m(x)$ for two firms on the same diagram. If one curve lies wholly above the other than we can say that the one group has a higher propensity to leave at all lengths of service. On the other hand if the two curves cross we cannot say, unequivocally, that propensity to leave is higher in one group than the other; at some lengths of service it is and at others it is not. The same type of argument can be carried through with the survivor function.

The foregoing argument shows that circumstances can arise in which one cannot unambiguously rank two groups according to their levels of wastage. Hence any attempt to force a ranking by constructing a single index of wastage is doomed to failure. Ideally, therefore, one ought to abandon the quest for indices of waste and concentrate instead on educating managers to interpret survivor functions or the like. However, most users have a strong preference for single indices and the statistician is therefore under some pressure to devise indices which are meaningful and with small risk of being seriously misleading.

We may obtain an idea of how to do this by looking at the problem from a slightly different point of view. Attempts to graduate CLS distributions by the one-parameter exponential distribution have been unsuccessful. The two- or three-parameter mixed exponentials and the two-parameter lognormal provided much better fits and this suggests that at least two parameters are needed to summarize the curves adequately. In looking for a single index we are therefore looking for the single most relevant parameter for our purposes.

Stated in this form the problem is a familiar one to statisticians. The object is to summarize a frequency distribution in a single number. The traditional way is to construct measures of location, dispersion, and so forth. In our case it is clearly the location of the distribution which is most relevant for our purposes. A measure of location will give us, in some sense, the 'typical' length of stay. If this is low it will indicate a situation of high turnover, and conversely if it is high. Our conclusion is, then, that if we have to use a single measure of wastage it should be some measure of location of the CLS distribution. We shall therefore proceed to discuss the merits of the two most commonly used measures, the mean and the median.

## The expected length of service (the mean)

Lane and Andrew (1955) first proposed the mean length of service as a measure of stability. Most managers are familiar with the idea of an average and it therefore has advantages for communication as well as some useful statistical properties. It will not be possible to calculate the mean directly unless we are in the unusual position of having a complete cohort distribution. In Chapter 2 we described a method of estimating the mean, $\mu$, using the fact that

$$\mu = D_0 = \int_0^{x_{k+1}} G(x)\,\mathrm{d}x$$

but as we noted this presents difficulties with a cohort because $G(x)$ is only estimated for part of the range. An alternative is to fit a theoretical distribution and then to find its mean. The means of the distributions that we discussed in Section 3.2 are as follows.

(a) Two-term mixed exponential: $\theta/\lambda_1 + (1-\theta)/\lambda_2$.

(b) Type XI: $c/(q-1)$.

(c) Lognormal: $\exp(\omega + \frac{1}{2}\sigma^2)$.

We have fitted each of these distributions to the survivor function of Example 3.1. Our estimates of the mean length of service derived from these results are shown in Table 3.2.

The figure for the two-term exponential is out of line with the others and this arises from the fact that the fitted distribution overestimated the number of survivors in the upper tail. (The single exponential of Example 3.1 was, in fact, better with an estimated mean of 2.39.) This feature illustrates one of the practical drawbacks of the expectation of service. The mean of a highly skewed distribution depends strongly on the upper tail of the distribution. In manpower work empirical evidence in this region is either lacking or sparse and we have to rely on the extrapolation of a fitted curve. It is perfectly possible, as the example shows, for a curve to fit the early part of the distribution well and yet to be rather poor in the tail. This means that our estimate of the mean depends critically on the form of that part of the distribution about which we have least information.

***Table 3.2***

| *Two-term exponential* | *Type XI* | *Lognormal* |
|---|---|---|
| 3.16 | 2.77 | 2.71 |
| (Example 3.4) | (Example 3.5) | (Example 3.3) |

A related point is that the fitted distribution may have a significant part of its tail area lying beyond the maximum possible length of service. This would certainly be the case, for example, with the distribution shown on Figure 3.1 where the maximum length of service might be 40 or 45 years. To cope with this difficulty Lane and Andrew (1955) supposed that there was a fixed maximum length of service which they denoted by $p$. They modified the lognormal distribution by assuming that everyone who, according to the distribution, should have had a longer length of service, had exactly $p$ years of completed service. Thus they used a distribution made up of a lognormal distribution censored at $p$ and a mass of probability equal to the tail area beyond $p$ located at $p$. This would be a reasonable procedure if everyone joined at the same age and hence had the same maximum length of service. However, this will rarely be the case as individual maxima are likely to vary considerably and this makes the method somewhat arbitrary.

## The half-life (median)

The use of the median of the CLS distribution as a measure of stability appears to have been first proposed by Silcock (1954) and it has many advantages. Being the expected time for half a cohort to leave, it has an obvious practical meaning which can easily be appreciated by managers. It has the further advantage that it can be estimated directly from cohort data provided that at least 50 per cent of the cohort have left and even when this is not the case the extent of the extrapolation needed is much less than is required for the mean.

The fact that the median does not depend so much on the tails of the distribution makes it less susceptible to the errors arising from lack of data in that region which we noted with the mean. Ease of calculation and interpretation thus combine to make the median, or half-life, one of the most useful indices of stability.

There is also a further theoretical point in its favour which arises from the connection between the mean and median of a lognormal distribution. For the lognormal the median is $e^{\omega}$ and hence

$$\text{mean} = \text{median} \times e^{\frac{1}{2}\sigma^2}. \qquad (3.27)$$

It often seems to be the case that the parameter $\sigma^2$ is fairly constant for a given type of employee. This phenomenon can be observed in the fact that many of the lognormal plots in Lane and Andrew (1955) appear as nearly parallel lines (recalling that the slope is proportional to $\sigma$). Where this is the case (3.27) shows that the mean is proportional to the median and hence the two measures are equivalent.

If the median has to be estimated from a cohort where fewer than 50 per cent have left or from census data, it will be necessary to fit a curve to the estimated survivor function and then to determine its median. If the lognormal distribution provides a satisfactory fit this can be done most easily by plotting the estimated survivor function on lognormal probability paper and then reading off the 50 per cent point. We did this for the data in Example 3.3 where we estimated the median to be 1.15 years. If it is desired to estimate the median from estimates of the parameters of a fitted distribution the following formulae for the median will be required. The numerical values for the data of Example 3.3 are given in brackets.

(a) Type XI: $c(2^{1/q} - 1)$ $(= 1.17)$
(b) Lognormal: $e^{\omega}$ $(= 1.15)$
(c) Two-term exponential: the root of $\theta\, e^{-\lambda_1 x} + (1 - \theta)\, e^{\lambda_2 x} = \frac{1}{2}$ $(= 1.16)$.

In the last case a non-linear equation has to be solved whereas a simple direct formula is available in the other two cases. It will be seen that the agreement between the three methods is very much better than in the case of the mean.

### The proportion who survive for a given time

One disadvantage shared by the mean and the median is that their sampling distributions are difficult to handle in small samples. For many purposes this is not a relevant consideration but sometimes we may wish to compare two groups and decide whether the difference between them is significantly different. An alternative measure which is very easy to handle in this respect is the proportion who survive for a fixed time. Thus for the cohort used for illustration throughout this chapter the proportion who survive for one year is 0.545. Its sampling distribution is binomial with estimated standard error

$\sqrt{0.545(1 - 0.545)/800)} = 0.018$. With census data the estimated proportion for given $x$ is $G(x)$ and we have shown in Chapter 2 how to find the standard error of this if $x$ coincides with one of the group boundaries. The arbitrariness in the choice of $x$ is an unsatisfactory feature of this measure. We would aim to choose $x$ so that the proportion surviving was not too close to 0 or 1 and this element of choice allows a subjective element to enter the calculation.

## Other methods

We have already drawn attention to the way in which crude wastage rates depend on the length-of-service structure of the group and this feature serves to disqualify them as indicators of wastage. This point may be reinforced further by an example from Lane and Andrew (1955) who gave figures for two firms as shown on Table 3.3.

***Table 3.3***

| | *Crude rate* | *Expected length of service* |
|---|---|---|
| Firm C | 8.8% | 5 years |
| Firm D | 17.3% | $7\frac{1}{2}$ years |

This apparently paradoxical result is explained by the fact that Firm D had a much larger proportion of short-service employees and so in spite of the higher stability reflected in the expectation of service it had a higher rate of leaving than Firm C. This is admittedly an extreme case but the dangers to which it points are real ones. Only if the system is in a steady state (Chapter 7) will the crude wastage rate give an unambiguous indication of propensity to leave. Standardized wastage rates avoid the difficulties but as we noted at the beginning of Chapter 2, they do not appear to have been used by manpower planners.

Recognizing the shortcomings of the crude rate, van der Merwe and Miller (1971) suggested that it should be supplemented by a second measure and that the two should be interpreted in conjunction. This is in line with our own argument that two measures are needed for a complete description of the process. These authors proposed that the second measure should be the median length of service of those who leave in a given interval (year). This is not the same as the half-life discussed above because, in general, the *current* leavers distribution depends on the length-of-service structure. In fact the whole point of introducing the census analysis in Chapter 2 was to eliminate the effects of this dependence. Thus they proposed that two measures, both affected by length-of-service structure, should be used. It may be possible to interpret one in the light of the other but it is not altogether clear how this is to be done.

Bowey (1974) has also proposed measures intended to bring out facets of leaving behaviour which the crude rate conceals. Thus, from a practical point of view a high level of wastage will be less serious if it is almost entirely among short-service people than if it is among those with long service in whom a substantial amount of training and experience have been invested. Hence she proposes, for example, to use a crude rate calculated only for those with, say, at least two years of service. Such an index is, in effect, a very special case of a standardized rate and may well be a useful index.

## Comparisons

In Chapter 2 we discussed statistical tests of whether two or more estimated survivor functions were products of the same underlying distribution. These can also be used in the present context as a means of judging whether levels of wastage in different organizations differ significantly. The comparison of two proportions mentioned above is, of course, a special case of the contingency table test.

Having fitted a distribution to a survivor function, there is now the possibility of comparing different functions by looking at differences between the estimates of their parameters. In the lognormal case, for example, we might base a test on $\hat{\omega}_1 - \hat{\omega}_2$, where the subscripts denote the two populations. The sampling distribution will depend on the method of fitting and, in most

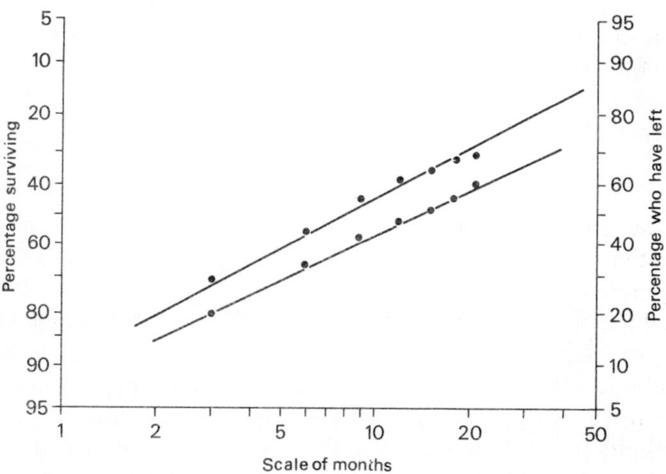


Figure 3.7 Survivor functions of the two distributions shown in Figure 3.2 plotted on lognormal probability paper

cases, is only available asymptotically. In the case of distributions fitted by eye such distributions could only be determined experimentally. We have therefore not thought it worth while to provide tests of this kind for this purpose, as those given in Section 2.4 will serve most practical purposes. It is worthwhile, however, pointing out that a graphical analysis will often be sufficient to make a formal test superfluous. In Figure 3.7 we have plotted the two distributions shown earlier in Figure 3.2 on the same diagram. It is at once clear that the scatter about the lines is much less than between the lines and we would therefore have little hesitation in concluding that the apparent difference is a real one.

## 3.6 COMPLEMENTS

Since most empirical CLS distributions come in grouped form it might seem more natural to develop models for the leaving process in discrete time. This is possible but the theory of the continuous distributions we have discussed is much better developed than that of their discrete analogues so there is no advantage to be gained. For the reader who wishes to explore what is involved in a discrete-time treatment we give the following introduction. We start with the discrete analogues of $f(x)$, $G(x)$, and $q(x)$ defined as follows:

$$f_i = \Pr\{x = i\} \qquad (i = 1, 2, \ldots)$$

$$G_i = \Pr\{x \geqslant i\} = \sum_{j=i}^{\infty} f_j \qquad (i = 1, 2, \ldots)$$

$$q_i = \Pr\{\text{service terminates at } x = i \mid \text{survival to } i\} = f_i/G_i$$

Note that this is slightly different from the treatment given in Chapter 2 in that we now assume that $x = 1, 2, 3, \ldots$ are the only times at which losses can occur.

The constant-risk model takes the form $q_i = \pi$, say, whence

$$G_i = (1 - \pi)^{i-1} \quad \text{and} \quad f_i = \pi(1 - \pi)^{i-1} \qquad (i = 1, 2, \ldots)$$

This is the geometric distribution and it plays the same role in discrete theory as the exponential does in this chapter. Mixed geometric distributions can be generated by giving $\pi$ a probability distribution; $\pi$ varies between 0 and 1 and if it is given a beta distribution we arrive at the discrete analogue of the Type XI distribution (see Bartholomew 1963a, Section 5). Network models also lead to mixtures of geometric distributions and the analysis can be developed using the theory of absorbing Markov chains (see Chapter 4 and Bartholomew (1973a, Section 6.6)). It is also shown in this last reference that it is possible to devise a model in which the logarithm of length of service has a binomial distribution, thus producing a discrete analogue of the log-normal distribution.

We have seen that the lognormal distribution often provides a good fit to CLS distributions but that the theoretical arguments underlying this derivation do not provide a very satisfying explanation of its occurrence. This leads one to wonder whether there may not be some other distribution, of similar shape, for which a more convincing explanation can be found. An interesting step in this direction has been taken by Cronin (1977) who investigated the log-logistic distribution. It is well known that the logistic distribution is similar in shape to the normal and this fact has been used to good effect in other fields such as dosage–response analysis. Hence we may expect the log-logistic distribution to be close to the lognormal in shape. A random variable $y$ with a logistic distribution has distribution function

$$F(y) = 1 - \left\{\exp\left(\frac{y-\omega}{\theta}\right) + 1\right\}^{-1}, \qquad -\infty < y < \infty$$

Hence, if $y = \ln x$, $x$ has distribution function of the form

$$F(x) = 1 - \{1 + Ax^{1/\theta}\}^{-1}, \qquad 0 \leqslant x < \infty$$

This is a special case of a family of distributions studied by Burr (1942). One practical advantage that $F(x)$ has over the lognormal is that its survivor function and propensity-to-leave function have very simple forms. On the modelling side Cronin has shown that the log-logistic can arise as a mixture of Weibull distributions which, in turn, arise in extreme-value theory. For small and large $x$ the log-logistic distribution is similar to the Pareto (or Type XI) distribution. It will be interesting to see whether these somewhat disconnected results can be made to yield a plausible explanation of the leaving process.

Hyman (1970) gives an interesting analysis of survival data in which, among other things, lognormal distributions are successfully fitted. His paper also contains a discussion of the problem of measuring wastage and includes a critical review of some earlier work. In particular he discusses another stability index proposed by Bowey (1969) which is essentially the same as the average length of service of the current stock. This may be a useful description of the current level of experience in the organization but it is not a measure of wastage or stability in the sense that those terms have been used here.

The theoretical properties of the crude wastage rate and, in particular, its dependence on the length-of-service structure of the system have been investigated in Bartholomew (1959 and 1973a, Chapter 7) (see also Exercise 5.1). This work shows how the wastage rate can be expected to decline as a system ages. The difference between the initial wastage rate and that eventually reached in the steady state can easily involve a factor of two or three, which provides further warning of the hazards of using crude rates as measures of propensity to leave.

## 3.7 EXERCISES AND SOLUTIONS

### Exercise 3.1

Find $G(x)$ and $m(x)$ for the mixed exponential distribution of equation (3.4) and comment on the form of these two functions.

### Exercise 3.2

Suppose that $h(\lambda)$ of (3.5) is given by

$$h(\lambda) = \theta\beta_1 \mathrm{e}^{-\beta_1\lambda_1} + (1-\theta)\beta_2 \mathrm{e}^{-\beta_2\lambda_2} \qquad (\beta_1, \beta_2 > 0, 0 \leqslant \theta < 1)$$

Find $f(x)$, $G(x)$, and $m(x)$ and comment on the results.

### Exercise 3.3

Investigate the form of (3.9) in the following special cases: (a) $\lambda_1 = 0$; (b) $\lambda_2 = 0$; (c) $\lambda_3 = 0$; (d) $\lambda_2 = \lambda_3$; (e) $\lambda_1 = \lambda_2 = \lambda_3$. Attempt, first of all, to deduce the answer in each case by inspection of the network diagram and then check your conclusions by substitution in the formula.

### Exercise 3.4

Verify that the CLS distribution of a $k$-component mixed exponential distribution is decreasing for all $x$.

### Exercise 3.5

Compute and plot $m(x)$ for the lognormal distribution with typical parameters $\omega = 1, \sigma = 1.5$.

### Exercise 3.6

The following questions all relate to the cohort data given in Exercise 2.1 and represent further analyses which can be carried out.

(a) Plot the logarithm of the estimated survivor function against $x$ as described in Section 3.3 and comment on the result.

(b) Plot $G(x)$ on lognormal probability paper as in Example 3.2 and comment on the result. (If lognormal paper is not available make the appropriate transformations and use ordinary graph paper.)

(c) Estimate $\omega$ and $\sigma$ from a line fitted by eye.

(d) Estimate the expectation of service from your results obtained in (c). Compare the expectation with the estimate, $D_0$, obtained in Exercise 2.1.

**Exercise 3.7**

Fit a two-term mixed exponential distribution to the distribution used in Example 3.1 by the method of percentage points. Take $x_1 = 1$, $x_2 = 2$, and $x_3 = 4$. Compare your result with that obtained in Example 3.4 and comment on the difference illustrating your comments with relevant calculations. Estimate the mean and median of the distribution using your fitted distribution and compare with the estimates in the text.

**Exercise 3.8**

Fit a Pearson Type XI curve to the data of Exercise 2.1 by the method of percentage points and estimate its mean and median. Compare your results with those of Exercise 3.6.

**Exercise 3.9**

Make a forecast of how many members of the cohort of Exercise 2.1 will remain after (a) 10, (b) 15 years.

**Solution to Exercise 3.1**

If

$$f(x) = \theta\lambda_1 \,\mathrm{e}^{-\lambda_1 x} + (1 - \theta)\lambda_2 \,\mathrm{e}^{-\lambda_2 x}$$

then

$$G(x) = \int_x^\infty f(u)\,\mathrm{d}u = \theta \,\mathrm{e}^{-\lambda_1 x} + (1 - \theta)\,\mathrm{e}^{-\lambda_2 x}$$

and

$$m(x) = f(x)/G(x) = \{\theta\lambda_1 \,\mathrm{e}^{-\lambda_1 x} + (1 - \theta)\lambda_2 \,\mathrm{e}^{-\lambda_2 x}\}/\{\theta \,\mathrm{e}^{-\lambda_1 x} + (1 - \theta)\,\mathrm{e}^{-\lambda_2 x}\}$$

The survivor function is a simple weighted average of two exponentials. It is less easy to see what form $m(x)$ has without further analysis. First assume that $\lambda_1 > \lambda_2$ and write

$$m(x) = \frac{\theta\lambda_1 \,\mathrm{e}^{-(\lambda_1 - \lambda_2)x} + (1 - \theta)\lambda_2}{\theta \,\mathrm{e}^{-(\lambda_1 - \lambda_2)x} + (1 - \theta)}$$

Then when $x = 0$, $m(0) = \theta\lambda_1 + (1 - \theta)\lambda_2$ and as $x \to \infty$, $m(x) \to \lambda_2$ so, clearly $m(0) > m(\infty)$. Next, consider the derivative of $m(x)$.

$$\frac{\mathrm{d}m(x)}{\mathrm{d}x} = \frac{-\theta(1 - \theta)(\lambda_1 - \lambda_2)^2 \,\mathrm{e}^{-(\lambda_1 + \lambda_2)x}}{\{\theta \,\mathrm{e}^{-\lambda_1 x} + (1 - \theta)\,\mathrm{e}^{-\lambda_2 x}\}^2}$$

and, obviously, this is negative for all values of $x$. We have thus shown that propensity to leave decreases monotonically from an initial value of $\theta\lambda_1 + (1-\theta)\lambda_2$ to an ultimate value of $\lambda_2$. This case is therefore rather like that where $h(\lambda)$ has the gamma distribution given by (3.6). The difference is that there $m(x)$ tends to zero with increasing $x$ whereas here $m(x)$ approaches a positive constant. This arises, of course, from the fact that after a sufficiently long time almost all those remaining in a cohort will belong to the group with the smaller $\lambda$, namely $\lambda_2$.

**Solution to Exercise 3.2**

$$f(x) = \int_0^\infty \lambda\, e^{-\lambda x} h(\lambda)\, d\lambda$$

$$= \theta\beta_1 \int_0^\infty \lambda\, e^{-\lambda(\beta_1 + x)}\, d\lambda + (1-\theta)\beta_2 \int_0^\infty \lambda\, e^{-\lambda(\beta_2 + x)}\, d\lambda$$

$$= \frac{\theta\beta_1}{(\beta_1 + x)^2} + \frac{(1-\theta)\beta_2}{(\beta_2 + x)^2}$$

$$G(x) = \int_x^\infty f(u)\, du = \frac{\theta\beta_1}{(\beta_1 + x)} + \frac{(1-\theta)\beta_2}{(\beta_2 + x)}$$

and

$$m(x) = \left\{\frac{\theta\beta_1}{(\beta_1 + x)^2} + \frac{(1-\theta)\beta_2}{(\beta_2 + x)^2}\right\} \Big/ \left\{\frac{\theta\beta_1}{\beta_1 + x} + \frac{(1-\theta)\beta_2}{\beta_2 + x}\right\}$$

It is clear that $f(x)$ is a decreasing density and it may be thought of as a mixture of two Type XI densities of the form given by (3.5) and (3.6) when $q = 1$. For this reason one would expect $m(x)$ to be a monotonically decreasing function and this is confirmed by inspection of the expression for $m(x)$ given above. Both numerator and denominator are monotonically decreasing functions of $x$ but the numerator decreases faster than the denominator. Hence the ratio is decreasing for all $x$. The same conclusion can be reached by considering the derivative of $x$ but the algebra involved is tedious.

**Solution to Exercise 3.3**

$$G(x) = \frac{\lambda_2 - \lambda_3}{\lambda_1 + \lambda_2 - \lambda_3} e^{-(\lambda_1 + \lambda_2)x} + \frac{\lambda_1}{\lambda_1 + \lambda_2 - \lambda_3} e^{-\lambda_3 x}$$

(a) Setting $\lambda_1 = 0$ means that no one becomes committed and hence that everyone is subject to a constant propensity to leave. The survivor function must therefore be $G(x) = e^{-\lambda_2 x}$ which is confirmed by putting $\lambda_1 = 0$ in $G(x)$ above.

(b) Putting $\lambda_2 = 0$ means that everyone becomes committed before being exposed to the risk of leaving. Their total CLS must therefore be the sum of two (independent) exponential random variables, the first being the time before entering the 'committed' state and the second the length of stay in that state. One could find the form of this distribution using standard techniques and the result is often known as an Erlang distribution. Its survivor function obtained by substituting in the formula above is

$$G(x) = \frac{\lambda_3}{\lambda_3 - \lambda_1} e^{-\lambda_1 x} - \frac{\lambda_1}{\lambda_3 - \lambda_1} e^{-\lambda_3 x}$$

Notice that this is a mixture of two exponentials but although the coefficients sum to 1, one is positive and the other negative; $G(x)$, however, is never negative.

(c) When $\lambda_3 = 0$ those who become committed never leave. This means that the survivor function should approach some value greater than zero as $x$ becomes large. Making the substitution

$$G(x) = \frac{\lambda_2}{\lambda_1 + \lambda_2} e^{-(\lambda_1 + \lambda_2)x} + \frac{\lambda_1}{\lambda_1 + \lambda_2}$$

The limiting value is $\lambda_1/(\lambda_1 + \lambda_2)$ and this may be interpreted as the proportion who become committed and therefore never leave.

(d) If $\lambda_2$ and $\lambda_3$ are equal then, as far as leaving is concerned, it does not matter which of the states 1 and 2 the individual is in. He is subject to a constant propensity to leave throughout his stay in the system and must therefore have an exponential survivor function. This may be verified by substituting $\lambda_2 = \lambda_3 = \lambda$, say, when we find $G(x) = e^{-\lambda x}$.

(e) If $\lambda_1 = \lambda_2 = \lambda_3$ we merely have a special case of (d) and since the value of $\lambda_1$ is irrelevant the conclusion is the same.

**Solution to Exercise 3.4**

For a $k$-component mixture

$$f(x) = \sum_{i=1}^{k} \theta_i \lambda_i e^{-\lambda_i x} \quad \text{with} \quad \sum_{i=1}^{k} \theta_i = 1 \quad \text{and} \quad \theta_i > 0 \text{ for all } i.$$

The result will be established if we show that the derivative of $f(x)$ is always negative.

$$\frac{df(x)}{dx} = -\sum_{i=1}^{k} \theta_i \lambda_i^2 e^{-\lambda_i x}$$

and since the $\theta_i$'s are non-negative the result follows at once.

**Solution to Exercise 3.5**

$$m(x) = \frac{1}{\sqrt{2\pi}\sigma x}\exp\left[-\tfrac{1}{2}\left(\frac{\ln x - 1}{1.5}\right)^2\right]\Big/\int_x^\infty \frac{1}{\sqrt{2\pi}\sigma u}\exp\left[-\tfrac{1}{2}\left(\frac{\ln u - 1}{1.5}\right)^2\right]du$$

$$= \frac{1}{\sqrt{2\pi}\sigma x}\exp(-\tfrac{1}{2}z^2)/\{1 - \Phi(z)\} \quad \text{where } z = (\ln x - 1)/1.5$$

$$= \frac{1}{1.5x}\frac{\phi(z)}{1 - \Phi(z)}$$

where $\phi(.)$ and $\Phi(.)$ are the density and distribution functions, respectively, of the standard normal distribution. The easiest way to tabulate the function $m(x)$ is to select values of $z$ and then compute $x$ and $m(x)$ as shown in Table 3.4. The calculation shows that $m(x)$ rises to a peak and then declines slowly over a long interval.

***Table 3.4***

| $z$ | $\phi(z)$ | $1 - \Phi(z)$ | $x = \exp(1.5z + 1)$ | $m(x)$ |
|---|---|---|---|---|
| −3 | 0.0044 | 0.9987 | 0.030 | 0.098 |
| −2.5 | 0.0175 | 0.9938 | 0.064 | 0.184 |
| −2 | 0.0540 | 0.9772 | 0.135 | 0.272 |
| −1.5 | 0.1295 | 0.9332 | 0.287 | 0.323 |
| −1 | 0.2420 | 0.8413 | 0.607 | 0.316 |
| −0.5 | 0.3521 | 0.6915 | 1.284 | 0.264 |
| 0 | 0.3989 | 0.5000 | 2.718 | 0.196 |
| 1 | 0.2420 | 0.1587 | 12.182 | 0.083 |
| 2 | 0.0540 | 0.0228 | 54.60 | 0.029 |
| 3 | 0.0044 | 0.0014 | 244.7 | 0.009 |

**Solution to Exercise 3.6**

(a) The plot is shown in Figure 3.8 and it exhibits the same kind of curvature that was evident in Example 3.1. We conclude that the exponential distribution does not provide a satisfactory fit to this distribution.

(b) The plot on Figure 3.9 shows that the points lie almost on a straight line showing that the lognormal fit is very good.

(c) The median read off from the graph is 1.3 so an estimate of $\omega$ is $\hat{\omega} = \ln 1.3 = 0.262$. For estimating $\hat{\sigma}$ we will take the same percentage points as in the text obtaining

$$\hat{\sigma} = \frac{\ln 9.0 - \ln 0.19}{2(1.645)} = 1.173$$

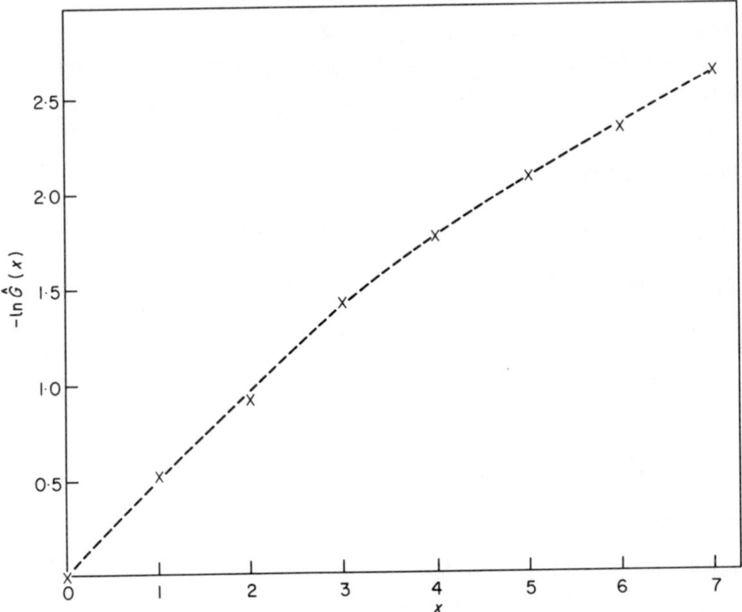


Figure 3.8 Estimated survivor function plotted on logarithmic paper for Exercise 3.6

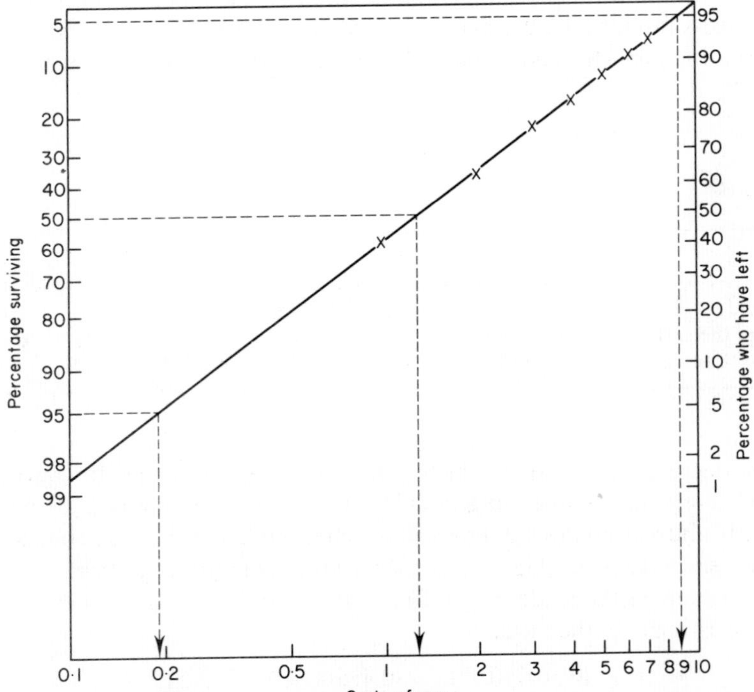


Figure 3.9 Estimated survivor function plotted on lognormal probability paper for Exercise 3.6

(d) The estimate of the mean is $\exp\{0.262 + \frac{1}{2}(1.173)^2\} = 2.59$ which compares with a value of about 3.4 estimated directly from the empirical distribution function with extrapolation on the assumption that $x_{k+1} = 35$.

**Solution to Exercise 3.7**

For the given values of $x$ we have, from the basic data:

$$G(x_1) = 0.545, \qquad G(x_2) = 0.342, \qquad G(x_3) = 0.170$$

$$U^2 = (0.170 - 0.342^2)/(0.342 - 0.545^2) = 1.179$$

$$U = 1.086, \qquad V = 0.545\,(1.086) - 0.342 = 0.250$$

$$X - Y = \{(X + Y)^2 - 4XY\}^{\frac{1}{2}} = \{1.179 - 4(0.250)\}^{\frac{1}{2}} = (0.1790)^{\frac{1}{2}} = 0.423$$

$$X = \tfrac{1}{2}(1.086 + 0.423) = 0.755$$

$$Y = \tfrac{1}{2}(1.086\text{–}0.423) = 0.332$$

$$\hat{\theta} = (0.545\text{–}0.332)/0.423 = 0.504$$

$$\hat{\lambda}_1 = 0.281, \qquad \hat{\lambda}_2 = 1.102$$

In this case both $\lambda$'s are larger than before and the mixing proportion has changed substantially. We may compare the performance of the two fits by computing the fitted and actual values of the survivor functions, as in Table 3.5.

***Table 3.5***

| $x_i$ | 1 | 2 | 3 | 4 | 5 | 6 |
|---|---|---|---|---|---|---|
| Actual | 0.545 | 0.342 | 0.245 | 0.170 | 0.131 | 1.105 |
| Fit in the text | 0.545* | 0.342* | 0.245* | 0.193 | 0.162 | 0.140 |
| Fit of Exercise 3.7 | 0.545* | 0.342* | 0.235 | 0.170* | 0.126 | 0.094 |

The asterisks denote values which were made to agree by the fitting process. Whereas the fit obtained in Example 3.4 overestimated the tail, the new fit slightly underestimates it but overall is somewhat better. This is as one would expect since the percentage points span a wider range of the distribution. As a further exercise the reader might like to try $x_1 = 2, x_2 = 4, x_3 = 6$.

The estimate of the mean is

$$(0.504)/(0.281) + (0.496)/(1.102) = 2.24$$

This compares with 2.71 for the lognormal and 2.77 for the Type XI distributions and is therefore now smaller by about the same amount as it previously exceeded them. This calculation provides further weight to our remarks about the difficulty of estimating the mean of a censored distribution. The median satisfies

$$0.504\,\mathrm{e}^{-0.281x} + 0.496\,\mathrm{e}^{-1.102x} = 0.500$$

From our earlier work we may expect the median to be about 1.16, for which the left-hand side is 0.502. Being slightly too large, we take $x = 1.17$ and find the left-hand side is 0.499. The median is therefore 1.17 correct to two figures and therefore virtually the same as for the other fitted distributions.

These calculations provide further illustration of the remark in the text that the median can be estimated much more reliably than the mean from censored data.

**Solution to Exercise 3.8**

We have a choice of percentage points. It seems reasonable to take $x_1 = 1$ because almost 40 per cent of the distribution lies to the left of this point and there is no other intermediate value of $x$ in this range. For the second point we need a value in the upper tail and we shall take $x_2 = 5$ but one might equally well take $x_2 = 4$ or 3. In our case

$$\hat{G}(x_1) = 0.597, \hat{G}(x_2) = 0.124$$

$$\kappa = \log 0.597/\log 0.124 = 0.2471$$

Hence $d$ is the root of the equation

$$\log(1 + d) - 0.2471 \log(1 + 5d) = 0$$

The non-zero root of this equation is $\hat{d} = 0.141$ and so $\hat{c} = 7.092$ and $q = -\log 0.595/\log 1.141 = 3.936$. Our estimate of the mean is 7.092/2.936 =2.42 years, and of the median $7.092\,(2^{1/3.936} - 1) = 1.37$. These values are reasonably close to those obtained from the lognormal fit of Exercise 3.6.

**Solution to Exercise 3.9**

We can extrapolate the line plotted on Figure 3.8 as far as 10 years, at which point the estimated percentage surviving is 4 per cent which with a cohort of 700 means about 28 people. To estimate the number surviving to 15 years the line would have to be re-plotted, either on 3-cycle paper, so that the plotted points appear in the middle cycle, or on 2-cycle paper as at present but with the first cycle used to span the range 1 to 10. An alternative method is to estimate the survivor function directly using tables of the normal integral. Thus

$$\hat{G}(15) = \int_{15}^{\infty} \frac{1}{\sqrt{2\pi}\sigma} \exp\left[-\tfrac{1}{2}\left(\frac{\ln x - \omega}{\sigma}\right)^2\right] dx = 1 - \Phi\left(\frac{\ln 15 - \omega}{\sigma}\right)$$

$$= 1 - \Phi\left(\frac{\ln 15 - 0.262}{1.173}\right) = 1 - \Phi(2.085)$$

$$= 0.019 \text{ or about 13 people.}$$

The reader should check that this method and the graphical method agree for $\hat{G}(10)$.

CHAPTER 4

# Transition Models Based on the Theory of Markov Chains

## 4.1 INTRODUCTION

The discussion of wastage in the previous chapters has been largely based on the assumption that we were dealing with a single homogeneous group of people. We now turn to the study of heterogeneous systems in which people are classified according to such things as grade, age, or location. This marks a return to the viewpoint of the first chapter in which a manpower system was regarded as a set of interconnected stocks and flows. Since manpower planning calls for both understanding and control of these quantities it requires a dynamic model of the system to investigate the effects of various strategies. Such a model must incorporate the restraints under which the organization operates together with assumptions about individual behaviour.

It is useful to begin thc discussion by distinguishing between those quantities which are fixed, or controlled, and those which are not. For example, in some organizations the number of jobs in each grade is fixed in relation to the work available. The stock numbers cannot therefore be treated as random variables and the need to maintain their values will place constraints upon the flows. Voluntary wastage on the other hand is the result of many individual decisions and must be treated as a random variable. When setting out to model a system it is thus necessary to begin by identifying which stocks and flows are predetermined and which are not. In this chapter we shall describe models in in which the stock numbers are free to vary and in Chapter 5 we shall turn to models where they are fixed.

The following list of questions gives some idea of the purposes which transition models may serve in a practical manpower planning situation.

(a) What will the grade (or age, or length-of-service) structure be at various dates in the future if present patterns of loss and promotion continue?
(b) What should the promotion rates and recruitment numbers be in order to achieve a desired structure in a specified time?
(c) What impact will expansion or contraction of the organization have on the promotion prospects or the grade structure? What can be done to anticipate and minimize the ill effects of such changes?
(d) Is there an 'ideal' age structure for a particular organization?

A model is essentially a description of the system together with a set of assumptions about the behaviour of the uncontrolled variables. Such assumptions may be based on two kinds of consideration which may be briefly described as *empirical* or *hypothetical.* By an empirical assumption we mean one derived from past observation of the system. A simple example, which figures prominently in what follows, is provided by wastage rates. If we examine records of leaving for people having similar length of service we often find that the number is approximately proportional to the stock of people with that length of service. Since this is consistent with the assumption that each individual has the same probability of leaving (not depending on calendar time) we might choose to incorporate this assumption in the model.

Such an assumption amounts to supposing that the pattern observed in the past will continue into the future. In many planning situations, however, we are less interested in projecting the past than in exploring a range of possible futures. That is, we are seeking the answers to 'what if...?' questions. In these circumstances we often find it useful to see the results of assuming different levels of the wastage probability because of the insight which it gives into the operation of the system. This would be an example of making hypothetical assumptions.

The empirical observation that flows are proportional to stocks is a common one which provides the basis for the widespread use of so-called Markov chain models which are the subject of this chapter. We begin, therefore, by investigating the class of models which arises when transition flows are, on average, proportional to the stocks from which they come. Generally, recruitment flows cannot be treated in this way since their source 'stocks' are indeterminate.

## 4.2 THE BASIC MARKOV CHAIN MODEL

The foregoing ideas, which have also been developed in Chapter 1 (particularly Section 1.3), can be formalized by assuming that each person has a given probability of making any particular transition. Thus, suppose that our system is divided into $k$ categories which, for brevity, but without loss of generality, will be described as grades. The *transition probabilities* between each of the grades may then be set out in an array as follows:

$$\begin{array}{cccc} p_{11} & p_{12} \cdots p_{1k} & w_1 \\ p_{21} & p_{22} \cdots p_{2k} & w_2 \\ \vdots & \vdots \quad\quad \vdots & \vdots \\ p_{k1} & p_{k2} \cdots p_{kk} & w_k \end{array}$$

where the element $p_{ij}$ is the probability that an individual in grade $i$ at the start of the time interval is in grade $j$ at the end; while $w_i$ is the probability that a member of grade $i$ at the start has left by the end of the interval. (As explained in Chapter 1 these are transition *probabilities* in the stochastic model, and transition *proportions* in the deterministic model.) The assumptions for the Markov chain are that individuals move independently and with identical probabilities which do not vary over time. Note that since each person must either stay where they are, move to another grade, or leave, the row sums

$$\sum_{j=1}^{k} p_{ij} + w_i = 1 \tag{4.1}$$

for all $i$. The matrix $\mathbf{P} = \{p_{ij}\}$ is called the *transition matrix* and the row vector $\mathbf{w} = (w_1, w_2, \ldots, w_k)$ is called the *wastage vector*. It is implicit in this specification that time is discrete and in practice the unit of time will typically be a year or a month.

The assumptions of this model are strong ones and not all of them are easy to test empirically. We shall discuss this point later but, as with all models, what matters is not whether the assumptions are precisely true but whether they are sufficiently realistic for the models to be useful in practice

The elements of $\mathbf{P}$ and $\mathbf{w}$ will have to be assigned numerical values in any application and this may be done by making hypothetical assumptions about them or by estimating the probabilities from past data. For the moment we shall by-pass this problem and return to it in Section 4.3.

The Markov specification given above is not complete because it says nothing about the recruitment flows. We begin by supposing that the total number of recruits at time $T$, denoted by $R(T)$, is given. Let these recruits be allocated to categories with probabilities $r_1, r_2, \ldots, r_k \left( \sum_{i=1}^{k} r_i = 1 \right)$ then $\mathbf{r} = \{r_i\}$ will be called the *recruitment vector*. Different versions of the Markov model will require different assumptions about the values to be given to $\{R(T)\}$.

## 4.3 THE BASIC EQUATION WHEN $\{R(T)\}$ IS FIXED

A full account of Markov models for manpower systems is given in Bartholomew (1973a). Here we shall give just enough of the theory to provide the tools for practical applications. Using the stock and flow notation introduced in Chapter 1 we have from the accounting equation (1.1a)

$$n_j(T) = \sum_{i=1}^{k} n_{ij}(T-1) + n_{0j}(T) \qquad (j = 1, 2, \ldots, k) \tag{4.2}$$

Recall that stocks relate to the point of time given by their argument and that flows, except for recruitment, refer to the unit interval whose starting-point corresponds to their argument. The transitions in the stochastic model are governed by probabilities so that the terms in (4.2) are random variables. However, since the equations are linear the same relationship will hold for their average or expected values so that

$$\bar{n}_j(T) = \sum_{i=1}^{k} \bar{n}_{ij}(T-1) + \bar{n}_{0j}(T) \qquad (j = 1, 2, \ldots, k) \tag{4.3}$$

where $\bar{n}$ denotes an expectation. Given the expected stock level at the start of the period, $\bar{n}_i(T-1)$, and the total recruitment, $R(T)$, it follows that the expected flows are

$$\left.\begin{aligned} \bar{n}_{0j}(T) &= R(T)r_j \\ \bar{n}_{ij}(T-1) &= \bar{n}_i(T-1)p_{ij} \end{aligned}\right\} (i, j = 1, 2, \ldots, k)$$

Hence

$$\bar{n}_j(T) = \sum_{i=1}^{k} \bar{n}_i(T-1)p_{ij} + R(T)r_j \qquad (j = 1, 2, \ldots, k) \tag{4.4a}$$

or, in matrix notation,

$$\bar{\mathbf{n}}(T) = \bar{\mathbf{n}}(T-1)\mathbf{P} + R(T)\mathbf{r} \tag{4.4b}$$

This is the basic prediction equation which occupies a prominent role in most applications of Markov chain models. A simple computer program named BASEQN (from BASic EQuatioN) for making computations with (4.4b) is given in Appendix A and we shall illustrate its use before proceeding.

The interested reader with access to a computer is recommended to set up this program and to work through the examples and the exercises in this chapter and the next.

*Example 4.1* This simple example concerns a three-grade management hierarchy in which the only internal transitions are promotions into the next higher grade. The initial stock vector is

$$\mathbf{n}(0) = [180, 145, 35]$$

and the parameters are

$$\mathbf{P} = \begin{bmatrix} 0.70 & 0.10 & 0 \\ 0 & 0.85 & 0.05 \\ 0 & 0 & 0.85 \end{bmatrix}$$

$$R(T)\mathbf{r} = [52, 15, 3]$$

Two things should be noted about this specification. First, we have not given (and, in general, will not give) the wastage vector since this can be deduced from $\mathbf{P}$ using (4.1). This reduces the input requirements of the program. The second point concerns recruitment. We defined the vector $\mathbf{r}$ as a set of probabilities. The vector $R(T)\mathbf{r}$ is thus the vector of expected numbers of recruits. As far as predicting expected stock numbers is concerned, it makes no difference whether we regard these as actual numbers or as expectations. In this example, therefore, we shall be investigating the effect of having the expected number of entrants at each level as given by this vector for each $T$. Printout 4.1 gives the results of repeated calculations using (4.4) up to $T = 99$. Full details of the program are given in Appendix A but the format is designed to be almost self-explanatory. The upper part of the printout (down to RUN) contains the parameters and specifications for the model and the lower part the table of results. The program is interactive and the appropriate parameter values required have to be typed in by the user after each question mark. The data prompts or input symbols printed by the computer before each question

```
  K =     ?3
  N =     ?180,145,35
  P =     ?.70,.10,0.     0,.85,.05,     0,0,.85
  R =     ?52,15,3,     1,0,0
T,% =     ?5,YES
<*>     ?RUN

  T           1           2           3           TOTAL        R

  0       180(50%)    145(40%)     35(10%)      360(100%)
  1       178(48%)    156(42%)     40(11%)      374(104%)     70
  2       177(46%)    166(43%)     45(12%)      387(108%)     70
  3       176(44%)    173(44%)     49(12%)      398(111%)     70
  4       175(43%)    180(44%)     54(13%)      409(113%)     70
  5       174(42%)    185(44%)     58(14%)      418(116%)     70
<*>     ?T=,10
 10       174(39%)    202(45%)     73(16%)      449(125%)     70
<*>     ?T=.20
 20       173(37%)    213(45%)     87(18%)      473(131%)     70
<*>     ?T=,99
 99       173(36%)    216(45%)     92(19%)      481(134%)     70
<*>     ?
```

Printout 4.1 For explanation see Example 4.1

mark are chosen as near as possible to those used in the development of the theory. Thus $K$ is the number of grades, $N$ is the initial stock vector (i.e. $\mathbf{n}(0)$), $P$ is the transition matrix, and $R$ refers to the recruitment term. The data for the latter may be entered in a variety of forms according to the version of the model in use, with the final three numbers in this line of data input specifying this form. Details of these forms are given in Appendix A, and will also be explained as necessary in the worked examples. In this case the use of $(1, 0, 0)$ indicates that the first $k = 3$ numbers in this line are the actual numbers $\{n_{0j}(T)\}$ to be recruited into each grade. Thus $R(T)\mathbf{r} = (52, 15, 3)$. The final data input line in response to the prompt '$T$, % =', must contain two pieces of information specifying, in this order, the length of time for which the projection is to run (this can easily be extended as in this example), and whether or not the stock numbers in the output should be expressed as percentages.

The '⟨*⟩' prompts indicate control points at which the user can give a number of different commands—again, full details can be found in Appendix A. The first command (RUN) used in Printout 4.1 instructs the program to print results up to and including the specified $T$ value which in this case is $T = 5$. The next three control commands shown request printout of the projected results at the indicated time points, while the final command ('END') terminates the run.

The output table thus gives the predicted stock vectors for each year for $T \leqslant 5$, while three values are given for larger $T$ to illustrate the long-run behaviour. The stock number in each grade is also shown expressed as a percentage of the total size. This latter appears in the 'Total' column, where the percentages in this column are obtained by comparing the current size with the initial size and are useful for seeing how the total size of the system is changing. The last column gives $R(T)$, which is superfluous in this particular example because it was directly specified in the input.

The calculations show an expected steady growth during the next 5 years, of 3 or 4 per cent per annum with a tendency for the top grade to grow at the expense of the bottom. The final three rows of the table show that the stock vector approaches a limit. This does not necessarily happen but in most practical applications such a limit does exist and in Chapter 7 we shall see that much can be learnt from the study of limiting steady-state behaviour. ◀

It follows from (4.4) that if there is a limiting vector, $\mathbf{n}$ say, it must satisfy

$$\mathbf{n} = \mathbf{nP} + R\mathbf{r} \tag{4.5a}$$

where $R$ is the limit of the sequence $\{R(T)\}$ which must be assumed to exist for this purpose. Equation (4.5a) can be solved directly to give

$$\mathbf{n} = R\mathbf{r}(\mathbf{I} - \mathbf{P})^{-1} \tag{4.5b}$$

or $\mathbf{n}$ can be calculated by repeated use of (4.4) as in the example. The inverse matrix in (4.5b) always exists for matrices $\mathbf{P}$ with row sums strictly less than one.

The following three examples illustrate something of the range and flexibility of the model as well as bringing to light other important characteristics of graded systems having constant transition rates.

*Example 4.2* The parameters are chosen to be typical of what might be expected in a five-grade hierarchical system. The calculations shown assume the following parameter values:

$$\mathbf{n}(0) = (400, 300, 150, 100, 50)$$

$$\mathbf{P} = \begin{bmatrix} 0.65 & 0.20 & 0 & 0 & 0 \\ 0 & 0.70 & 0.15 & 0 & 0 \\ 0 & 0 & 0.75 & 0.15 & 0 \\ 0 & 0 & 0 & 0.85 & 0.10 \\ 0 & 0 & 0 & 0 & 0.95 \end{bmatrix}$$

$$R(T) = 102 \text{ for all } T, \qquad \mathbf{r} = (0.75, 0.25, 0, 0, 0)$$

These show a system in which promotion prospects fall as one progresses up the hierarchy. Similarly the wastage rates fall from 0.15 in the lowest two grades to 0.05 at the top. Recruitment is into the bottom two grades only and the constant total input of 102 was selected to keep the total size roughly constant. The results of the projection are given in Printout 4.2.

```
=    ?5
=    ?400,300,150.100,50
=    ?.65,.20.0,0,0,   0,.7,.15,0,0,   0,0,.75,.15,0,   0,0,0,.85,.10
       0,0,0,0,.95
=    ?.75,.25,0,0,0,    102.+.0
=    ?5,YES
?RUN

          1           2           3           4           5          TOTAL       R

    400(40%)    300(30%)    150(15%)    100(10%)     50( 5%)    1000(100%)
    336(35%)    315(32%)    157(16%)    107(11%)     57( 6%)     975( 97%)     102
    295(31%)    314(33%)    165(17%)    115(12%)     65( 7%)     955( 95%)     102
    268(29%)    304(32%)    171(18%)    123(13%)     74( 8%)     940( 94%)     102
    251(27%)    292(31%)    174(19%)    130(14%)     82( 9%)     929( 93%)     102
    240(26%)    280(30%)    174(19%)    136(15%)     91(10%)     922( 92%)     102
?T=,10
    221(24%)    243(27%)    159(17%)    154(17%)    137(15%)     915( 91%)     102
?T=.15
    219(24%)    233(25%)    146(16%)    153(17%)    176(19%)     927( 93%)     102
?T=,20
    219(23%)    231(25%)    141(15%)    148(16%)    204(22%)     943( 94%)     102
?T=,25
    219(23%)    231(24%)    139(15%)    143(15%)    224(23%)     956( 96%)     102
?T=.30
    219(23%)    231(24%)    139(14%)    141(15%)    238(25%)     966( 97%)     102
?T=,99
    219(22%)    231(23%)    138(14%)    138(14%)    276(28%)    1002(100%)     102
?
```

Printout 4.2 For explanation see Example 4.2

Although the total size changes relatively little, the structure changes radically. As in the previous example there is a steady growth at the top accompanied by contraction at the bottom. In fact after 10 years the top grade is projected to treble in size and the system eventually becomes very top-heavy. This is a common occurrence in practice (see, for example, Young (1971)) and yet there is nothing very remarkable about the promotion and wastage probabilities of the example. Indeed a more detailed analysis would show that almost any 'reasonable' set of parameters would lead to similar top-heaviness. Such pressures for growth at the top, sometimes recognized under the term 'grade-drift', are often felt in organizations where they are sometimes cynically attributed to the operation of Parkinson's law. Calculations such as these show that they may arise simply because a tapering grade structure is virtually incompatible with constant transition rates of the levels regarded as desirable in practice. For a fuller theoretical discussion of this point the reader should consult Bartholomew (1973a, Chapter 3, particularly pp. 62–64). ◀

*Example 4.3* This example is based on data from Gani (1963) concerning the university system of Australia. The reader should now be familiar with the input format so the parameter values are not listed separately but will be found at the top of Printout 4.3.

```
  K =     ?4
  N =     ?600,400,330,70
  P =     ?.15,.61,0,0,   0,.11,.71,0,   0,0,.10,.20,   0,0,0,.05
  R =     ?500,0,0,0,    1,0.0
T,% =     ?3,YES
<*>    ?RUN

 T           1          2          3           4            TOTAL       R

 0       600(43%)   400(29%)   330(24%)    70( 5%)     1400(100%)
 1       590(43%)   410(30%)   317(23%)    69( 5%)     1387( 99%)     500
 2       588(43%)   405(29%)   323(23%)    67( 5%)     1383( 99%)     500
 3       588(43%)   404(29%)   320(23%)    68( 5%)     1380( 99%)     500
<*>    ?*R=, 700,0,0.0,    1,0,0,    2,YES
 4       788(50%)   403(26%)   318(20%)    67( 4%)     1577(113%)     700
 5       818(47%)   525(30%)   318(18%)    67( 4%)     1729(123%)     700
<*>    ?*R=, 500,0,0,0.    1.0,0,    5,YES
 6       623(38%)   557(34%)   405(25%)    67( 4%)     1651(118%)     500
 7       593(38%)   441(28%)   436(28%)    84( 5%)     1555(111%)     500
 8       589(41%)   411(28%)   357(25%)    91( 6%)     1448(103%)     500
 9       588(42%)   404(29%)   327(23%)    76( 5%)     1396(100%)     500
10       588(43%)   403(29%)   320(23%)    69( 5%)     1381( 99%)     500
<*>    ?T=,99
99       588(43%)   403(29%)   318(23%)    67( 5%)     1376( 98%)     500
<*>    ?
```

Prinout 4.3 For explanation see Example 4.3

In this example the 'grades' are years of study so that grade 1 is all first-year students, grade 2 consists of second-years, and so on; grade 4 is an honours year. In contrast to Example 4.2 the main diagonal elements of **P** are relatively small and 'promotion' probabilities are large. This is because promotion now refers to normal progression to the next level. Those not promoted either drop out or repeat the year. The calculations show the effect of an age 'bulge' passing through the system such as might result from a peak in the birth rate 18 years or so prior to entry. All admissions are into the lowest grade and the system starts in an approximate steady state (compare $\mathbf{n}(0)$ with $\mathbf{n}(99)$). The total rises from 500 for each of the first three years to 700 for two years before returning to the original level. The peak comes one year later at each level and it is progressively reduced in absolute size as it moves up the hierarchy although relative to the size of each grade it remains constant. Its effect also lasts for longer higher up the hierarchy.

Notice especially that there is no long-term distortion of the grade structure which remains remarkably stable. Also, the system returns to its steady state by $T = 10$, which is much quicker than in the other examples and is due in this case to the small diagonal elements in **P**. These cause people to be forced through the system relatively quickly. Compare this with the hierarchical grade structures with large diagonal elements in **P** and typical lifetimes within the system of up to 40 years. ◀

All of the examples so far have dealt with simple hierarchies in which individuals move up by one step at a time. The following example gives results for a more complicated network as illustrated in Figure 4.1.

*Example 4.4* A non-hierarchical system. (See Figure 4.1 and Printout 4.4.)◀

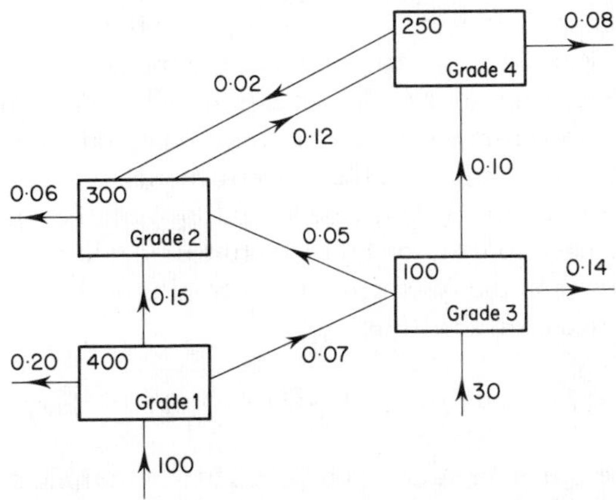


Figure 4.1 Diagram of the system for Example 4.4 showing stocks and flow rates

```
  K =    ?4
  N =    ?400,300,100,250
  P =    ?.58,.15,.07,0,   0,.82,0,.12,   0,.05,.71,.10,   0,.02,0,.90
  R =    ?100,0,30,0.    1,0,0
T,% =    ?5,YES
<*>    ?RUN

  T        1           2           3           4            TOTAL          R

  0     400(38%)    300(29%)    100(10%)    250(24%)    1050(100%)
  1     332(32%)    316(30%)    129(12%)    271(26%)    1048(100%)      130
  2     293(28%)    321(30%)    145(14%)    295(28%)    1053(100%)      130
  3     270(25%)    320(30%)    153(14%)    318(30%)    1061(101%)      130
  4     256(24%)    317(30%)    158(15%)    340(32%)    1071(102%)      130
  5     249(23%)    313(29%)    160(15%)    360(33%)    1082(103%)      130
<*>    ?T=,10
 10     239(21%)    299(26%)    161(14%)    429(38%)    1128(107%)      130
<*>    ?T=,15
 15     238(21%)    295(25%)    161(14%)    465(40%)    1159(110%)      130
<*>    ?T=,20
 20     238(20%)    296(25%)    161(14%)    486(41%)    1180(112%)      130
<*>    ?T=,99
 99     238(19%)    301(25%)    161(13%)    522(43%)    1222(116%)      130
<*>    ?
```

Printout 4.4 Showing the projected grade sizes for Example 4.4

## 4.4 THE FIXED-SIZE MODEL

The assumption of a fixed, or at any rate known, flow of recruits is a natural one when we are exploring the consequences of various recruitment strategies as in Example 4.3. In the case of a firm with a plentiful supply of labour, however, it is more likely that the number of recruits will be determined by the vacancies occurring in the system. In the following chapter we shall consider such a situation when all the grade sizes are fixed. First, however, we shall consider an intermediate case when the total size only is fixed. This is particularly relevant when the categories of the model are age or length-of-service groups but there can be other situations where individual grade sizes are allowed to vary within a fixed global total. When this is the case we shall speak of the *fixed-size* version of the Markov model.

The number of recruits will now be a random variable composed of two parts. The first part consists of those recruited to fill any new vacancies arising from growth in the system and the second those who replace leavers. The expected value of $R(T)$ is thus

$$\bar{R}(T) = N(T) - N(T-1) + \sum_{i=1}^{k} \bar{n}_i(T-1)w_i \tag{4.6}$$

where $N(T)$ is the size of the system at time $T$. There is nothing in the argument to prevent $N(T) - N(T-1)$ being negative, providing that $\bar{R}(T)$ is positive. If this is not the case it means that the fixed size can only be achieved by having redundancies (i.e. negative recruitment). This presents no problems algebraically although the model may need some re-interpretation.

Substituting (4.6) in (4.4b) we find

$$\bar{\mathbf{n}}(T) = \bar{\mathbf{n}}(T-1)\{\mathbf{P} + \mathbf{w}'\mathbf{r}\} + M(T)\mathbf{r} \tag{4.7}$$

where $M(T) = N(T) - N(T-1)$.

Each term in this equation can be identified in the following way:

$\bar{\mathbf{n}}(T-1)\mathbf{P}$ represents normal internal movements;
$\bar{\mathbf{n}}(T-1)\mathbf{w}'\mathbf{r}$ represents recruits who replace leavers;
$M(T)\mathbf{r}$ represents recruits filling new or created vacancies.

We shall write $\mathbf{Q} = \mathbf{P}+\mathbf{w}'\mathbf{r}$. This emphasizes that (4.7) has the same form as (4.4b), with $\mathbf{Q}$ as the 'transition' matrix whose typical element is $p_{ij} + w_i r_j$. The first term $p_{ij}$ corresponds to a real direct flow from $i$ to $j$, while $w_i r_j$ can be interpreted as a hypothetical indirect flow comprising that part of the wastage flow from $i$ which goes back to $j$ as recruitment.

Since the row sums of $\mathbf{Q}$ are all 1, $\mathbf{Q}$ is a stochastic matrix. Thus if $M(T) = 0$ for all $T$ we have a constant-size system and in this case (4.7) is essentially the same as the basic equation of Markov-chain theory. (An account of this theory can be found in Kemeny and Snell (1960).)

The other point to observe about (4.7) having the same form as (4.4b) is that the same computer program can be used for both, as in the following examples.

*Example 4.5a* If we take the transition matrix of Example 4.1 with $\mathbf{r} = (1, 0, 0)$ then

$$\mathbf{Q} = \mathbf{P} + \mathbf{w}'\mathbf{r}' = \begin{bmatrix} 0.70 & 0.10 & 0 \\ 0 & 0.85 & 0.05 \\ 0 & 0 & 0.85 \end{bmatrix} + \begin{bmatrix} 0.20 & 0 & 0 \\ 0.10 & 0 & 0 \\ 0.15 & 0 & 0 \end{bmatrix} = \begin{bmatrix} 0.90 & 0.10 & 0 \\ 0.10 & 0.85 & 0.05 \\ 0.15 & 0 & 0.85 \end{bmatrix}$$

A projection of the grade sizes under the assumption of no expansion or contraction is given in Printout 4.5 using the same program as in the earlier examples. Thus the matrix $\mathbf{Q}$ was entered in place of $\mathbf{P}$, while the data for $R$, which corresponds to the term $M(T)\mathbf{r}$ of (4.7), were set equal to zero. ◀

Using the program in this way requires the preliminary calculation of $\mathbf{Q}$. To avoid this the program BASEQN contains special recruitment options (see (d) and (e) in Appendix A, Section A3) in which the program requires the input $\mathbf{P}$ rather than $\mathbf{Q}$, and also calculates directly the values of $R(T)$ required to make the total size of the system expand or contract as desired. The use of these options and the specification of the expansion and contraction is made by means of the last three data items input in response to the '$R$'

```
    K =      ?3
    N =      ?180,145,35
    P =      ?.90,.10,0,     .10,.85,.05.     .15,0,.85
    R =      ?0,0,0,     1,0.0
  T,% =      ?5,YES
  <*>     ?RUN

    T            1              2             3             TOTAL        R

    0         180(50%)     145(40%)      35(10%)        360(100%)
    1         182(50%)     141(39%)      37(10%)        360(100%)       0
    2         183(51%)     138(38%)      39(11%)        360(100%)       0
    3         185(51%)     136(38%)      40(11%)        360(100%)       0
    4         186(52%)     134(37%)      40(11%)        360(100%)       0
    5         187(52%)     132(37%)      41(11%)        360(100%)       0
  <*>     ?T=,99
   99         191(53%)     127(35%)      42(12%)        360(100%)       0
  <*>     ?
```

Printout 4.5 For explanation see Example 4.5a

prompt. These options have the further advantage that the numbers output in the '*R*' column of the results table are the total recruitment numbers and not the numbers of additional vacancies, $M(T)$, as in Example 4.5a where the total recruitment numbers are not shown and would have to be calculated separately.

*Example 4.5b* Printout 4.6 shows a specimen calculation with this option using the same matrix as in Example 4.5a but with the historical recruitment vector (0.743, 0.214, 0.043) and with the total size of the system growing at $M(T) = 30$ for all $T$. Notice the slow but steady growth at the top and the continuing rise in the number of recruits required to support the expansion. ◀

```
    K =     ?3
    N =     ?180,145,35
    P =     ?.70,.10,0,     0,.85,.05.     0,0,.85
    R =     ?.743,.214,.043,      -1,+,30
  T,% =     ?5,YES
  <*>    ?RUN

    T            1              2             3             TOTAL        R

    0         180(50%)     145(40%)      35(10%)        360(100%)
    1         190(49%)     160(41%)      41(10%)        390(108%)      86
    2         200(48%)     174(41%)      46(11%)        420(117%)      90
    3         210(47%)     188(42%)      52(12%)        450(125%)      94
    4         220(46%)     202(42%)      58(12%)        480(133%)      99
    5         231(45%)     216(42%)      64(13%)        510(142%)     103
  <*>    ?T=,10
   10         284(43%)     283(43%)      93(14%)        660(183%)     125
  <*>    ?T=,20
   20         392(41%)     418(44%)     151(16%)        960(267%)     168
  <*>    ?T=,99
   99        1246(37%)    1480(44%)     604(18%)       3330(925%)     513
  <*>    ?
```

Printout 4.6 For explanation see Example 4.5b

*Example 4.6* As a further illustration of the fixed-size model we use data from the officers' system of one of the British women's services taken from Forbes (1971a). The calculation is designed to investigate the effect of having two periods of constant total size interrupted by a 10-year interval of linear growth. The results are given in Printout 4.7.

It is clear that the change in the recruitment flow has an impact at every level but that its effect on the actual grade sizes is much more marked in the

```
  K =    ?4
  N =    ?129,74,28,11
  P =    ?.728,.102,0,0,   0,.83,.046,0,   0,0,.867,.033,   0,0,0,.902
  R =    ?1,0,0,0,    -1,+,0
T,% =    ?0,YES
<*>    ?RUN

 T            1          2          3          4           TOTAL      R

 0        129(53%)    74(31%)    28(12%)    11( 5%)     242(100%)
<*>    ?T=,5
 5        129(53%)    76(31%)    27(11%)    10( 4%)     242(100%)    35
<*>    ?*R=, 1,0.0,0,    -1,+.24,    1,YES
 6        153(57%)    76(29%)    27(10%)    10( 4%)     266(110%)    59
<*>    ?T=,9
 9        212(63%)    89(26%)    27( 8%)    10( 3%)     338(140%)    71
<*>    ?T=,12
12        261(64%)   110(27%)    29( 7%)    10( 2%)     410(169%)    82
<*>    ?T=.15
15        304(63%)   134(28%)    33( 7%)    10( 2%)     482(199%)    93
<*>    ?*R=, 1.0.0,0,    -1,+.0,    1.YES
16        294(61%)   142(30%)    35( 7%)    10( 2%)     482(199%)    73
<*>    ?T=,20
20        272(56%)   157(33%)    42( 9%)    11( 2%)     482(199%)    71
<*>    ?T=,25
25        262(54%)   159(33%)    49(10%)    13( 3%)     482(199%)    70
<*>    ?T=,30
30        259(54%)   157(33%)    52(11%)    14( 3%)     482(199%)    70
<*>    ?T=,99
99        257(53%)   154(32%)    53(11%)    18( 4%)     482(199%)    70
<*>    ?
```

Printout 4.7 For explanation see Example 4.6

lower grades. What is, perhaps, more surprising is that the relative grade sizes do not show more change over the 30-year period. The forecast of the total recruitment required shows a very sudden drop immediately the expansion ceases. This is characteristic of the effect of abrupt changes in policy. If this were unacceptable in practice the program could be used to investigate the effect on recruitment of smoothing the transition from the 'growth' to 'no growth' situations. ◀

## 4.5 THE MARKOV MODEL APPLIED TO AGE AND LENGTH-OF-SERVICE DISTRIBUTIONS

Markov models have a particularly valuable role to play in the study of age and length-of-service distributions. Bulges in an age distribution can lead to promotion bottlenecks whereas lack of people in particular age groups may make it difficult to find suitable candidates for promotion. When a bulge reaches the retirement ages an excessive demand for recruits to fill the vacancies will lead to a new bulge at the recruitment ages and so perpetuate the problem. It is therefore important in practice to identify and anticipate irregularities in the age distribution. A model can help to do this and can also provide a means of evaluating the likely effects of strategies for alleviating problems which arise.

In this section we revert to a homogeneous system in which grades are not distinguished. This restriction will be removed later. First we consider a population in which people are classified according to length of service. Suppose also that each length-of-service category is the same width as the discrete time interval of the model. Under these circumstances there are only two possible transitions open to an individual; either he must leave or increase his length of service by one time unit. This means that the $k \times (k+1)$ array of transition probabilities will have the following form

$$\begin{matrix} 0 & p_{12} & 0 & \ldots 0 & w_1 \\ 0 & 0 & p_{23} & \ldots 0 & w_2 \\ \vdots & \vdots & \vdots & \vdots & \vdots \\ 0 & 0 & 0 & \ldots p_{k-1,k} & w_{k-1} \\ 0 & 0 & 0 & \ldots 0 & 1 \end{matrix}$$

The $w$'s are the length of service specific wastage rates and, of course,

$$p_{i,i+1} = 1 - w_i \qquad (i = 1, 2, \ldots, k-1)$$

$k$ is the maximum length of service after which everyone must leave. Sometimes it is necessary for the last category to include those with $k$ *or more* years service in which case the last two elements in the final row will be $p_{k,k}$ and $w_k = (1 - p_{k,k})$.

If the length-of-service specific wastage rates can be assumed constant through time then the Markov model can be used to project the length-of-service structure. In this application it is particularly important to know the steady-state structure associated with the system since this is, in a sense, the ideal structure. It has the property that once achieved, it will be maintained as long as the parameters are unchanged. This aspect will be developed in

Chapter 7 but it is easy to calculate the steady-state structure when **P** has the very special form given above. A simple argument from first principles, which is equivalent to (4.5a), depends on the fact that, in the steady state, the expected inflow and outflow from each category must be the same. Thus

$$\left.\begin{aligned} Rr_1 &= n_1 \\ n_{j-1}p_{j-1,j} + Rr_j &= n_j \qquad (j = 2, \ldots, k-1) \\ n_{k-1}p_{k-1,k} + Rr_k &= n_k w_k \end{aligned}\right\} \tag{4.8}$$

where we know all $p_{ij}$ and $r_i$. If the input is fixed, $R$ is also known so that starting with $n_1 = Rr_1$ the steady-state structure, $n_i$, can then be computed. The calculation is a little more complicated in the case of the fixed-size model because then $R = \sum_{i=1}^{k} n_i w_i$. However, $R$ can be taken as an arbitrary constant so long as the result is scaled to make $\sum_{i=1}^{k} n_i = N$ (see the solution to Exercise 4.5b).

*Example 4.7* Take a system in which the length-of-service classes and associated wastage rates are as follows:

| *Class* | 1 | 2 | 3 |
|---|---|---|---|
| Length of service (yrs) | 0–1 | 1–2 | 2 or more |
| Wastage probability | 0.4 | 0.2 | 0.1 |

This shows a declining propensity to leave with increasing length of service which we saw to be typical in Chapters 2 and 3. Let the recruitment be 100 per unit time (which must, of course, be into the lowest class) then the parameters are

$$\mathbf{P} = \begin{bmatrix} 0 & 0.6 & 0 \\ 0 & 0 & 0.8 \\ 0 & 0 & 0.9 \end{bmatrix}, \qquad R\mathbf{r} = (100, 0, 0)$$

The steady state, derived from (4.8) is

$$\mathbf{n} = (100\,(16\%),\ 60\,(9\%),\ 480\,(75\%)),\ \text{total size } 640 \tag{4.9}$$

It is interesting to see the effect of changes in the wastage rate on the size and structure. If the rates are all halved then with the same recruitment

$$\mathbf{n} = (100\,(6\%),\ 80\,(5\%),\ 1440\,(89\%)),\ \text{total size } 1620 \tag{4.10}$$

Thus the total size is more than doubled and the length-of-service structure has relatively more people at the upper end. To see the effect of halving the wastage rates and keeping the total size fixed we must use the fixed-size version of the model which gives

$$\mathbf{n} = (40\ (6\%),\ 32\ (5\%),\ 569\ (89\%)),\ \text{total size } 640 \tag{4.11}$$

The structure here is obviously the same as in the previous case but the main point of interest is that the annual recruitment is now only 40.

These calculations tell us only about what will happen eventually if a change in wastage takes place. There will, of course, be a transitional phase which can be studied using the BASEQN program. Taking the initial structure as in (4.9), the **P** associated with the reduced wastage rates, and the recruitment level corresponding to the steady state shown in (4.10), we obtain the results in Printout 4.8.

```
  K =    ?3
  N =    ?100,60.480
  P =    ?0,.80,0,    0,0,.90,    0,0,.95
  R =    ?100,0,0,    1.0.0
T,% =    ?5,YES
<*>    ?RUN

  T           1           2           3            TOTAL        R

  0       100(16%)     60( 9%)   480(75%)     640(100%)
  1       100(14%)     80(12%)   510(74%)     690(108%)      100
  2       100(14%)     80(11%)   556(76%)     737(115%)      100
  3       100(13%)     80(10%)   601(77%)     781(122%)      100
  4       100(12%)     80(10%)   643(78%)     823(129%)      100
  5       100(12%)     80( 9%)   683(79%)     863(135%)      100
<*>    ?T=,10
 10       100(10%)     80( 8%)   854(83%)    1034(162%)      100
<*>    ?T=,20
 20       100( 8%)     80( 6%)  1089(86%)    1269(198%)      100
<*>    ?T=,30
 30       100( 7%)     80( 6%)  1230(87%)    1410(220%)      100
<*>    ?T=,99
 99       100( 6%)     80( 5%)  1434(89%)    1614(252%)      100
<*>    ?
```

Printout 4.8 For explanation see Example 4.7

The speed at which each class approaches its steady-state value varies markedly. It is rapid in the middle class but rather slow at the top where after 10 years only 40 per cent of the eventual change has taken place.

If the total size is to be held constant the fixed-size version of the program should be used leading to the result shown in Printout 4.9.

Here the new equilibrium is reached much more quickly because the initial and final structures are much closer. Notice too the big drop in expected recruitment from the present level of 100. This feature of the example serves to illustrate a very practical point: even quite small changes in growth can imply relatively large changes in the recruitment levels.

```
     K =    ?3
     N =    ?100,60,480
     P =    ?0,.8,0,     0,0,.9,      0,0,.95
     R =    ?1,0,0,      -1,+,0
   T,% =    ?5,YES
   <*>    ? RUN

     T            1            2            3            TOTAL        R

     0       100(16%)     60( 9%)    480(75%)       640(100%)
     1        50( 8%)     80(12%)    510(80%)       640(100%)       50
     2        44( 7%)     40( 6%)    556(87%)       640(100%)       44
     3        41( 6%)     35( 5%)    565(88%)       640(100%)       41
     4        40( 6%)     32( 5%)    568(89%)       640(100%)       40
     5        40( 6%)     32( 5%)    569(89%)       640(100%)       40
   <*>    ?
```

Printout 4.9 For explanation see Example 4.7

Comparing Printouts 4.8 and 4.9 we see that a drop in the expansion rate from approximately 7 to 0 per cent implies in this system a drop in the level of recruitment from 100 to 40: a factor of $\frac{2}{3}$. ◀

It is not necessary that the class intervals should be the same as the time unit adopted for the model. When projecting an age distribution, for example, one might wish to adopt a five-year grouping for age but make projections at one-year intervals. In these circumstances three transitions will be possible to each individual—to stay in the same class, move up one class or leave. The matrix **P** will thus have non-zero entries on the main diagonal as well as on the superdiagonal. Apart from the problem of specifying the probabilities of moving up one class (again these are usually based on the observed historical proportions) there are no practical difficulties in using a Markov model. There are, however, theoretical difficulties arising from the fact that if the states of a Markov chain are amalgamated the resulting process is not, in general, Markovian. In our experience this problem is best dealt with empirically using the kind of considerations discussed in Section 4.3.

*Example 4.8* Consider a projection of an age distribution with three classes and the following parameters:

| *Age class* | 1 | 2 | 3 |
|---|---|---|---|
| Age (years) | 20–30 | 30–55 | 55 + |
| Wastage probability | 0.25 | 0.05 | 0.20 |

$$\mathbf{n}(0) = (357, 153, 15)$$

$$\mathbf{P} = \begin{bmatrix} 0.72 & 0.03 & 0 \\ 0 & 0.93 & 0.02 \\ 0 & 0 & 0.80 \end{bmatrix}$$

Printout 4.10 shows what happens when this system starts in a stationary state, passes through a five-year period of expansion at 15 per cent annum, and thereafter remains at a constant size. Recruitment is calculated by the program using one of the special options, to give 15 per cent expansion per annum for the first 5 years (i.e. $T \leqslant 5$) and a constant size thereafter (i.e. $T > 5$). Unfortunately, the rather coarse grouping partly obscures the growth and evolution of the age bulge. Note particularly how the demand for recruits increases rapidly during the expansion phase and then immediately drops when expansion ceases. The input data and output are shown in Printout 4.10. ◀

```
    K =     ?3
    N =     ?357.153,15
    P =     ?.72,.03.0,     0,.93,.02,      0,0,.80
    R =     ?1.0,0,     -1.*.1.15
  T,% =     ?5,YES
  <*>    ? RUN

    T             1            2           3            TOTAL      R

    0      357(68%)     153(29%)     15( 3%)      525(100%)
    1      436(72%)     153(25%)     15( 2%)      604(115%)     179
    2      524(75%)     155(22%)     15( 2%)      694(132%)     210
    3      623(78%)     160(20%)     15( 2%)      798(152%)     246
    4      735(80%)     168(18%)     15( 2%)      918(175%)     287
    5      862(82%)     178(17%)     16( 1%)     1056(201%)     333
  <*>    ?*R=, 1,0,0,     -1.*.1.0,      5,YES
    6      848(80%)     191(18%)     16( 2%)     1056(201%)     228
    7      836(79%)     203(19%)     17( 2%)     1056(201%)     225
    8      824(78%)     214(20%)     17( 2%)     1056(201%)     222
    9      814(77%)     224(21%)     18( 2%)     1056(201%)     220
   10      804(76%)     233(22%)     19( 2%)     1056(201%)     218
  <*>    ?T=,15
   15      768(73%)     265(25%)     23( 2%)     1056(201%)     211
  <*>    ?T=,20
   20      747(71%)     283(27%)     26( 2%)     1056(201%)     207
  <*>    ?T=.30
   30      727(69%)     300(28%)     29( 3%)     1056(201%)     203
  <*>    ?T=,99
   99      718(68%)     308(29%)     31( 3%)     1056(201%)     201
  <*>    ?
```

Printout 4.10 For explanation see Example 4.8

## 4.6 THE MARKOV MODEL WITH GRADE- AND SENIORITY-SPECIFIC TRANSITION RATES

In the foregoing sections we have modelled manpower systems in which wastage was grade specific and length-of-service (or age) specific but not both. In practice, propensity to leave certainly depends on length of service or seniority and it is quite likely to depend on grade also. Likewise, promotion probabilities often depend on the length of time a person has spent in their current grade. In the interests of realism it is thus desirable to define the classes of the model so as to allow for these variations. It is one of the great practical advantages of the Markov model that the classes can be defined in whatever way is most appropriate for the problem in hand. Indeed, the selection of appropriate classes is one of the most important steps in any application and we shall offer some advice on how to do this in the following section. The next example illustrates an application to a three-grade system in which the transition probabilities are functions of both grade and seniority within the grade.

*Example 4.9* The basic data are given on Figure 4.2 where the various transition probabilities are shown against arrows marking the flows and the seniority bands are given in brackets. The initial structure is assumed to be

$$\mathbf{n}(0) = (150, 160, 170 \mid 100, 100, 100 \mid 200)$$

where the order of the elements corresponds to the numbering of the classes in Figure 4.2. The various transition probabilities may be collected together to give

$$\mathbf{P} = \left[\begin{array}{ccc|ccc|c} 0.70 & 0.10 & 0 & 0 & 0 & 0 & 0 \\ 0 & 0.78 & 0.02 & 0.10 & 0 & 0 & 0 \\ 0 & 0 & 0.95 & 0 & 0 & 0 & 0 \\ \hline 0 & 0 & 0 & 0.88 & 0.07 & 0 & 0 \\ 0 & 0 & 0 & 0 & 0.83 & 0.05 & 0.10 \\ 0 & 0 & 0 & 0 & 0 & 0.95 & 0 \\ \hline 0 & 0 & 0 & 0 & 0 & 0 & 0.90 \end{array}\right]$$

If the total size remains constant the projected grade sizes are as in Printout 4.11.

The total numbers in the grades change relatively slightly but there are

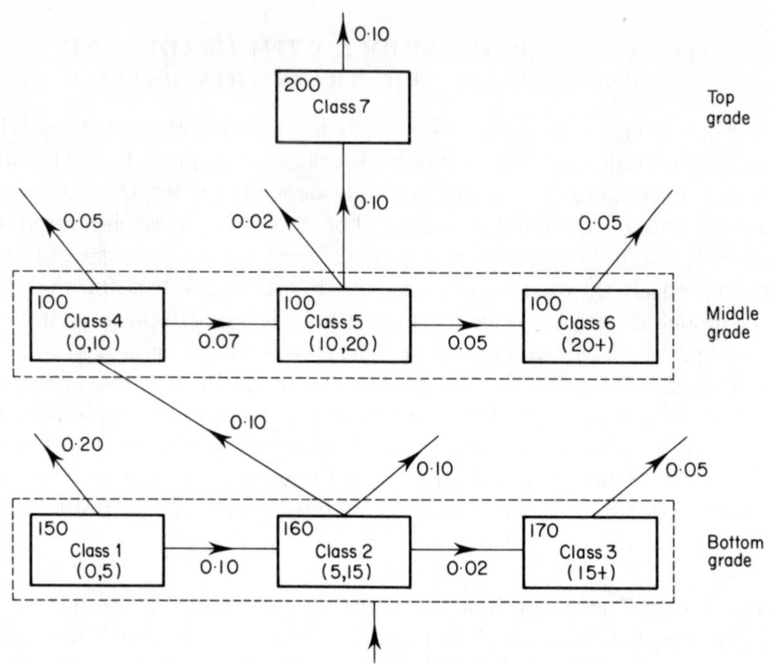


Figure 4.2 Depicting the system of Example 4.9

```
  K =    ? 7
  N =    ?  150,   160,   170,   100,   100,   100,   200
  P =    ?  .70,   .10,     0,     0,     0,     0,     0
?             0,   .78,   .02,   .10,     0,     0,     0
?             0,     0,   .95,     0,     0,     0,     0
?             0,     0,     0,   .88,   .07,     0,     0
?             0,     0,     0,     0,   .83,   .05,   .10
?             0,     0,     0,     0,     0,   .95,     0
?             0,     0,     0,     0,     0,     0,   .90
  R =    ? 1, 0, 0, 0, 0, 0, 0,        -1,+,0
T,% =    ?5,NO
<*>    ? RUN

  T          1      2      3      4      5      6      7          TOTAL      R

  0        150    160    170    100    100    100    200      980(100%)
  1        192    140    165    104     90    100    190      980(100%)     87
  2        226    128    159    106     82     99    180      980(100%)     92
  3        254    123    154    106     75     99    170      980(100%)     96
  4        277    121    149    105     70     97    161      980(100%)     99
  5        296    122    144    105     65     96    152      980(100%)    103
<*>    ?T=,10
 10        356    142    123    106     52     88    113      980(100%)    112
<*>    ?T=,20
 20        391    171     99    126     49     72     71      980(100%)    118
<*>    ?T=,99
 99        395    179     72    149     62     61     61      980(100%)    118
<*>    ?
```

Printout 4.11 For explanation see Example 4.9

big changes in the seniority structure within the bottom two grades. There is a growing concentration in the shorter seniority groups and this accounts for the rising number of recruits needed since wastage rates tend to be higher in the less senior classes. ◀

## 4.7 PROBLEMS OF IMPLEMENTATION

### Estimation

In all the work on Markov models so far, we have taken the parameters of the model as given. In any practical application numerical values have to be assigned either by making hypothetical assumptions or by estimating their current values. Even when it is desired to explore the effects of a range of parameter values the current values will often provide a useful baseline from which to start.

If the Markov assumptions hold, it is easy to obtain point estimates of the transition probabilities from historical data by the method of maximum likelihood. For this we need complete stock and flow data. If $n_{ij}(T)$ is the observed number in $i$ at $T$ who are in $j$ at $T + 1$ and if $n_i(T)$ is the stock at the beginning of this interval, then the estimate of $p_{ij}$ is

$$\hat{p}_{ij} = n_{ij}(T)/n_i(T) \qquad (i, j = 1, 2, \ldots, k) \tag{4.12}$$

If stock and flow data are available over several time intervals for which the rates can be assumed to be the same then

$$\hat{p}_{ij} = \sum_T n_{ij}(T)/\sum_T n_i(T) \qquad (i, j = 1, 2, \ldots, k) \tag{4.13}$$

where the summation is taken over the periods for which data are available. The wastage probabilities are estimated in an identical manner by using the appropriate wastage flow in the numerator. Similarly, the recruitment probabilities $\mathbf{r}$ can be estimated by the observed proportions.

*Table 4.1 Flows for a university system for the two academic years 1970–72*

| 1970–71 | 1 | 2 | 3 | *Left* | *Total* |
|---|---|---|---|---|---|
| 1. Lecturer | 158 | 8 | 0 | 9 | 175 |
| 2. Senior Lecturer & Reader | 0 | 62 | 2 | 3 | 67 |
| 3. Professor | 0 | 0 | 55 | 1 | 56 |
| **1971–72** | **1** | **2** | **3** | ***Left*** | ***Total*** |
| 1. Lecturer | 168 | 3 | 0 | 8 | 179 |
| 2. Senior Lecturer & Reader | 0 | 64 | 2 | 7 | 73 |
| 3. Professor | 0 | 0 | 58 | 1 | 59 |

*Example 4.10* The data in Table 4.1 relate to the three grades, professor, reader and senior lecturer, and lecturer, at a university in the two academic years shown. The entries in the body of the table are the flows between the grades indicated by the row and column in which the value appears. The penultimate column gives the number of leavers.

The stocks at the beginning of each interval can be obtained by summing the rows. Thus the basic information for estimating the transition probabilities is as follows.

$$\hat{p}_{11} = \frac{158 + 168}{175 + 179} = 0.921 \quad \hat{p}_{12} = \frac{8 + 3}{175 + 179} = 0.031 \quad \hat{w}_1 = \frac{9 + 8}{175 + 179} = 0.048$$

$$\hat{p}_{22} = \frac{62 + 64}{67 + 73} = 0.900 \quad \hat{p}_{23} = \frac{2 + 2}{67 + 63} = 0.029 \quad \hat{w}_2 = \frac{3 + 7}{67 + 63} = 0.071$$

$$\hat{p}_{33} = \frac{55 + 58}{56 + 59} = 0.983 \quad - \quad \hat{w}_3 = \frac{1 + 1}{56 + 59} = 0.017$$

It is clear that these estimators can only be computed if complete flow data are available. This is not always the case but some progress can be made if a sequence of stock vectors is available. The idea is to estimate **P** by choosing that value which would have most nearly (in some sense) produced the observed stocks. The fullest account of the theory is contained in Lee, Judge, and Zellner (1970). A much briefer account with explicit formulae for some special cases will be found in Bartholomew (1977b); applications to manpower planning were made in Teather (1971), and to the estimation of voter transition matrices in McCarthy and Ryan (1977). ◀

### Validation of the model

It is perfectly possible to fit Markov models to stock and flow data in a purely mechanical way whether or not the underlying assumptions are correct. If there are gross violations of the assumptions any forecasts made with the model are likely to be wide of the mark, although even in this situation the parameter estimates may still be useful as a description of the system. Three courses are open to the analyst which separately or in combination can help avoid the worst errors.

(a) To carry out statistical tests of the assumptions.
(b) To compare the predictions of the model with actual outcomes (usually on historical data).
(c) To design the model, especially with regard to the choice of categories, so as to make the assumptions as nearly correct as possible.

We shall deal with each of these in turn.

The two principal assumptions which one might aim to test are as follows:

(a) The transition probabilities do not depend on time.

(b) The transition probabilities are the same for all individuals within a class. In addition, of course, there are the assumptions that individuals behave independently and that transition probabilities are functions of the current state only. These are more difficult to test and, in our experience, are not likely to be as serious in their effects as assumptions (a) and (b).

As a first step it is worthwhile to compute the standard errors of the quantities estimated. According to the assumptions of the Markov model, the flows originating in any one category are jointly distributed in the multinomial form and hence the standard errors of the flow proportions may be written down from standard theory. Thus assuming assumption (b) is true the standard error of $\hat{p}_{ij}$ given by (4.12) is

$$\{p_{ij}(1 - p_{ij})/n_i(T)\}^{\frac{1}{2}}$$

which can be estimated by replacing $p_{ij}$ by $\hat{p}_{ij}$. Forbes (1971a), Sales (1971), and others have compared rates by plotting the estimates with limits set at one standard error on either side. Any marked deviation from constancy can then easily be picked out by eye. If the probabilities are estimated over several time periods as in (4.13), $n_i(T)$ is replaced by $\sum_T n_i(T)$ in the denominator of the standard error.

A more formal test can be made using the $\chi^2$ statistic for the comparison of several multinomial samples. Examples of such tests on manpower matrices will be found in Forbes (1971a) and Sales (1971).

Both approaches can be used in testing either of assumptions (a) or (b) above. In the first case we compare flow numbers at different times; in the second the comparison will be between different categories of person at the same time. For example, we might suspect that the two sexes had different transition probabilities, in which case one would make comparisons grade by grade between the sexes. Notice that all of these methods involve looking at each grade separately and hence one might conclude that the assumptions were true in some grades but not in others.

Although we have followed the orthodox statistical approach in discussing hypotheses which might be tested, it is arguable that such tests are of marginal relevance. The objects of fitting a model to data (in this context) are to provide insights into the dynamics of the system and to make projections. A model should be judged therefore on how successfully it achieves these objects. It may not matter very much whether there is heterogeneity in the data or dependencies between one person's behaviour and another if the eventual effect on the predicted stocks is negligible. An inadequate model may well be good enough for the purpose in hand and is not to be dismissed on the grounds that, in principle, one could do better. It is this philosophy which underlies the widespread use of Markov models in many fields of application where the assumptions, strictly interpreted, are certainly false. Indeed a simple model is often more effective in the sense that its results may be more likely to be heeded simply because management find it easier to understand and therefore more acceptable.

One very useful way of validating a model is by testing its performance on historical data. If a sufficient run of data is available one can predict the later part of the series using the earlier part. This tells us how successful we would have been had we used the method in the past. A bad performance in the past would cast doubt on the model's usefulness in the future while success in the past would inspire some measure of confidence about future use.

*Example 4.11* A test of the Markov model on the Scientific Officer Class of the British Civil Service (from Sales (1971)). This is a six-grade hierarchical system and the annual transition matrix estimated from flow data for the three years 1963–65 was as follows.

$$\mathbf{P} = \begin{bmatrix} 0.727 & 0.178 & 0.023 & 0.004 & 0.004 & 0.002 \\ 0 & 0.826 & 0.139 & 0.001 & 0.001 & 0 \\ 0 & 0.008 & 0.920 & 0.033 & 0 & 0 \\ 0 & 0.002 & 0.008 & 0.927 & 0.033 & 0 \\ 0 & 0 & 0.007 & 0.003 & 0.884 & 0.030 \\ 0 & 0 & 0.007 & 0 & 0.007 & 0.918 \end{bmatrix}$$

(The very small off-diagonal probabilities are almost certainly due to errors in the data but their effect on the calculation is negligible.) The stock vector in 1965 was

$$\mathbf{n}(1965) = (231, 793, 1413, 503, 154, 72)$$

and the average number of recruits per year over the period of estimation was 154.5. Using this information the stocks for 1966–68 were predicted assuming a fixed input equal to the average historical level. The results are compared with the actual outcome in Table 4.2.

From knowledge of how this particular system works some, at least, of the assumptions of the model are false. Nevertheless it performs quite well in the short term and one might well be satisfied with the degree of accuracy achieved. ◀

It is often possible to improve the performance of a model by setting it up in such a way that the assumptions are more nearly satisfied. For example, if we were able to establish that the transition probabilities for men and women were different then it would clearly be better to model each sex separately. The main opportunity for improving the fit of the model lies in the choice of grades or classes. We saw in Example 4.9 how this could be done to advantage by incorporating seniority into the grade specification. Had we treated this as a three-grade system, ignoring the effect of seniority, we would have been attributing the same promotion probability to each member of the grade. If it is known *a priori* that seniority affects the chance of promotion then we may form subgrades, as in that example, so that the chance of promotion is more nearly constant within those subgrades. This device can be used with any other attribute which is expected to influence the transition rates. The price to be paid is that the size and complexity of the model is increased but this, of itself, creates no problems of principle. The multiplication of grades in this manner does, however, have an important drawback. As the grades become more numerous the stocks and flows became smaller until a point is reached where the probabilities cannot be estimated with adequate

***Table 4.2 Comparison of observed and predicted stocks for the Markov model***

| | 1966 | | 1967 | | 1968 | |
|---|---|---|---|---|---|---|
| *Grade* | *Observed* | *Predicted* | *Observed* | *Predicted* | *Observed* | *Predicted* |
| 1 | 217 | 226 | 225 | 222 | 220 | 219 |
| 2 | 753 | 780 | 790 | 768 | 761 | 758 |
| 3 | 1411 | 1439 | 1444 | 1461 | 1461 | 1480 |
| 4 | 515 | 520 | 526 | 537 | 555 | 553 |
| 5 | 154 | 157 | 166 | 159 | 174 | 162 |
| 6 | 77 | 72 | 78 | 72 | 74 | 73 |
| *Total* | 3127 | 3194 | 3229 | 3219 | 3245 | 3245 |

precision. We thus have to balance realism against loss of accuracy. In a particular application the consequences can be investigated empirically by adopting alternative classifications and comparing the effects which they have on the predictions. In our experience, it is better to err on the side of too few rather than too many groups.

## 4.8 COMPLEMENTS

In this chapter we have concentrated on the straightforward use of Markov chain models for prediction. We shall take this discussion further in Chapters 6 and 7 where career prospects and steady-state aspects are considered in more detail. However, this is still leaves many topics untouched and many problems unsolved.

One of the chief limitations of our treatment is that it has been almost entirely deterministic, in the sense that we have dealt with expected values.

This is both a strength and a weakness. The weakness lies in the fact that we have presented an incomplete picture by ignoring the variation which is an integral part of the stochastic models. There is a considerable body of theory concerned with the distribution of projected stock numbers stemming from the work of Pollard (1966). It was expressed in terms suitable for use in manpower planning in Bartholomew (1973a) and examples of its use in this connection will be found in Forbes (1971a) and Sales (1971). This work shows that the errors in forecasts are likely to be quite large—the variances of the predictions being of the same order as the predicted values themselves. On top of this there is a further source of error arising from the fact that in most applications some, at least, of the parameters have to be estimated. The effects of this have been investigated by Bartholomew (1975a) who shows that this source can give rise to errors of a similar magnitude to the random error arising from the stochastic assumptions of the model. This takes no account of the uncertainties of yet another kind arising from changes in the parameters which may occur during the forecast period. The whole question of how to cope with uncertainty in manpower planning is a complex one and the interested reader should consult Bartholomew, Hopes, and Smith (1976).

The strength of our emphasis on expected values is that at this level the models are more robust. There is a certain amount of empirical evidence, direct and indirect, that the stochastic aspects of the Markov model are less realistic than when it is viewed deterministically. For example, while it is often an empirical fact that flow rates remain nearly constant over a period of time it is less likely to be the case that the distribution of the numbers moving have the multinomial form as required by the theory. In fact it often seems that organizational constraints limit the amount of variation which can occur so that the stochastic theory overestimates the variation which actually occurs in practice. There thus seem to be grounds for placing greater reliance on the deterministic aspects of the model. This interpretation has been emphasized by some writers (e.g. Grinold and Stanford (1974) and Grinold and Marshall (1977)) who have coined the term 'fractional flow model' for what we have termed a Markov model.

An obvious defect of the Markov model in the eyes of many practitioners is its failure to incorporate the feedback mechanisms which are widely believed to exist in practice. For example, if promotion opportunities contract one might expect the wastage from the grades affected to increase and vice versa. There is no difficulty, in principle, in including lagged relationships between parameters in the model and some reasearch on these lines is in progress. Young (1971) and Young and Vassiliou (1974) have explored a non-linear model with the same end in view and in the latter paper have demonstrated that their model gives an improved fit to data from a system which underwent substantial organizational change during the period over which the model was tested. While improvements are to be looked for in this direction, the basic linear Markov model is likely to continue to give good, if somewhat less than perfect, service.

We have already remarked on the possibility of heterogeneity in the system and advised that, where possible, the system should be divided into more homogeneous subsystems. This depends on our ability to identify some attributes on the basis of which the division can be made. Very often heterogeneity may be suspected without knowing its nature. One might therefore try to generalize the model to allow for possible variation of the parameters in the population in rather the same manner as when we were treating models for wastage in Chapter 3. Various attempts to do this have been made and some indication of the possibilities will be found in Bartholomew (1973a). The simplest is the 'mover–stayer' model in which only part of the population move while the remainder always stay in the same category. This model does not appear to have been used in manpower planning at the institutional level and it is doubtful whether it is ever appropriate. The difficulty with more complicated models which allow for heterogeneity is that the number of parameters and the problems of estimating them make their practical application very difficult. Once again it seems that it may be better to accept the limitations of the very much simpler Markov model than attempt to achieve greater realism at unacceptable cost.

The decision to treat time as discrete was primarily dictated by convenience though it is often an adequate approximation even when not strictly accurate. In real life where movements can occur at any time, it may seen to be more realistic to treat time as continuous. This takes us into the realm of continuous-time Markov processes and, from a theoretical point of view, this offers many advantages. An account of the theory is given in Bartholomew (1973a) and a discussion of problems of estimation in Bartholomew (1977b). For most applied work we have found that the arithmetical advantages of the discrete-time model and the fact that data are often only available at discrete intervals outweigh any small gain in realism which might result from the use of the continuous version.

One of the main areas of current research is in the theory of control of manpower systems. We shall treat this in an elementary way in Chapter 7 but the subject is much larger than this might indicate. It is rarely sufficient to predict changes. Such prediction is almost always followed by the question of how to steer the system in the desired direction. Questions of this kind have been tackled on a trial-and-error basis in some of the exercises of this chapter but clearly a theoretical formulation is required for an adequate development of control strategies. This is an active field of research some of the early results of which will be found in Chapter 7 of this book, and in Bartholomew (1973a, Chapter 4), Grinold and Stanford (1974), Vajda (1975), Bartholomew (1975b) and (1977c), Davies (1973), and Grinold and Marshall (1977).

We have concentrated on the stock numbers but there are other characteristics of a manpower system which may be relevant for manpower mangement. In Chapter 6 we shall consider certain probabilities and waiting times. Glen (1977) has shown how to calculate the length-of-service distribution for each grade and has given formulae for means and variances. Such quantities are useful for estimating the degree of experience available to the organization.

Many applications of Markov models have been made, a few of which have been referred to in this chapter. Few of these use recent data because the subject has long ceased to have the novelty required for publication. Some examples can be found in theses but most have never been written up. The following is a short list of papers in which the reader will find applications to manpower planning described: Young and Almond (1961), Gani (1963), Dawson and Denton (1974), Forbes (1971a), Hopkins (1974), Mahoney and Milkovich (1971). Nielson and Young (1973), Rowland and Soverign (1969), Sales (1971), Stewman (1975), Thonstad (1969), and Vassiliou (1976).

## 4.9 EXERCISES AND SOLUTIONS

### Exercise 4.1

In Example 4.1, suppose that the promotion rates from grades 1 and 2 are reduced to 0.05 and 0.04 respectively but that the wastage rates remain unchanged. The recruitment is reduced to 22 per year into grade 1 only. Write down the new **P** and **r** and use BASEQN to project the present grade sizes for 10 years.

### Exercise 4.2

In the five grade hierarchy of Example 4.2, assume that the grade sizes shown for $T = 3$ years, i.e. $\mathbf{n} = [268, 304, 171, 123, 74]$, are now the present ones. Keeping the same recruitment numbers and distribution, guess what constant set of promotion rates will bring the grades back to $\mathbf{n} = [376, 282, 141, 94, 47]$. This is the original structure but with the total size at its present level of 940. Use the program to test your guesses.

### Exercise 4.3

For the university system of Example 4.3 and assuming a constant inflow of 500 into the first year (i.e. no age 'bulge'), use the program to investigate the effect of the following contingencies:

(a) the pass rate for first-year students increases to 70 per cent while the repeat rate and fail rate decrease to 10 per cent and 20 per cent respectively;

(b) the proportion going on to the fourth year increases to 30 per cent, with the repeat rate remaining at 10 per cent;

(c) both (a) and (b) happen simultaneously.

**Exercise 4.4**

For the non-hierarchical system of Example 4.4:

(a) calculate the stationary grade sizes for the system as at present and check these with the forecast ones for $T = 99$ in Printout 4.4;

(b) if the wastage rates in grades 1, 2, 3, and 4 changed to 0.15, 0.03, 0.10, 0.04, respectively, the other rates out of the grades being unchanged, calculate the stationary grade sizes, and use the program to project the present grade sizes; check the stationary values with $T = 99$.

**Exercise 4.5**

For the officers' system of Example 4.6:

(a) what would have been the effect on the grade sizes of expanding to double the total size in one step? Compare the grade sizes in this case with those of Example 4.6.

(b) calculate the stationary grade sizes, and check these with those forecast for large $T$.

**Exercise 4.6**

Set up the transition matrices for the system with wastage rates as in Exercise 2.3. Hence, project the stock distribution for at least 5 years ahead under the assumption of 170 recruits during each year. Repeat the projection when the total size is fixed at 680.

**Exercise 4.7**

For the age model of Example 4.8 the expansion produced a marked shift in the age distribution towards younger ages. In an attempt to avoid this, consider the effect of recruiting older people (i.e. over the age of 30) during the expansion. In particular, set the recruitment distribution $\mathbf{r} = [0.80, 0.20, 0]$ during the first 5 years.

**Exercise 4.8**

For the data of Table 4.3, estimate the transition rates and their errors, and

***Table 4.3 Male and Female transition data for 1964 and 1965***

| | Male 1964 | | | | Male 1965 | | | |
|---|---|---|---|---|---|---|---|---|
| *Grade* | S | P | L | R | S | P | L | R |
| 1 | 4530 | 169 | 168 | 241 | 4434 | 146 | 121 | 434 |
| 2 | 1998 | 90 | 89 | 33 | 2021 | 92 | 69 | 22 |
| 3 | 737 | 32 | 21 | 0 | 774 | 65 | 14 | 1 |
| 4 | 674 | — | 21 | 4 | 690 | — | 24 | 1 |
| *Total* | 7939 | | | 278 | 7919 | | | 458 |

| | Female 1964 | | | | Female 1965 | | | |
|---|---|---|---|---|---|---|---|---|
| *Grade* | S | P | L | R | S | P | L | R |
| 1 | 1272 | 11 | 154 | 93 | 1195 | 7 | 128 | 131 |
| 2 | 234 | 7 | 9 | 5 | 234 | 5 | 3 | 8 |
| 3 | 56 | 1 | 3 | 0 | 59 | 1 | 1 | 0 |
| 4 | 12 | — | 0 | 0 | 13 | — | 0 | 0 |
| *Total* | 1574 | | | 98 | 1501 | | | 139 |

hence consider whether for a transition model with push flows the Male and Female subsystems should be modelled separately or could be combined. Consider also whether the transition probabilities changed from 1964 to 1965 and hence whether each year's data should be amalgamated to estimate the parameters of the model. In the table, S are the stocks at the beginning of the year, and P, L, and R are respectively the appropriate promotion, leaving, and recruitment transition flows as defined in Section 1.2.

**Exercise 4.9**

This exercise involves the fitting, the simple validation and the projection of a push model for the system whose data are given in Table 4.4.

(a) *Inspection of the data*: Plot the total and individual grade sizes over the whole period. Also calculate, and if necessary plot, the promotion rates, the leaving rates and the total inflow. Comment on whether these can reasonably be estimated in the push model by constant rates.

(b) *Validation of the model*—I: By averaging over the whole period, estimate a transition matrix (**P**) and the inflow numbers (**R**). Using these parameters and starting from the actual 1961 grade sizes, project values for the years 1962 to 1971. Hence test the model by comparing plots of these projections and the actual grade sizes.

**Table 4.4 Stock and flow data for the system of Exercise 4.9**

| | | 1961 | | 1962 | | 1963 | | 1964 | | 1965 | | 1966 | | 1967 | | 1968 | | 1969 | | 1970 | | 1971 | | 1972 |
|---|---|---|---|---|---|---|---|---|---|---|---|---|---|---|---|---|---|---|---|---|---|---|---|---|
| *Stocks in grade* | 1 | 123 | | 119 | | 124 | | 120 | | 137 | | 151 | | 153 | | 161 | | 169 | | 175 | | 179 | | 181 |
| | 2 | 54 | | 56 | | 58 | | 70 | | 65 | | 71 | | 70 | | 68 | | 67 | | 67 | | 73 | | 67 |
| | 3 | 38 | | 39 | | 40 | | 47 | | 50 | | 56 | | 58 | | 54 | | 56 | | 56 | | 59 | | 61 |
| | *Total* | 215 | | 214 | | 222 | | 237 | | 252 | | 278 | | 281 | | 283 | | 292 | | 298 | | 311 | | 309 |
| *Leavers from grade* | 1 | | 5 | | 6 | | 9 | | 8 | | 6 | | 10 | | 9 | | 17 | | 18 | | 9 | | 8 | |
| | 2 | | 1 | | 2 | | 8 | | 7 | | 1 | | 7 | | 6 | | 1 | | 5 | | 3 | | 7 | |
| | 3 | | 4 | | 2 | | 2 | | 2 | | 0 | | 1 | | 7 | | 5 | | 3 | | 1 | | 1 | |
| *Entrants to grade* | 1 | | 8 | | 16 | | 32 | | 30 | | 28 | | 18 | | 21 | | 28 | | 29 | | 21 | | 13 | |
| | 2 | | 1 | | 2 | | 0 | | 1 | | 2 | | 0 | | 2 | | 0 | | 1 | | 3 | | 0 | |
| | 3 | | 0 | | 0 | | 2 | | 1 | | 3 | | 3 | | 1 | | 4 | | 2 | | 2 | | 1 | |
| | *Total* | | 9 | | 18 | | 34 | | 32 | | 33 | | 21 | | 24 | | 32 | | 32 | | 26 | | 14 | |
| *Promotion from grade* | 1 | | 7 | | 5 | | 27 | | 5 | | 8 | | 6 | | 4 | | 3 | | 5 | | 8 | | 3 | |
| | 2 | | 5 | | 3 | | 7 | | 4 | | 3 | | 0 | | 2 | | 3 | | 1 | | 2 | | 2 | |

The stock numbers relate to the beginning of the year shown and the flows are as defined in Section 1.2. The flows are shown positioned between the stock numbers to emphasize that the flows occur in the period between the points in time at which the stocks are counted

(c) *Validation of the model*—II: Estimate **P** and **R** using the data for 1961–67 as you consider appropriate. Then starting from the actual 1967 grade sizes and using these parameters, project grade sizes for 1968–72. Compare the plots of these projections and the actual grade sizes.

(In the sense that (b) uses observed values from the period of projection to estimate the parameters it is more a test of the model's ability to *simulate* or *describe* the system over this period. By contrast (c) is a truer *forecasting* test since it shows how well we could have forecast 1968–72 using the model in 1967. Note that for this test the parameter values need not be exclusively based on historical 1961–67 data: for example, we should incorporate any prior knowledge that we might have had in 1967 about changes in the parameters for the period ahead.)

(d) *Projection*: Project the grade sizes up to 1977 starting from the actual 1972 values and using whatever values you consider are appropriate for **P** and **R**. Hence comment on the long-term situation if these parameter values continue to hold.

## Solution to Exercise 4.1

The parameters and projections of the Markov-chain model are shown in Printout 4.12.

```
   K =      ?3
   N =      ?180,145.35
   P =      ?.75,.05,0.    0,.86,.04,    0,0,.85
   R =      ?22,0,0,    1,0,0
 T,% =      ?5,YES
 <*>      ? RUN

   T              1            2            3           TOTAL       R

   0         180(50%)    145(40%)     35(10%)       360(100%)
   1         157(48%)    134(41%)     36(11%)       326( 91%)      22
   2         140(47%)    123(41%)     36(12%)       298( 83%)      22
   3         127(46%)    113(41%)     35(13%)       275( 76%)      22
   4         117(46%)    103(41%)     34(13%)       255( 71%)      22
   5         110(46%)     95(40%)     33(14%)       238( 66%)      22
 <*>      ?T=,10
  10          93(51%)     63(35%)     26(14%)       183( 51%)      22
 <*>      ?T=,20
  20          88(62%)     39(27%)     15(11%)       142( 40%)      22
 <*>      ?T=,99
  99          88(69%)     31(25%)      8( 7%)       128( 36%)      22
 <*>      ?
```

Printout 4.12 Showing the projected grade sizes for Exercise 4.1

**Solution to Exercise 4.2**

By first calculating the promotion into the top grade necessary to maintain the required size for that grade and then similarly for the next grade down, and so on, it is easy to calculate that the promotion rates which will maintain the target grade sizes once they have been reached are

$$p_{12} = 0.090, \quad p_{23} = 0.075, \quad p_{34} = 0.050, \quad p_{45} = 0.025$$

These promotion rates are the steady-state ones and will always move the grade sizes towards and eventually attain their steady-state values. As a control policy they have the advantage that they need never be changed (and are therefore historically 'fair'). However, they generally move the system towards the target rather slowly. This is the case for this system as Printout 4.13 shows.

```
    K =    ?5
    N =    ?   268,   304,   171,   123,    74
    P =    ?   .76,  .090,     0,     0,     0
 ?               0,  .775,  .075,     0,     0
 ?               0,     0,  .850,  .050,     0
 ?               0,     0,     0,  .925,  .025
 ?               0,     0,     0,     0,  .950
    R =    ? .750,  .250,     0,     0,     0,     -1,+,0
  T,% =    ?5,NO
  <*>   ? RUN

    T          1      2      3      4      5          TOTAL       R

    0        268    304    171    123     74      940(100%)
    1        288    288    168    122     73      940(100%)     113
    2        304    277    165    122     73      940(100%)     113
    3        316    271    161    121     72      940(100%)     113
    4        325    267    157    120     72      940(100%)     114
    5        333    264    153    119     71      940(100%)     114
  <*>   ?T=,10
   10        353    265    141    112     68      940(100%)     116
  <*>   ?T=,20
   20        366    275    137    101     62      940(100%)     118
  <*>   ?T=,30
   30        370    279    138     96     57      940(100%)     119
  <*>   ?T=,99
   99        374    283    141     94     47      940(100%)     120
  <*>   ?
```

Printout 4.13 Showing the effect of the steady-state promotion rates in Exercise 4.2

Notice that it is at least 10 years before the bottom three grades are at all close to their required sizes which are $\mathbf{n} = [376, 282, 141, 94, 47]$. For the higher grades it takes much longer to bring them down to their targets. As Printout 4.14 shows even the extremely drastic policy of no promotion would require at least 5 years before the top two grades were near their targets. (Dynamic control policies that may change each year can also be evaluated using this model with very little extra effort.)

```
  K =    ? 5
  N =    ?  268,   304,   171,   123,    74
  P =    ? .05,     0,     0,     0,     0
?           0,   .85,     0,     0,     0
?           0,     0,   .90,     0,     0
?           0,     0,     0,   .95,     0
?           0,      0,    0,     0,   .95
  R =    ? .75,   .25,     0,     0,     0,    -1,+,0
T,% =    ? 5,NO
<*>    ?  RUN

  T        1      2      3      4      5        TOTAL        R

  0       268    304    171    123    74      940(100%)
  1       312    287    154    117    70      940(100%)    113
  2       351    272    139    111    67      940(100%)    115
  3       386    260    125    105    63      940(100%)    116
  4       416    251    112    100    60      940(100%)    118
  5       444    243    101     95    57      940(100%)    119
<*>    ?
```

Printout 4.14 Showing the effect of no promotion at all (see solution to Exercise 4.2)

## Solution to Exercise 4.3

(a) If the pass rate for first-year students increases from 61 to 70 per cent and the repeat and fail rates decrease from 15 to 10 per cent and 24 to 20 per cent, the transition matrix becomes that shown in Printout 4.15. From this printout we see that the effect of this would be a decrease in the size of the first year, and an increase in the size of the other years, with the total size remaining almost constant.

(b) If the proportion going on to the fourth year increased from 20 to 30 per cent, the number in the fourth year would increase from 70 to 100 as shown in Printout 4.16. Note that if our new projection is to follow on immediately after Printout 4.15 there is no need to input all the parameters a second time. We can use the facilities available in BASEQN to change only those parameters which are to be given new values. The data input and commands to do this are shown in Printout 4.16.

(c) If both (a) and (b) happen simultaneously the result as we might expect would be a combination of (a) and (b) as is shown in Printout 4.17. (The printout is assumed to follow on from Printout 4.16 so that only those parameters which are different need be specified in the data input.)

```
  K =      ? 4
  N =      ? 600,400,330,70
  P =      ? .1,.7,0,0,    0,.11,.71,0,    0,0,.1,.2,    0,0,0,.05
  R =      ? 500,0,0,0,    1,0,0
T,% =      ?3,YES
<*>      ?RUN

  T              1           2           3           4          TOTAL       R

  0         600(43%)    400(29%)    330(24%)     70( 5%)     1400(100%)
  1         560(40%)    464(33%)    317(22%)     69( 5%)     1411(101%)    500
  2         556(39%)    443(31%)    361(25%)     67( 5%)     1427(102%)    500
  3         556(39%)    438(31%)    351(25%)     76( 5%)     1420(101%)    500
<*>      ?T=,5
  5         556(39%)    437(31%)    345(24%)     73( 5%)     1410(101%)    500
<*>      ?T=,10
 10         556(39%)    437(31%)    345(24%)     73( 5%)     1410(101%)    500
<*>      ?
```

Printout 4.15 Showing the effect of the changes (a) in the first-year transition rates (see solution to Exercise 4.3)

```
<*>      ?N=N(0)
<*>      ?P=, .15,.61,0,0,   0,.11,.71,0,   0,0,.1,.3,   0,0,0,.05
<*>      ?T%=, 3,YES
<*>      ?RUN

  T              1           2           3           4          TOTAL       R

  0         600(43%)    400(29%)    330(24%)     70( 5%)     1400(100%)
  1         590(42%)    410(29%)    317(22%)    103( 7%)     1420(101%)    500
  2         588(42%)    405(29%)    323(23%)    100( 7%)     1417(101%)    500
  3         588(42%)    404(29%)    320(23%)    102( 7%)     1413(101%)    500
<*>      ?T=,5
  5         588(42%)    403(29%)    318(23%)    101( 7%)     1410(101%)    500
<*>      ?T=,10
 10         588(42%)    403(29%)    318(23%)    100( 7%)     1410(101%)    500
<*>      ?
```

Printout 4.16 Showing the effect of the changes (b) in the fourth-year transition rates (see solution to Exercise 4.3)

```
<*>      ?N=N(0)
<*>      ?P=, .1,.7,0,0,    0,.11,.71,0,    0,0,.1,.3,    0,0,0,.05
<*>      ?T%=, 3,YES
<*>      ?RUN

  T              1           2           3           4          TOTAL       R

  0         600(43%)    400(29%)    330(24%)     70( 5%)     1400(100%)
  1         560(39%)    464(32%)    317(22%)    103( 7%)     1444(103%)    500
  2         556(38%)    443(30%)    361(25%)    100( 7%)     1460(104%)    500
  3         556(38%)    438(30%)    351(24%)    113( 8%)     1458(104%)    500
<*>      ?T=,5
  5         556(38%)    437(30%)    345(24%)    109( 8%)     1447(103%)    500
<*>      ?T=,10
 10         556(38%)    437(30%)    345(24%)    109( 8%)     1446(103%)    500
<*>      ?
```

Printout 4.17 Showing the effect of both changes (a) and (b) (see solution to Exercise 4.3)

The main difference between the results shown in this printout and a simple summing of the effects shown in Printouts 4.15 and 4.16 is that the number of fourth-year students would increase to 109 due to the increase in third-year numbers. This is an interaction effect.

**Solution to Exercise 4.4**

(a) Assuming the stationary grade sizes have been reached, and equating the inflows and outflows for each grade, we have:

grade 1: $n_1 \times 0.42 = 100$ $\qquad n_1 = 238$

grade 3: $n_3 \times 0.29 = 30 + 238 \times 0.07$ $\qquad n_3 = 161$

grade 2: $n_2 \times 0.18 = 238 \times 0.15 + 161 \times 0.05 + n_4 \times 0.02$ $\qquad n_2 = 301$

grade 4: $n_4 \times 0.10 = 161 \times 0.10 + n_2 \times 0.12$ $\qquad n_4 = 522$

These check with the grade sizes for $T = 99$ in Printout 4.4.

(b) When the wastage decreases as specified, the stationary grade sizes are:

grade 1: $n_1 \times 0.37 = 100$ $\qquad n_1 = 270$

grade 3: $n_3 \times 0.25 = 30 + 270 \times 0.07$ $\qquad n_3 = 196$

grade 2: $n_2 \times 0.15 = n_4 \times 0.02 + 270 \times 0.15 + 196 \times 0.05$ $\qquad n_2 = 517$

grade 4: $n_4 \times 0.06 = n_2 \times 0.12 + 196 \times 0.05$ $\qquad n_4 = 1360$

Printout 4.18 contains a projection of the present grade sizes using the decreased wastage rates. This shows that with the decreased wastage it now takes longer to reach the corresponding steady-state grade sizes despite the fact that they are closer to the initial ones. This is an example of the general result that the lower the wastage the slower a system can change. Notice that the value for $T = 99$ does not provide a check for the steady-state grade sizes in this case because they have not yet been attained: a higher value for $T$ would be needed.

**Solution to Exercise 4.5**

(a) The consequences of expanding the system to twice the size are shown in Printout 4.19 and in Figure 4.3 by the broken lines.

(b) Assume the system is stationary and that there are, say, 100 recruits each year. Then equating the inflows and outflows for each grade and starting

```
K =     ? 4
N =     ? 400,300,100,250
P =     ? .63,.15,.07,0,    0,.85,0,.12,    0,.05,.75,.1,    0,.02,0,.94
R =     ? 100,0,30,0,    1,0,0
,% =    ? 5,YES
*>    ?    RUN

T          1          2          3          4          TOTAL       R

0      400(38%)  300(29%)  100(10%)  250(24%)    1050(100%)
1      352(32%)  325(30%)  133(12%)  281(26%)    1091(104%)    130
2      322(28%)  341(30%)  154(14%)  316(28%)    1134(108%)    130
3      303(26%)  352(30%)  168(14%)  354(30%)    1177(112%)    130
4      291(24%)  360(30%)  177(15%)  392(32%)    1220(116%)    130
5      283(22%)  367(29%)  183(15%)  429(34%)    1263(120%)    130
*>    ?T=,10
10     272(19%)  389(27%)  194(13%)  600(41%)    1454(138%)    130
*>    ?T=,15
15     270(17%)  409(25%)  195(12%)  738(46%)    1613(154%)    130
*>    ?T=,20
20     270(16%)  427(24%)  196(11%)  850(49%)    1743(166%)    130
*>    ?T=,30
30     270(14%)  455(24%)  196(10%) 1016(52%)    1937(185%)    130
*>    ?T=,99
99     270(12%)  513(22%)  196( 8%) 1337(58%)    2315(221%)    130
*>    ?
```

Printout 4.18 A projection using decreased wastage (see solution to Exercise 4.4)

```
K =     ? 4
N =     ? 129,74,28,11
P =     ? .728,.102,0,0,    0,.83,.046,0,    0,0,.867,.033,    0,0,0,.902
R =     ? 1,0,0,0,    -1,+,0
,% =    ?0,YES
*>    ?    RUN

T          1          2          3          4          TOTAL       R

0      129(53%)   74(31%)   28(12%)   11( 5%)     242(100%)
*>    ?T=,5
5      129(53%)   76(31%)   27(11%)   10( 4%)     242(100%)     35
*>    ?*R=, 1,0,0,0,    -1,*,2.0,    1,YES
6      371(77%)   76(16%)   27( 6%)   10( 2%)     484(200%)    277
*>    ?*R=, 1,0,0,0,    -1,*,1.0,    2,YES
7      346(72%)  101(21%)   27( 6%)   10( 2%)     484(200%)     76
8      327(68%)  119(25%)   28( 6%)   10( 2%)     484(200%)     75
*>    ?T=,10
10     301(62%)  142(29%)   32( 7%)   10( 2%)     484(200%)     73
*>    ?T=,15
15     272(56%)  159(33%)   42( 9%)   11( 2%)     484(200%)     71
*>    ?T=,20
20     263(54%)  160(33%)   49(10%)   13( 3%)     484(200%)     71
*>    ?T=,25
25     260(54%)  158(33%)   52(11%)   14( 3%)     484(200%)     70
*>    ?T=,30
30     259(53%)  156(32%)   53(11%)   16( 3%)     484(200%)     70
*>    ?
```

Printout 4.19 The projected grade sizes for an all-at-once expansion (see solution to Exercise 4.5)

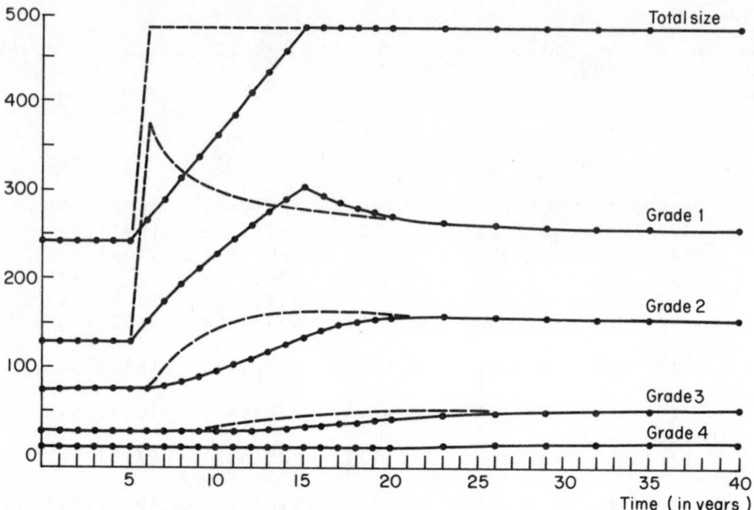


Figure 4.3 Showing the consequences of the immediate expansion (broken line) and the more gradual expansion

with the lowest grade, we have:

$$
\begin{aligned}
100 &= n_1(0.102 + 0.170) & n_1 &= 367.6 \quad (53\%) \\
n_1 \times 0.102 &= n_2(0.046 + 0.124) & n_2 &= 220.6 \quad (32\%) \\
n_2 \times 0.046 &= n_3(0.033 + 0.100) & n_3 &= 76.3 \quad (11\%) \\
n_3 \times 0.033 &= n_4(0.098) & n_4 &= 25.7 \quad (4\%) \\
& & \sum n_i &= 690.2 \quad (100\%)
\end{aligned}
$$

These are the stationary grade sizes for an inflow of 100. But whatever the inflow, the stationary structure (i.e. the proportion in each grade) will always be the same. Hence for a total size of 484 the required stationary grade sizes are

$$\mathbf{n}(\infty) = (257, 154, 53, 18)$$

which are very close to those forecast for $T = 30$.

**Solution to Exercise 4.6**

The appropriate model to set up is one in which the 'grades' correspond to each of the single-year length-of-service classes. The projection is shown in Printout 4.20. The recruits in this model, however, must be the number of people recruited during the year who remain in the stocks at the year end-point (see Section 2.3). We usually refer to these as the 'net' recruits to distinguish them from the actual or 'gross' recruits. Since the gross number is 170 and the survival rate for half a year is 0.67 we can estimate the required net recruits as $170 \times 0.67 = 114$.

```
 K  =    ? 7
 N  =    ? 100,   75,   65,   55,   70,   45, 270
 P  =    ?   0, .75,    0,    0,    0,    0,   0
?            0,   0,  .81,    0,    0,    0,   0
?            0,   0,    0,  .85,    0,    0,   0
?            0,   0,    0,    0,  .88,    0,   0
?            0,   0,    0,    0,    0,  .90,   0
?            0,   0,    0,    0,    0,    0, .90
?            0,   0,    0,    0,    0,    0, .90
 R  =    ?   1,   0,    0,    0,    0,    0,   0,     114,+,13
T,% =    ? 5,NO
<*>    ?   RUN

 T         1      2      3      4      5      6      7          TOTAL      R

 0       100     75     65     55     70     45    270      680(100%)
 1       114     75     61     55     48     63    283      700(103%)    114
 2       127     86     61     52     49     44    312      729(107%)    127
 3       140     95     69     52     45     44    320      765(113%)    140
 4       153    105     77     59     45     41    327      808(119%)    153
 5       166    115     85     66     52     41    331      855(126%)    166
<*>    ?T=,10
 10      231    164    125     99     81     68    391     1158(170%)    231
<*>    ?T=,20
 20      361    261    204    166    140    121    701     1954(287%)    361
<*>    ?
```

Printout 4.20 Showing the projection of a length-of-service distribution (see solution to Exercise 4.6)

For a constant total size Printout 4.21 shows the projected length-of-service distribution and the required recruitment levels. Since this is the net recruitment it must be divided by 0.67 to give the gross level.

```
<*>     ?N=N(0)
<*>     ?R=, 1,0,0,0,0,0,0,    -1,+,0
<*>     ?T%=, 5,NO
<*>     ?RUN

 T         1      2      3      4      5      6      7          TOTAL      R

 0       100     75     65     55     70     45    270      680(100%)
 1        94     75     61     55     48     63    283      680(100%)     94
 2        93     71     61     52     49     44    312      680(100%)     93
 3        92     70     57     52     45     44    320      680(100%)     92
 4        92     69     57     49     45     41    327      680(100%)     92
 5        92     69     56     48     43     41    331      680(100%)     92
<*>     ?
```

Printout 4.21 Following on from Printout 4.20 and showing the effect of holding the total size constant (see solution to Exercise 4.6)

The problem of 'net' and 'gross' recruitment can be avoided by setting up an extra class which takes in the gross recruits, one year ahead, and then moves them on the following year to the 0–1 class with transition rate 0.67. Note too that if there is no need to differentiate the last three length-of-service classes then the model can be simplified by amalgamating these classes into one final class since they all have the same wastage rate.

**Solution to Exercise 4.7**

Printout 4.22 shows that if, during the expansion, 20 per cent of recruits were over the age of 30 then the distribution would now shift towards the older ages, suggesting that this policy recruits rather too many older people if it is required to keep the age distribution fixed. (The interested reader is recommended to try and determine by trial and error (or otherwise) what recruitment policy during expansion would maintain the present age distribution.) It is also interesting to note from Printout 4.22 that significantly fewer recruits will be required both during and after the expansion: this is a direct consequence of older people having lower wastage rates. In practice

```
    K =     ? 3
    N =     ? 357,153,15
    P =     ? .72,.03,0,     0,.93,.02,     0,0,.80
    R =     ? .80,.20,0,     -1,*,1.15
  T,% =     ? 5,YES
  <*>    ?    RUN

    T            1            2            3            TOTAL       R

    0       357(68%)    153(29%)     15( 3%)      525(100%)
    1       400(66%)    189(31%)     15( 2%)      604(115%)     179
    2       450(65%)    228(33%)     16( 2%)      694(132%)     203
    3       509(64%)    272(34%)     17( 2%)      798(152%)     231
    4       578(63%)    321(35%)     19( 2%)      918(175%)     264
    5       658(62%)    376(36%)     22( 2%)     1056(201%)     302
  <*>    ?*R=,   1,0,0,     -1,*,1.0,     5,YES
    6       661(63%)    370(35%)     25( 2%)     1056(201%)     188
    7       665(63%)    364(34%)     27( 3%)     1056(201%)     189
    8       669(63%)    358(34%)     29( 3%)     1056(201%)     190
    9       672(64%)    353(33%)     30( 3%)     1056(201%)     191
   10       676(64%)    349(33%)     31( 3%)     1056(201%)     192
  <*>    ?T=,15
   15       691(65%)    332(31%)     33( 3%)     1056(201%)     195
  <*>    ?T=,20
   20       702(66%)    321(30%)     33( 3%)     1056(201%)     198
  <*>    ?T=,30
   30       712(67%)    312(30%)     32( 3%)     1056(201%)     200
  <*>    ?T=,99
   99       718(68%)    308(29%)     31( 3%)     1056(201%)     201
  <*>    ?
```

Printout 4.22 Showing the effect of taking older recruits (see solution to Exercise 4.7)

this effect may not be so simple since wastage is also likely to be related to length of service, so that recruitment into the 30–35 class may significantly alter the overall wastage rate for this class.

## Solution to Exercise 4.8

The estimated transition rates (and their standard errors) are shown in Table 4.5.

***Table 4.5***

| | | M64 | M65 | F64 | F65 |
|---|---|---|---|---|---|
| *Promotion rates from grade* | 1 | 0.037 (0.003) | 0.033 (0.003) | 0.009 (0.003) | 0.006 (0.002) |
| | 2 | 0.045 (0.005) | 0.046 (0.005) | 0.030 (0.011) | 0.021 (0.009) |
| | 3 | 0.043 (0.008) | 0.084 (0.010) | 0.018 (0.018) | 0.017 (0.017) |
| *Leaving rates from grade* | 1 | 0.037 (0.003) | 0.027 (0.002) | 0.121 (0.009) | 0.107 (0.009) |
| | 2 | 0.045 (0.005) | 0.034 (0.004) | 0.038 (0.013) | 0.013 (0.007) |
| | 3 | 0.028 (0.006) | 0.018 (0.005) | 0.054 (0.030) | 0.017 (0.017) |
| | 4 | 0.031 (0.007) | 0.035 (0.007) | 0 | 0 |
| *Recruitment distribution over grade* | 1 | 0.867 (0.020) | 0.948 (0.010) | 0.949 (0.022) | 0.942 (0.020) |
| | 2 | 0.119 (0.019) | 0.048 (0.010) | 0.051 (0.022) | 0.058 (0.020) |
| | 3 | 0 | 0.002 (0.002) | 0 | 0 |
| | 4 | 0.014 (0.007) | 0.002 (0.002) | 0 | 0 |
| *Total* | | 1.000 | 1.000 | 1.000 | 1.000 |

Comparing the Male and Female data there would appear to be a significant difference between the promotion rates and leaving rates in grades 1 and 4. This would suggest modelling these subsystems separately.

For Males the promotion rates in grades 1 and 2 appear constant over time while in grade 3 there is almost certainly a significant change. The leaving rates for all grades except grade 4 show a similar and significant decrease of the order of 70 per cent. The recruitment distribution also appears to change significantly (although note that the 1965 distribution is similar to the 1964 and 1965 Female distribution).

For Females there appears to be no change in the promotion rates or the recruitment distribution, while the wastage rates again show a decrease which is probably real, at least in grades 1 and 2.

Overall therefore the differences between the two years implies that a simple push model is probably not appropriate.

**Solution to Exercise 4.9**

(a) The unbroken lines in Figure 4.4 show how the actual grade sizes and total size have changed over the whole period 1961–72. The only conclusions we can reasonably draw from this figure at this stage are that overall these

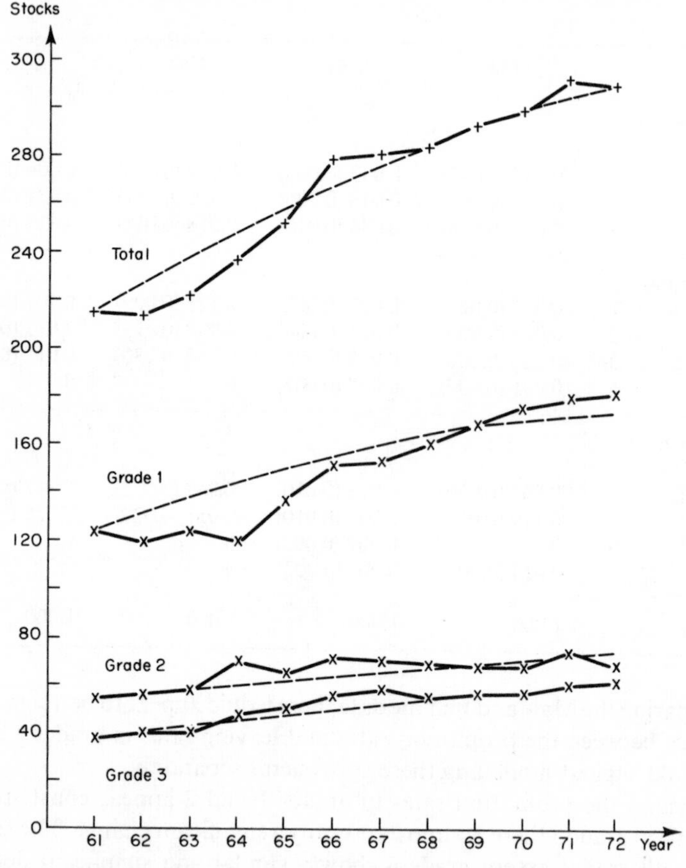


Figure 4.4 The actual and projected grade sizes 1961–72

show an increasing trend, with the total size in particular suggesting there may have been two periods of expansion corresponding approximately to the first and second halves of the period.

Table 4.6 contains the leaving and promotion rates over the whole period

***Table 4.6 Leaving and promotion rates (per cent) for 1961–72—see solution to Exercise 4.9***

| | | 61/2 | 62/3 | 63/4 | 64/5 | 65/6 | 66/7 | 67/8 | 68/9 | 69/70 | 70/1 | 71/2 | 1961–72 overall† |
|---|---|---|---|---|---|---|---|---|---|---|---|---|---|
| *Leaving rates from grade* | 1 | 4.1 | 5.1 | 7.3 | 6.7 | 4.4 | 6.6 | 5.9 | 10.6 | 10.7 | 5.1 | 4.5 | 6.5 (2.0) |
| | 2 | 1.9 | 3.6 | 13.8 | 10.0 | 1.5 | 9.9 | 8.6 | 1.5 | 7.5 | 4.5 | 9.6 | 6.7 (3.1) |
| | 3 | 10.5 | 5.1 | 5.0 | 4.3 | 0 | 1.8 | 12.1 | 9.3 | 5.4 | 1.8 | 1.7 | 5.1 (3.1) |
| *Promotion rates from grade* | 1 | 5.7 | 4.2 | 21.8 | 4.2 | 5.8 | 4.0 | 2.6 | 1.9 | 3.0 | 4.6 | 1.7 | 5.0 (1.8) |
| | 2 | 9.3 | 5.4 | 12.1 | 5.7 | 4.6 | 0 | 2.9 | 4.4 | 1.5 | 3.0 | 2.7 | 4.5 (2.6) |

† the numbers in brackets are the 'average' errors–see solution to Exercise 4.9.

1961–72. As we have seen in Chapter 2 it is often difficult to judge whether the fluctuations observed in the flow rates over time are likely to be due to chance or to more systematic factors. A simple visual method suggested in Chapter 2 was to plot these rates as a time series together with confidence intervals based on their individual standard errors. A quicker but less rigorous method is used in Figure 4.5 where only the point estimates are

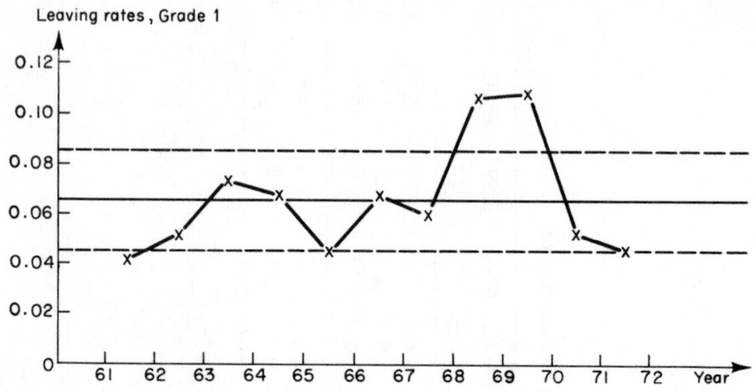


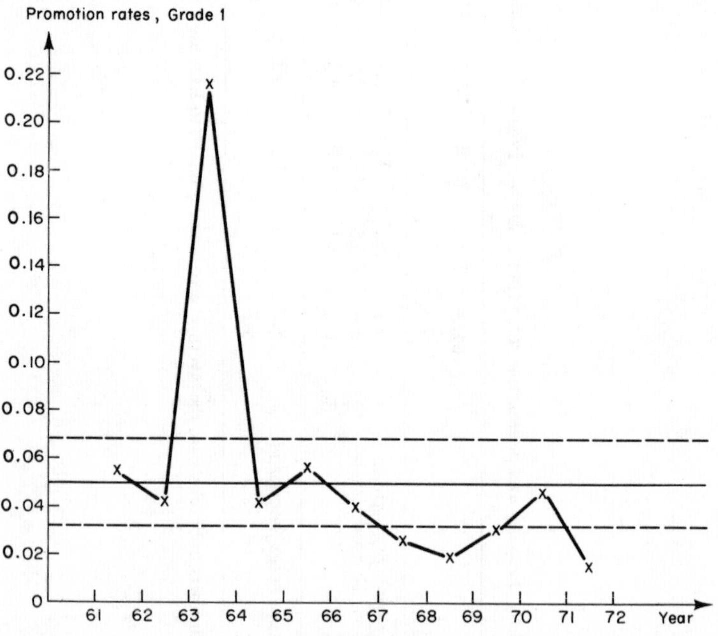


Figure 4.5 Leaving and promotion rates from grade 1 for the system of Exercise 4.9

shown together with an error band (see broken lines) based on the overall estimate and a typical or average error, which is calculated as follows. From (4.13) the estimate for the overall flow rate is

$$\hat{p}_{ij} = \sum_T n_{ij}(T) / \sum_T n_i(T)$$

An 'average' estimate of error is then

$$[\hat{p}_{ij}(1 - \hat{p}_{ij}) / \{k^{-1} \sum_T n_i(T)\}]^{\frac{1}{2}}$$

where $k$ is the number of years of data, so that the denominator is the average stock level. This error assumes a binomial model which implies people behave independently and are all subject to the same flow rate. These error bands would not be appropriate if the stock levels change markedly from year to year since in this case the individual point estimates would have widely different errors. The bands in Figure 4.5 have been drawn as $\hat{p}_{ij} \pm$ one 'average' error.

The graph of the leaving rates in Figure 4.5 is typical of those for the other grades. There appears to be no systematic trend and approximately 70 per cent of the points lie within the band, as we would expect. We can therefore conclude that subject to our assumptions there is no strong evidence to suggest that the rates have not remained constant over the period. The spread of the observed promotion rates are also as expected except for 1963/4. In fact it turns out that there was some re-grading in the system during this year which accounts for this exceptional promotion rate.

***Table 4.7***

| *Grade* | *Stocks at beginning* | *Promotions from* | *Leavers from* | *Entrants to* |
|---|---|---|---|---|
| 1 | 1611 | 81 | 105 | 244 |
| 2 | 719 | 32 | 48 | 12 |
| 3 | 553 | — | 28 | 20 |

(b) Summing the data over the whole period 1961–71 gives the figures in Table 4.7 which in turn give the estimated parameters

$$\mathbf{P} = \begin{bmatrix} 0.8845 & 0.0503 & 0 \\ 0 & 0.8887 & 0.0445 \\ 0 & 0 & 0.9494 \end{bmatrix}, \quad \mathbf{R}' = \begin{bmatrix} 22.2 \\ 1.1 \\ 1.8 \end{bmatrix}$$

The projections for 1961–72 using these parameters are shown by the broken lines in Figure 4.4. Although in some cases these projections do not give good predictions of individual values they do give good overall predictions of the trends. Furthermore, at the end of the period the projected and actual values agree reasonably well as we might hope. The total size and the size of grade 1 will inevitably be particularly sensitive to the assumed level of inflow, so we could expect the model to simulate the data more closely if we put **R** equal to the actual inflow (the interested reader is recommended to do this).

(c) In deciding how to use the 1961–67 data to estimate parameters to project for 1967–72 it seems reasonable to exclude the obviously untypical promotion experience for 1963/4. Furthermore, in view of the re-grading during this year it would seem appropriate to exclude all that year's data, so that the data on which our estimates are to be based are as shown in Table 4.8.

***Table 4.8***

| *Grade* | *Stocks at beginning* | *Promotions from* | *Leavers from* | *Entrants to* |
|---|---|---|---|---|
| 1 | 650 | 31 | 35 | 100 |
| 2 | 316 | 15 | 18 | 6 |
| 3 | 230 | — | 9 | 7 |

giving as estimates of the parameters

$$\mathbf{P} = \begin{bmatrix} 0.8985 & 0.0477 & 0 \\ 0 & 0.8956 & 0.0475 \\ 0 & 0 & 0.9609 \end{bmatrix}$$

$$\mathbf{R} = [20.0 \quad 1.2 \quad 1.4]$$

The projections for 1967–71 using these parameters are shown in Figure 4.6. From an inspection of the data in Tables 4.4 and 4.5 it appears that the leaving rate from grade 3 used in this projection is lower than that actually experienced during 1967–71, particularly during the earlier years of the projection period. This is likely to have caused most of the discrepancy in the forecast for grade 3, although the promotion rate from grade 2 may also have contributed since that used in the forecast appears rather higher than the one actually experienced. The grade 1 discrepancy is probably due to mis-specification of the inflow numbers. This can be checked by re-running the forecast with the actual inflow levels and again the interested reader is recommended to do this.

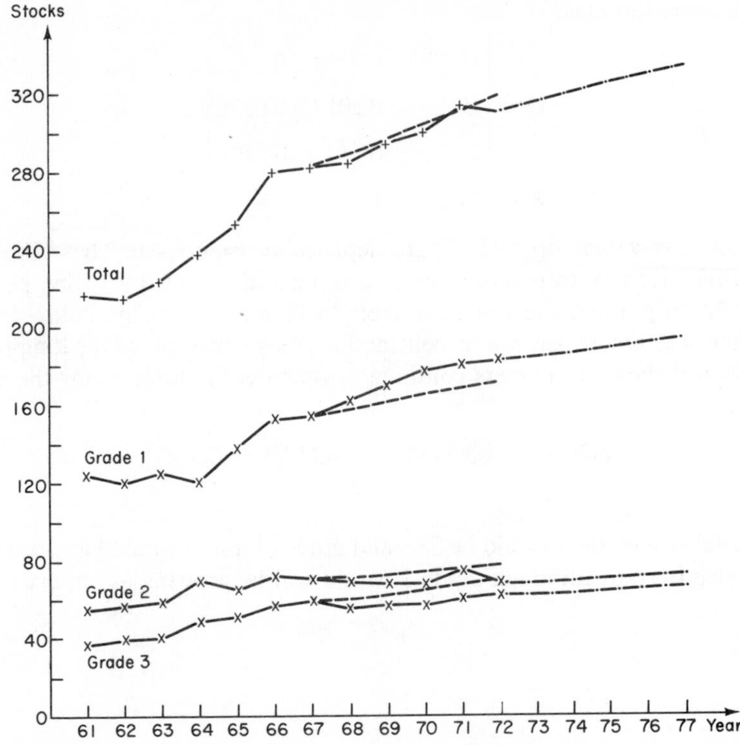


Figure 4.6 Projections for 1967–72 [Solution to Exercise 4.9c], and 1972–77 [Solution to Exercise 4.9d]

(d) For a projection from 1972 onwards it would appear reasonable to estimate the parameters using all the data from 1961–72 excluding the year 1963/4. This gives the data of Table 4.9.

***Table 4.9***

| *Grade* | *Stocks at beginning* | *Promotions from* | *Leavers from* | *Entrants to* |
|---|---|---|---|---|
| 1 | 1487 | 54 | 97 | 212 |
| 2 | 661 | 25 | 40 | 12 |
| 3 | 513 | — | 26 | 18 |

which give estimates

$$\mathbf{P} = \begin{bmatrix} 0.8985 & 0.0363 & 0 \\ 0 & 0.9017 & 0.0378 \\ 0 & 0 & 0.9493 \end{bmatrix}$$

$$\mathbf{R} = [21.2 \quad 1.2 \quad 1.8]$$

The forecast values up to 1977 are depicted in Figure 4.6. These show a straightforward extrapolation of the trends observed over the period 1961–72. In practice the system is likely to be sensitive to the inflow levels and hence to any expansion or contraction. As an example of the long-term situation if these parameters continue to operate, the forecast for the year 2000 is

$$\mathbf{n}(2000) = [207\,(54\%), 86\,(23\%), 88\,(23\%)]$$

The total size by then would be 381, and grades 1 and 2 would have reached their steady state, although grade 3 would still be growing at almost 1 per year.

CHAPTER 5

# Transition Models Based on Renewal Theory

## 5.1 INTRODUCTION

In the Markov family of models, wastage and promotion flows are treated as 'push' flows determined by fixed transition probabilities. The main use of these models in the planning process is to investigate how the grade sizes would change under the operation of constant average flow rates. In many organizations, however, we know that the grade sizes are not free to vary in the way postulated by these models but are constrained by the amount of work to be done or the finance available. The grade sizes are thus known in advance or, at least, can be predicted on the basis of manpower demand studies as described in Chapter 8. Under these circumstances promotion and recruitment can only take place to fill vacancies as they occur. They are thus what we described in Chapter 1 as 'pull' flows and models which operate on these principles are known as *renewal models*. In a renewal model the main object is not to predict the stocks, which are given, but the flows. This exercise hinges on the wastage flow which may be said to 'drive' the system by creating vacancies to be filled by promotion or recruitment. The aim of this chapter is to describe some of the simpler renewal models and to illustrate their use. For reasons which will become apparent later, renewal models are apt to possess awkward mathematical features which make it difficult to give a well-rounded and simple account of the theory. However, the models are no more difficult to handle computationally than their Markovian counterparts and are often much more realistic.

## 5.2 ONE-GRADE SYSTEMS

The simplest renewal situation is one, either where there is no internal differentiation into grades, or where we choose to ignore such categorizations. The only flow in this case, apart from wastage, is that of recruitment into the system. If the size of the organization is fixed, recruitment and wastage must always be equal, so predicting one is equivalent to predicting the other. If the size is changing the only modification required is to add (or subtract) the change in the number of posts from the predicted wastage.

The central assumption of a renewal model will thus be about wastage and our earlier discussions suggest it should be made dependent on length of service. In discrete time this can be done by specifying a set of length-of-service specific wastage rates as in Chapter 2. The total wastage flow can then be predicted by applying these wastage rates to the stocks in each length-of-service category. This is precisely what we were doing in the last chapter where we used the fixed-size Markov model to predict a length-of-service structure. Using that version of the Markov model there is thus no difficulty in projecting both the recruitment needs and the length-of-service structure. An example illustrating the use of the model in this way was provided by Example 4.7.

The foregoing argument shows that a simple renewal system can be handled within the framework of Markov chain theory. However, there are advantages in approaching the problem directly using standard renewal theory arguments. The problem can be treated in discrete or continuous time and we shall give formulae for both cases. In the discrete case one might assume that all movements take place at the time points 0, 1, 2, . . . . However, we find it more natural to proceed by defining a discrete approximation to the underlying continuous process in which losses can occur at any time. This is in line with our treatment of wastage in Chapter 2 and is also consistent with the analysis using the Markov model referred to in the last paragraph. We begin therefore with the definitions of the discrete analogues of the functions $f(x)$, $m(x)$, and $G(x)$ as follows (see, also, Sections 2.2 and 3.6 where slightly different notations were used).

$$\left.\begin{aligned} f_i &= \Pr\{\text{entrant leaves with length of service in } (i, i+1)\} \\ w_i &= \Pr\{\text{person with length of service } i \text{ leaves in } (i, i+1)\} \\ G_i &= \Pr\{\text{entrant's length of service is } i \text{ or more}\} \end{aligned}\right\} (i = 0, 1, 2, \ldots)$$

The relationships between these three functions are easily derived by elementary probability arguments. Thus

$$\left.\begin{aligned} G_i &= (1 - w_0)(1 - w_1)\ldots(1 - w_{i-1}) \\ \text{and} \qquad f_i &= G_i w_i \end{aligned}\right\} \tag{5.1}$$

We make the further assumption that those who leave during an interval are replaced at the end of that interval. (This means that the system will only be at full strength at times which are multiples of the basic time interval. We regard this as a reasonable, and not unrealistic, approximation to the true situation.)

Let $h_j$ denote the replacement rate at time $j$ so that, if the size of the system is $N$, the expected number of replacements will be $Nh_j$. Then if all initial members of the system start with zero length of service we shall show that

$$h_1 = f_0$$

$$h_j = f_{j-1} + h_1 f_{j-2} + h_2 f_{j-3} + \ldots + h_{j-1} f_0 \qquad (j = 2, 3, \ldots) \tag{5.2}$$

The first part of this equation follows from the fact that the number to be replaced at time 1 is precisely equal to the number out of the initial stock who leave with length of service in (0,1). The second part computes the total replacement rate by classifying leavers according to their time of joining and then adding them up. Thus, the expected number joining at time $i$ is $Nh_i$ and of these an expected proportion $f_{j-i-1}$ will leave in the interval $(j-1, j)$ with length of service between $j-i-1$ and $j-i$. This group of joiners therefore contributes $Nh_i f_{j-i-1}$ to the total loss in that interval. Summing these contributions and equating the result to $Nh_j$ gives (5.2). The computations are easily carried out as illustrated in the following example.

*Example 5.1* Consider the system with the following leaving functions which necessarily satisfy the relationships of (5.1):

| $i$ | 0 | 1 | 2 | 3 | 4 | 5 | 6 | 7 |
|---|---|---|---|---|---|---|---|---|
| $G_i$ | 1.00 | 0.60 | 0.45 | 0.38 | 0.34 | 0.32 | 0.30 | 0.29 |
| $f_i$ | 0.40 | 0.15 | 0.07 | 0.04 | 0.02 | 0.02 | 0.01 | ... |
| $w_i$ | 0.40 | 0.25 | 0.15 | 0.10 | 0.07 | 0.05 | 0.05 | ... |

Equation (5.2) gives the following renewal or replacement rates when the system remains at constant size and all initial members of the population have zero length of service.

$$
\begin{aligned}
h_1 &= f_0 && = 0.40 && = 0.40\\
h_2 &= f_1 + h_1 f_0 && = 0.15 + (0.40)(0.40) && = 0.31\\
h_3 &= f_2 + (h_1 f_1 + h_2 f_0) && = 0.07 + 0.18 && = 0.25\\
h_4 &= f_3 + (h_1 f_2 + h_2 f_1 + h_3 f_0) && = 0.04 + 0.17 && = 0.21\\
h_5 &= f_4 + (h_1 f_3 + \ldots + h_4 f_0 && = 0.02 + 0.16 && = 0.18\\
h_6 &= f_5 + (h_1 f_4 + \ldots + h_5 f_0) && = 0.02 + 0.14 && = 0.16\\
h_7 &= f_6 + (h_1 f_5 + \ldots + h_6 f_0) && = 0.01 + 0.13 && = 0.14
\end{aligned}
$$

etc.

During the first 6 years, therefore, there will be a considerable drop in wastage, and hence in the level of recruitment required. This is due to the ageing of the population, since wastage in this example decreases with length of service. This calculation also illustrates the point about the dependence of crude wastage (=recruitment) rates on the age structure of the system made in Section 3.5. As this system ages its loss rate declines substantially. ◀

The theory may be extended to cover the case where the members of the initial stock do not have zero length of service. Only the first term, $f_{j-1}$, of (5.2) relates to the initial stock so all we have to do is to replace it by a term giving the expected number of the initial stock who leave at calendar time $j-1$. Suppose that, initially, the proportion of the initial stock with length of service $i$ is $S_i$. The probability that a person with length of service $i$ will leave at time $j-1$ ($j > i$) is

$$f_{i+j-1}/G_i$$

where $G_i$ is the probability of survival to $i$. Hence the expected proportion of losses at time $j-1$ arising from the original stock is

$$\sum_i S_i f_{i+j-1}/G_i \tag{5.3}$$

and this must replace $f_{j-1}$ in (5.2).

*Example 5.2* Suppose in Example 5.1 that the initial population has the following length-of-service distribution $\{S_i\}$:

| $i$ | 0 | 1 | 2 | 3 | 4 | 5 | 6 | 7 | |
|---|---|---|---|---|---|---|---|---|---|
| $S_i$ | 0.50 | 0.30 | 0.10 | 0.10 | 0 | 0 | 0 | 0 | ... |
| $S_i/G_i$ | 0.50 | 0.50 | 0.22 | 0.26 | 0 | 0 | 0 | 0 | ... |

Since members of the initial population do not now all have zero lengths of service the first term $f_{j-1}$ in the expression for the renewal rate $h_j$ is replaced by (5.3) so that

$$h_1 = (\sum_i f_i S_i/G_i) \qquad = 0.30 \qquad = 0.30$$

$$h_2 = (\sum_i f_{i+1} S_i/G_i) + h_1 f_0 \qquad = 0.12 + 0.16 = 0.28$$

$$h_3 = (\sum_i f_{i+2} S_i/G_i) + (h_1 f_1 + h_2 f_0) \qquad = 0.06 + 0.18 = 0.24$$

$$h_4 = (\sum_i f_{i+3} S_i/G_i) + (h_1 f_2 + h_2 f_1 + h_3 f_0) = 0.04 + 0.17 = 0.21$$

etc.

The initial distribution in this example has therefore had the effect of reducing the number of recruits required, at least during the first 4 years. ◀

The limiting, or steady-state, behaviour of the recruitment flow can be found by an appeal to renewal theory where it is shown, under very general conditions that

$$\lim_{j\to\infty} h_j = \mu^{-1} \tag{5.4}$$

where $\mu = \sum_{j=1}^{\infty} jf_{j-1}$. (This is not quite the mean of the CLS distribution because if we assume that those who leave in the interval $(j-1, j)$ do so, on average, half-way through then the mean CLS would be $\sum_{j=1}^{\infty} (j - \frac{1}{2})f_{j-1} = \mu - \frac{1}{2}$. In other words our model treats leavers as if they had remained in post until the time at which they were replaced.) The limiting result of (5.4) corresponds to a rule of thumb which is often used in manpower planning. For example, if the average length of service is 4 years one would expect 25 per cent of the stock to turn over each year. The result of (5.4) shows that this is only the case in the long run. In the short term the actual wastage rate may be very different as the calculations of Example 5.1 show. Further calculations given in Bartholomew (1973a, Chapter 7) confirm this and suggest that crude rates may possibly differ from their steady-state values by a factor of as much as two or three. It is for this reason that the crude wastage rate can be so misleading as an indicator of organizational stability, as already noted in Chapters 2 and 3.

The restriction that the size of the organization be fixed can easily be removed in the case of an expanding organization. Let the initial size be $N(0)$ and suppose that the size is increased by an increment $M(j)$ at time $j$ $(j = 1, 2, \ldots)$. The total size at time $j$ is thus

$$N(j) = N(0) + \sum_{i=1}^{j} M(i) \qquad (j = 1, 2, \ldots)$$

Because the size is changing it is easier to work with the expected number of replacements rather than the replacement rate and so we define $R_j$ as the expected number of recruits required at time $j$. This number can be found by summing the numbers required by the initial population and by each of the renewal processes associated with each of the increments $M(1), M(2), \ldots$. The process which began at time $i$ with $M(i)$ new places will require $M(i)h_{j-i}$ recruits at time $j$. In total, therefore, we shall require

$$R_j = N(0)h_j + \sum_{i=1}^{j-1} M(i)h_{j-i} + M(j) \qquad (j = 1, 2, \ldots) \tag{5.5}$$

recruits at time $j$. When the growth pattern is given, the calculation thus only requires a knowledge of the renewal function $h_j$ for the fixed-size system.

*Example 5.3* Following on from Example 5.2, suppose that the initial total size of the system is $N(0) = 100$ and that during the first 4 years this increases by increments $M(1) = 10$, $M(2) = 15$, $M(3) = 20$ and $M(4) = 25$. Then from (5.5) the numbers of recruits required and the replacement rates are

$$R_1 = 100h_1 + M(1) = 40, R_1/N(0) = 0.40$$
$$R_2 = 100h_2 + M(1)h_1 + M(2) = 46, R_2/N(1) = 0.42$$
$$R_3 = 100h_3 + (M(1)h_2 + M(2)h_1) + M(3) = 51, R_3/N(2) = 0.41$$
$$R_4 = 100h_4 + (M(1)h_3 + M(2)h_2 + M(3)h_1) + M(4) = 59, R_4/N(3) = 0.44$$

Expansion of the system has caused the required *number* of recruits to increase in contrast to the decreasing requirements shown in Examples 5.1 and 5.2. The replacement rate, however, is almost constant because the increase caused by the expansion of the system has almost exactly balanced the decrease resulting from the ageing of the initial population. ◀

Most textbook treatments of renewal theory are in continuous time and this offers many advantages as far as the mathematical analysis of renewal processes is concerned. For numerical purposes we find it more convenient to work in discrete time but even then there are sometimes advantages in changing to a continuous-time formulation. For example, if we have fitted a continuous CLS distribution to wastage data as described in Chapter 3 it is rather tiresome to convert it back into a discrete distribution in order to use the theory of this chapter. A full discussion of continuous-time renewal theory applied to manpower systems will be found in Bartholomew (1973a, Chapters 7 and 8). Here we merely review a few basic results which are useful in practical manpower planning work.

Let $f(x)$ denote the probability density function of CLS and let $h(T)$ be the *renewal density* at time $T$. The latter is the continuous analogue of $h_j$ and it is defined so that $h(T)\delta T$ is the probability that a randomly chosen individual will require replacement in the interval $(T, T + \delta T)$. Hence it may be thought of as the expected proportion of replacements needed in that interval. The continuous form of the renewal equation when the initial stock all have zero length of service is then

$$h(T) = f(T) + \int_0^T h(x)f(T - x)\,dx \tag{5.6}$$

Special techniques, involving the use of the Laplace transform, exist for solving this equation but they do not always work as, for example, in the case when $f(x)$ is lognormal. In such cases the following simple approximation (Bartholomew (1973a) and (1963b)) is often useful:

$$h(T) \approx f(T) + F^2(T)/\int_0^T G(x)\,dx \tag{5.7}$$

where

$$F(T) = \int_0^T f(x)\,dx \quad \text{and} \quad G(T) = \int_T^\infty f(x)\,dx$$

A plot of $h(T)$ will show how the rate of recruitment changes with time but for most applied purposes what is required is the expected number of recruits needed in an interval of time. This can be obtained at once by integrating $h(T)$ between the appropriate limits.

All of the foregoing theory relates to expected values. In reality the numbers of leavers and recruits will be random variables and it would often be useful in planning work to know something about the distribution of our predictions. The required result for discrete time follows from the fact that the total number of leavers at time $j$ can be expressed in the form

$$\sum_{i=1}^{N} X_i$$

where $X_i = 1$ if the person in the $i$th job leaves and $X_i = 0$ otherwise. Now $X_i$ is a binomial random variable with $\Pr\{X_i = 1 \text{ at time } j\} = h_j$. Hence, if people behave independently the total number of leavers will have a binomial distribution with parameters $N$ and $h_j$. If $h_j$ is small—as it usually will be—then the distribution will be approximately Poisson with mean $Nh_j$. This approximation is in fact more general and will apply even if the $h_j$'s are different, as they would be in an expanding organization or where the propensity to leave varies between jobs. Provided that all of the $h_j$'s are small it remains true that

$$\sum_{i=1}^{N} X_i$$

has a Poisson distribution approximately. Having forecast that the recruitment needs at some future time will be $R_j$, say, it is thus reasonable to suppose that the actual requirement, all other things remaining constant, will be within $R_j \pm 2\sqrt{R_j}$.

It may be helpful at this point to indicate how the prediction of recruitment (and hence wastage) discussed in this section relates to the earlier discussion of the same topic in Section 3.4. There we are concerned with predicting the future leaving experience of a given cohort or of all the cohorts making up the present stock of an organization. Here we have extended the scope of the exercise to include wastage arising from future cohorts recruited to replace losses from the current cohorts. This problem is complicated by the fact that the sizes of these new cohorts are determined by the numbers who leave in the forecast period. It is this last feature which renewal theory enables us to deal with.

## 5.3 A SIMPLE HIERARCHICAL MODEL WITH INSTANTANEOUS FILLING OF VACANCIES

The approach to multi-grade systems is best made in easy stages so that the complications which arise can be properly appreciated. We start therefore with a simple hierarchy of $k$ grades in which promotion occurs only in to the next higher grade; recruitment may occur at any level. Let the (fixed) numbers in the grades be $n_1, n_2, \ldots, n_k$. Wastage is assumed to be grade-specific with $w_i$ being the probability of leaving for an individual in grade $i$. In this respect the renewal model makes exactly the same assumption as the Markov model. The difference arises with promotions which can now only take place when vacancies occur. If a vacancy occurs in grade $i$ it may be filled either by direct recruitment from outside or by promotion from the grade below. Suppose promotion occurs with probability $s_i$ and recruitment with probability $1 - s_i$. In order for the model to be applicable it is necessary for these probabilities to be the same for all posts in a given grade and to be constant over time. Finally, we assume that the time to fill a vacancy is so small compared with the discrete interval used as a basis for the calculation that it can be neglected.

We are now in a position to calculate the expected promotion and recruitment flows. Let $P_{i,i+1}$ denote the expected number of promotions between grade $i$ and grade $i + 1$ over one time interval. Beginning at the top, $P_{k-1,k}$ must be equal to that number of vacancies arising at that level which are filled by promotion. The expected number of vacancies is $N_k w_k$ and the proportion to be filled by promotion is $s_k$; hence

$$P_{k-1,k} = n_k w_k s_k \tag{5.8}$$

Turning next to $P_{k-2,k-1}$ we have

$$\begin{aligned} P_{k-2,k-1} &= (\text{Number of vacancies in } k - 1)s_{k-1} \\ &= (n_{k-1}w_{k-1} + P_{k-1,k})s_{k-1} \end{aligned} \tag{5.9}$$

In general, the same mode of argument gives

$$P_{i,i+1} = \{n_{i+1}w_{i+1} + P_{i+1,i+2}\}s_{i+1} \qquad (i = 1, 2, \ldots, k-2)$$

or

$$P_{i,i+1} = \sum_{j=i+1}^{k} n_j w_j \prod_{h=i+1}^{j} s_h \qquad (i = 1, 2, \ldots, k-1) \tag{5.10}$$

The recruitment numbers, $R_i$, are obviously given by

$$\begin{aligned} R_i &= \text{total loss from grade } i - \text{total input to grade } i \text{ by promotion} \\ &= n_i w_i + P_{i,i+1} - P_{i-1,i} \end{aligned} \tag{5.11}$$

where $P_{i,i+1} - 0$ if $i = k$ and $P_{i-1,i} = 0$ if $i = 1$.

The expected promotion numbers can be expressed as proportions of the stocks from which they originate. Note that these proportions will be constant over time for this model and hence in an average sense it is indistinguishable in this respect from a Markov model.

The argument used above may break down and it is important and instructive to consider how this might come about. Consider, again the calculation of $P_{k-2,k-1}$ leading to (5.9). The number of vacancies arising in grade $k-1$ was computed as the sum of the number of leavers and the number of promotions to the higher grade. This involves the implicit assumption that wastage occurs first and that the promotees are drawn from those who remain. It could happen that more promotees are called for by the higher grade than are available, in which case the model breaks down and would need to be modified. In practice, promotees usually constitute only a small proportion of the available stock and so this situation is unlikely to arise. However, in making the calculations one should check that the total number of vacancies in a grade at no time exceeds the total stock. This is an example of the sort of difficulty which complicates the mathematical analysis of discrete-time renewal processes. Many of these difficulties vanish if the derivation is conducted in continuous time but they cannot be avoided in practice if discrete-time computations have to be made.

(An alternative way of specifying the model would be to suppose that wastage occurs only among those not selected for promotion. This would create other problems which the reader is invited to explore, but it would still not completely avoid the possibility of a greater demand for promotees than the grade could supply.)

The practical implementation of the model requires values to be chosen for the parameters $\{w_i\}$ and $\{s_i\}$. If one wishes to make a projection using current rates these can be estimated from flow rates in the obvious way. The $w_i$'s are estimated from observed leaving rates exactly as with the Markov model; the estimation of $s_i$'s requires a record of the filling of each vacancy so that the proportions filled from inside can be calculated.

*Example 5.4* A three-grade hierarchical system has fixed grade sizes with $\mathbf{n} = (140, 105, 35)$, $\mathbf{w} = (0.20, 0.10, 0.15)$, $\mathbf{s} = (0, 0.60, 0.60)$. Thus all vacancies at the lowest level are filled by recruitment and 40 per cent at the two higher levels likewise. We will calculate the promotion and recruitment flows required to maintain the grade sizes. The two promotion flows are

$$P_{12} = n_2 w_2 s_2 + n_3 w_3 s_2 s_3 = 6.30 + 1.89 = 8.19$$

$$P_{23} = n_3 w_3 s_3 = 3.15$$

Expressed as rates (i.e. proportions of the stock from which they originate) they are 0.059 and 0.030 respectively. The expected recruitment numbers are

$$R_1 = n_1 w_1 + P_{12} = 28.00 + 8.19 = 36.19$$

$$R_2 = n_2 w_2 + P_{23} - P_{12} = 10.50 + 3.15 - 8.19 = 5.46$$

$$R_3 = n_3 w_3 - P_{23} = 5.25 - 3.15 = 2.10$$

Alternative expressions for $R_2$ and $R_3$ are

$$R_2 = n_2 w_2(1 - s_2) + n_3 w_3 s_3(1 - s_2), \quad R_3 = n_3 w_3(1 - s_3)$$

from which we conclude that the model can never lead to negative values for recruitment. ◀

## 5.4 VACANCY CHAIN MODELS: NON-HIERARCHICAL SYSTEMS WITH INSTANTANEOUS FILLING OF VACANCIES

A natural extension of the foregoing model is to relax the requirement that a vacancy filled from inside the system should only be filled from the grade below. This is not without its problems but before attempting to generalize the model we shall introduce a new way of looking at manpower systems due to White (1970). The idea is to look at the flow of vacancies instead of people. Whenever a person moves from one location to another in a renewal system a vacancy moves in the opposite direction. Thus a loss can be viewed as the recruitment of a vacancy and the recruitment of a person is the loss of a vacancy. There is thus a dual relationship between the flows of people and the flows of vacancies. White applies these ideas to the study of the mobility of clergymen in several US churches and Stewman (1975) has applied them to a state police force. Both authors were primarily interested in the sociological aspects of the mobility process. Our treatment here is directed to manpower planning problems and, in some respects, goes further than the earlier work by including results on stocks of vacancies.

The vacancy chains arising in the model of Section 5.3 were of a particularly simple kind. When a vacancy entered a particular grade as a result of promotion or wastage it travelled instantaneously down the system, making its exit at one of the lower levels. We could set up the transition table governing the path of the vacancy chain in that case and it would have the following form:

$$\begin{matrix} 0 & 0 & 0 \ldots 0 & 0 & 1 \\ s_2 & 0 & 0 \ldots 0 & 0 & 1 - s_2 \\ 0 & s_3 & 0 \ldots 0 & 0 & 1 - s_3 \\ \vdots & & & \vdots & \vdots \\ 0 & 0 & 0 \ldots s_k & 0 & 1 - s_k \end{matrix}$$

in the $i$th row of this array we have the probabilities governing the movement of a vacancy when in grade $i$. In particular the last column contains the probabilities that it leaves the system at each level; in the $j$th column we have the probabilities that it moves to grade $j$—these are zero unless $j = i - 1$.

The more general version of the vacancy chain model allows non-zero elements to ccur anywhere in the transition table. Generalizing the notation slightly this leads to the array

$$\begin{array}{cccc|c} s_{11} & s_{12} & \cdots & s_{1k} & s_{1,k+1} \\ s_{21} & s_{22} & \cdots & s_{2k} & s_{2,k+1} \\ \vdots & & & \vdots & \vdots \\ s_{k1} & \cdots & \cdots & s_{kk} & s_{k,k+1} \end{array} \equiv \mathbf{S} \,\vdots\, \mathbf{s}'_{k+1}, \text{ say,}$$

where $s_{ij}$ is the probability of a vacancy moving from $i$ to $j$ and $s_{i,k+1}$ replaces the former $1 - s_i$ as the probability of loss. This array should be compared with that at the beginning of Section 4.2 when it will be recognized that we have a Markov-chain model for the instantaneous movement of vacancies. Its usefulness in manpower applications depends on whether the assumptions of constancy and independence are satisfied. White and Stewman have shown in their applications that these assumptions were acceptable but they should always be tested if possible.

If the elements $s_{ii}$ on the diagonal of $\mathbf{S}$ are non-zero it means that vacancies can be filled from within the same stratum. Such moves, of course, contribute nothing to the promotion and recruitment flows between strata.

When we try to apply the more general model in practice a question of realism arises. In the model of Section 5.3 the length of a vacancy chain could be no more than $k - 1$, which would happen if the vacancy entered at the top and moved to the bottom. With a general $\mathbf{S}$ the length of the chain could be indefinitely long since vacancies could move around within the system for a very long time before leaving. This makes it more difficult to postulate that the whole sequence of moves must be completed within the span of one time period. With hierarchical or fairly simple structures the chance of an unreasonably long chain occurring will be negligible and the difficulty can then be ignored.

Accepting the assumption of instantaneous filling, we are now faced with the problem of finding the expected number of transitions along every path in the system which will be generated by each round of leavers. We expect $n_i w_i$ vacancies to enter grade $i$ as a result of the wastage of individuals from that grade. These vacancies will move through the system until finally they make their exit. The required probabilities can easily be found from the fundamental matrix (see Chapter 6) for the chain with transition matrix $\mathbf{S}$. We shall see in that chapter that $d_{ij}$, the $(i, j)$th element of $(\mathbf{I} - \mathbf{S})^{-1}$, is the expected length of time a vacancy which enters at $i$ will spend in $j$. It may, equivalently, be interpreted as the number of visits paid to $j$. On each visit to $j$ the probability of the vacancy moving to $h$ is $s_{jh}$ so the expected number of transitions from $j$ to $h$ for a vacancy entering at $i$ will be $d_{ij}s_{jh}$. There is a similar contribution for each vacancy entering $i$ of which there are $n_i w_i$ in all. Finally, as vacancies can arise at all levels, the product has to be summed over $i$ giving a total number of vacancy transitions between $j$ and $h$ equal to

$$\sum_{i=1}^{k} n_i w_i d_{ij} s_{jh}$$

Since each movement of a vacancy in one direction implies the movement of a person in the opposite direction it follows that the expected flows generated by each round of leavers are

$$P_{hj} = s_{jh} \sum_{i=1}^{k} n_i w_i d_{ij} \qquad \begin{Bmatrix} j = 1, 2, \ldots, k \\ h = 1, 2, \ldots, k+1 \end{Bmatrix} \tag{5.12}$$

When $h = k + 1$, $P_{k+1,j}$ must be interpreted as the recruitment flow into $j$. The reader should check that (5.12) includes (5.10) as a special case.

To illustrate the calculations we take the transition matrix used by Stewman (1975) for a police force with five grades.

*Example 5.5* Table 5.1 gives the estimated transition matrix, **S**, for the movement of vacancies. The grades are Trooper (1), Corporal (2), Sergeant (3), Lieutenant (4), and Colonel (5).

***Table 5.1*†**

| | | *Destination* | | | | | |
|---|---|---|---|---|---|---|---|
| | | 1 | 2 | 3 | 4 | 5 | *Out* |
| ***Origin*** | 1 | 0·0494 | 0·0031 | 0·0004 | 0 | 0 | 0·9471 |
| | 2 | 0·6125 | 0·2811 | 0·0024 | 0 | 0 | 0·1040 |
| | 3 | 0·0015 | 0·5539 | 0·3443 | 0 | 0 | 0·1003 |
| | 4 | 0 | 0·0169 | 0·7022 | 0·1404 | 0·0056 | 0·1349 |
| | 5 | 0 | 0 | 0 | 0·6635 | 0·2115 | 0·1250 |

† Table 5.1 and the matrix **D** (below) are reproduced from Stewman (1975) by permission of Gordon and Breach Science Publishers Ltd.

The fundamental matrix $\mathbf{D} = (\mathbf{I} - \mathbf{S})^{-1}$ is as follows:

$$\begin{array}{c|ccccc} & 1 & 2 & 3 & 4 & 5 \\ \hline 1 & 1.0552 & 0.0051 & 0.0007 & 0 & 0 \\ 2 & 0.9016 & 1.3993 & 0.0057 & 0 & 0 \\ 3 & 0.7640 & 1.1820 & 1.5229 & 0 & 0 \\ 4 & 0.6454 & 0.9986 & 1.2567 & 1.1697 & 0.0083 \\ 5 & 0.5431 & 0.8403 & 1.0575 & 0.9843 & 1.2752 \end{array}$$

We may now calculated the set of expected flows for any given stocks and wastage rates on the assumptions of the model. Suppose that the stocks and wastage rates are as follows:

| $i$ | 1 | 2 | 3 | 4 | 5 |
|---|---|---|---|---|---|
| $n_i$ | 2500 | 1500 | 1000 | 500 | 150 |
| $w_i$ | 0.20 | 0.15 | 0.10 | 0.10 | 0.05 |

Then Table 5.2 shows the expected flows during each year.

***Table 5.2***

| | | Destination | | | | | |
|---|---|---|---|---|---|---|---|
| | | 1 | 2 | 3 | 4 | 5 | *Outside* |
| *Origin* | 1 | 42 | 301 | 0 | — | — | 500 |
| | 2 | 3 | 138 | 124 | 1 | — | 225 |
| | 3 | 0 | 1 | 77 | 46 | — | 100 |
| | 4 | — | — | — | 9 | 7 | 50 |
| | 5 | — | — | — | 0 | 2 | 8 |
| | *Outside* | 800 | 51 | 23 | 9 | 1 | — |

These are rounded to the nearest whole number. Zeros signify possible flows which round to zero and dashes denote impossible flows. Apart from rounding errors the inflows equal the outflows for each grade. ◀

Notice again that the calculations are made on the assumption that there will always be sufficient people at any level to meet the demand for promotees to fill grades at higher levels. It is easily checked that this requirement is met by adding up the rows in the $P_{hj}$ table and checking that they do not exceed the corresponding $n_h$.

The theory can easily be extended to cope with expanding grade sizes. For example, if at some time the size of grade $i$ increases to $n_i + m_i$ then the entry of vacancies will become $n_i w_i + m_i$ and the consequential promotion between $h$ and $j$ will be given by substituting this for $n_i w_i$ in (5.12).

As with the simpler model discussed in the last section, the expected flow numbers are not time-dependent when the grade sizes are fixed. At first sight this may appear surprising since the Markov model for flows of people showed a time dependence in the predicted stocks and hence in the flow numbers. The explanation of the difference lies in the fact that the transient phase of the vacancy chain model occurs between successive points on our discrete time scale as a consequence of the assumption that vacancies are filled instantaneously. In many applications this is unrealistic as vacancies often take some time to be filled and therefore, at any time, there will be a stock of unfilled vacancies. We, therefore, go on to consider a model in which a vacancy can only move once per time interval. By choosing the time interval to be equal to the average time to fill a vacancy this should give a more realistic description of the process.

## 5.5 VACANCY CHAIN MODELS: A MODEL IN WHICH VACANCIES MOVE ONE STEP AT A TIME

The parallel between this vacancy flow model and the one for the flow of people is closer than in the earlier case. Just as then a person was assumed to make only one transition per unit time, so now we assume that the associated vacancy does likewise. The immediate consequence of adopting this position is that there are always vacancies in the system. We therefore introduce $v_j(T)$ to denote the expected number (or stock) of vacancies in grade $j$ at time $T$ with $\mathbf{v}(T)$ denoting the vector $\{v_j(T)\}$. We continue to use the notation $s_{ij}$ for the transition probability of a vacancy from $i$ to $j$ but there is an important distinction to be drawn between its present and earlier meaning. Since we are assuming vacancies can make only one movement per time period, $s_{ij}$ is now the probability that a vacancy existing in grade $i$ at the beginning of the time period will be filled from grade $j$, by the end of the period. The difference between the present use and that of the last section is that in the earlier model time did not appear in the specification. With this distinction in mind the difference equation for expected vacancies can be written

$$v_j(T) = \sum_{i=1}^{k} v_i(T-1)s_{ij} + (\text{new vacancies entering } j) \quad (j = 1, 2, \ldots, k) \quad (5.13)$$

If the grade sizes (i.e. number of posts or establishment) are $n_1, n_2, \ldots, n_k$ the expected number of new vacancies in grade $j$ at time $T$ will be

$$e_j(T) = \{n_j - v_j(T-1)\}w_j$$

Notice that this is not $n_j w_j$ because vacancies arising from losses can only occur in places already occupied. In matrix notation

$$\mathbf{v}(T) = \mathbf{v}(T-1)\mathbf{S} + \mathbf{e}(T) \quad (5.14)$$

This is of exactly the same form as the equation for the person-flow model and calculations can be made using the same computer program. However, remembering that $\mathbf{e}(T)$ depends on $\mathbf{v}(T-1)$, it is more convenient to rewrite the equation

$$\mathbf{v}(T) = \mathbf{v}(T-1)\mathbf{S}^* + \mathbf{R} \quad (5.15)$$

where $\mathbf{S}^* = \mathbf{S} - \mathbf{W}$ and $\mathbf{R} = \mathbf{n}\mathbf{W}$, $\mathbf{W}$ being a diagonal matrix formed from the wastage probabilities. **S*** is not now a stochastic matrix but this does not invalidate the use of BASEQN for calculation. The vector **R** now plays the role of a fixed recruitment vector. (There is no point in having a fixed size vacancy model unless there is some arbitrary rule requiring a fixed number of posts to be kept vacant.)

In practice we are more likely to want to know the expected flows than the numbers of vacancies. These are easily found as follows:

$$\begin{aligned}&\text{expected recruitment flow to } j \text{ at } T = v_j(T-1)s_{j,k+1} = P_{k+1,j}\\&\text{expected transfer flow from } i \text{ to } j = v_j(T-1)s_{ji} = P_{ij}\end{aligned} \qquad (5.16)$$

*Example 5.6* Suppose we have three grades with

| $i$ | 1 | 2 | 3 |
|---|---|---|---|
| $n_i$ | 300 | 200 | 100 |
| $w_i$ | 0.20 | 0.05 | 0.10 |

If (a) recruitment occurs only into the lowest grade, and (b) vacancies occurring in the higher two grades are filled by promotion from the next lower grade, then

$$\mathbf{S} = \begin{bmatrix} 0 & 0 & 0 \\ 1 & 0 & 0 \\ 0 & 1 & 0 \end{bmatrix}, \qquad \mathbf{S}^* = \begin{bmatrix} -0.20 & 0 & 0 \\ 1 & -0.05 & 0 \\ 0 & 1 & -0.10 \end{bmatrix}$$

and **R** = (60, 10, 10). Using the BASEQN program and starting with zero stocks of vacancies we obtain the result shown in Printout 5.1.

```
    K =     ? 3
    N =     ? 0,0,0
    P =     ? -.2,0,0,     1,-.05,0,     0,1,-.1
    R =     ? 60,10,10,   1,0,0
  T,% =     ? 5,NO
  <*>    ?    RUN

    T           1       2       3          TOTAL        R

    0           0       0       0        0( 10%)
    1          60      10      10       80(***%)       80
    2          58      20       9       87(***%)       80
    3          68      18       9       95(***%)       80
    4          64      18       9       92(***%)       80
    5          65      18       9       93(***%)       80
  <*>    ?
```

Printout 5.1 For explanation see Example 5.6

The number of vacancies soon reaches a steady state in which, for example, there will be 18 promotions from grade 1 to grade 2. ◀

As we pointed out earlier, this model assumes that the average time to fill a vacancy is the same as the time interval adopted for the projection. The model can be modified to cope with more general situations as the following examples show.

*Example 5.7* Suppose that wastage rates are available on an annual basis but that vacancies take, on average, 3 months to fill. We can use a quarterly accounting interval if we are able to estimate quarterly wastage rates from the annual ones available. As a first approximation the quarterly rates may be taken as a quarter of the annual rates. Using the data of Example 5.6 with a time interval of one-quarter of a year

$$\mathbf{S}^* = \begin{bmatrix} -0.05 & 0 & 0 \\ 1 & -0.0125 & 0 \\ 0 & 1 & -0.025 \end{bmatrix}, \qquad \mathbf{R} = (15, 2.5, 2.5)$$

Printout 5.2 gives the results of projecting the stock vector.

```
    K =     ?3
    N =     ?0,0,0
    P =     ?-.05,0,0,     1,-.0125,0,     0,1,-.025
    R =     ?15,2.5,2.5,     1,0,0
  T,% =     ?5,NO
  <*>     ?RUN

    T          1       2       3           TOTAL        R

    0          0       0       0          0( 10%)
    1         15       3       3         20(***%)      20
    2         17       5       2         24(***%)      20
    3         19       5       2         26(***%)      20
    4         19       5       2         26(***%)      20
    5         19       5       2         26(***%)      20
  <*>    ?
```

Printout 5.2 For explanation see Example 5.7

As there are eventually 5 vacancies per quarter in grade 2 there will be that number of promotions from grade 1, that is 20 a year, on average. ◀

The model can also cope with a system in which the grade sizes are expanding. Expansion implies the addition of new vacancies over and above those created by wastage. If every grade expands at a rate of $100\alpha$ per cent per annum then we must add $\alpha\mathbf{n}$ to the right-hand side of (5.15). We can also deal with contraction by this method using a negative $\alpha$ provided that we check that the stock vectors have elements which are never negative.

*Example 5.8* If the system discussed in the previous two examples expands at a rate of 10 per cent per annum and vacancies again take 1 year to fill, then the vacancy difference equation is

$$\mathbf{v}(T)=\mathbf{v}(T-1)\begin{bmatrix}0&0&0\\1&0&0\\0&1&0\end{bmatrix}+\{\mathbf{n}(T-1)-\mathbf{v}(T-1)\}\begin{bmatrix}0.20&0&0\\0&0.05&0\\0&0&0.10\end{bmatrix}+0.1\,\mathbf{n}(T-1)$$

$$=\mathbf{v}(T-1)\begin{bmatrix}-0.20&0&0\\1&-0.05&0\\0&1&-0.10\end{bmatrix}+\mathbf{n}(T-1)\begin{bmatrix}0.30&0&0\\0&0.15&0\\0&0&0.20\end{bmatrix} \tag{5.17}$$

In other words, we add the expansion rates to the wastage rates in the **W** component of **R** (see equation (5.15)). To evaluate this equation using our program we note that the number of posts

$$\mathbf{n}(T-1)=(1.1)^{T-1}\,[300, 200, 100]$$

so that **R** in (5.15) becomes

$$(1.1)^{T-1}\,[90, 30, 20]$$

BASEQN has an option to deal with recruitment expressed in this form and using it the vacancy projection is as shown in Printout 5.3. ◀

```
     K  =    ?3
     N  =    ?0,0.0
     P  =    ?-.2,0,0,      1,-.05,0,     0,1,-.1
     R  =    ?90,30,20,      1,*,1.1
   T,%  =    ?5,NO
   <*>    ?RUN

     T          1        2        3           TOTAL         R

     0          0        0        0          0( 10%)
     1         90       30       20        140(***%)        1
     2        111       52       20        183(***%)        1
     3        138       54       22        214(***%)        1
     4        146       59       24        230(***%)        1
     5        162       65       27        254(***%)        1
   <*>    ?
```

Printout 5.3 For explanation see Example 5.8

The **S**-matrix used in the foregoing examples had a very simple structure as recruitment was only at the lowest level and all other vacancies were filled by promotion. In the following two examples we illustrate the calculations required in slightly more complicated situations.

*Example 5.9* Suppose

$$\mathbf{S} = \begin{bmatrix} 0.30 & 0 & 0 \\ 0.80 & 0.20 & 0 \\ 0 & 0.90 & 0.10 \end{bmatrix}$$

The presence of non-zero diagonal elements could be given either of two interpretations. They could be taken to mean that some vacancies are filled by sideways movements in the same grade. This would mean, for example, that 30 per cent of vacancies in grade 1 would be filled from within that grade. Alternatively, it could be interpreted to mean that 30 per cent of vacancies are deliberately kept open in that grade. In either case the calculation of the expected future stock vectors is the same. Thus using the **R** of Example 5.6 we have the results of Printout 5.4. ◀

```
      K =      ? 3
      N =      ? 0,0,0
      P =      ? .1,0,0,    .8,.15,0,    0,.9,0
      R =      ? 60,10,10,  1,0,0
    T,% =      ? 5,NO
    <*>     ?    RUN

      T          1      2      3        TOTAL        R

      0          0      0      0       0( 10%)
      1         60     10     10      80(***%)      80
      2         74     20     10     105(***%)      80
      3         84     22     10     116(***%)      80
      4         86     22     10     118(***%)      80
      5         86     22     10     119(***%)      80
    <*>     ?
```

Printout 5.4 For explanation see Example 5.9

*Example 5.10* Here we suppose recruits enter at all levels with 20 per cent of vacancies in grade 3, 40 per cent in grade 2, and 100 per cent in grade 1 being filled by this means. The only change required to Example 5.6 is to make

$$\mathbf{S} = \begin{bmatrix} 0 & 0 & \\ 0.60 & 0 & 0 \\ 0 & 0.80 & \end{bmatrix}$$

giving the results of Printout 5.5 ◀

The various modifications illustrated in these examples could, of course, be applied simultaneously.

```
    K =    ? 3
    N =    ? 0,0,0
    P =    ? -.2,0,0,    .6,-.05,0,    0,.8,-.1
    R =    ? 60,10,10,    1,0,0
  T,% =    ? 5,NO
  <*>    ?   RUN

    T          1      2      3          TOTAL       R
    0          0      0      0        0( 10%)
    1         60     10     10       80(***%)      80
    2         54     18      9       81(***%)      80
    3         60     16      9       85(***%)      80
    4         58     16      9       83(***%)      80
    5         58     16      9       84(***%)      80
  <*>    ?
```

Printout 5.5 For explanation see Example 5.10

## 5.6 MODELS WITH LENGTH-OF-SERVICE SPECIFIC WASTAGE RATES

The main shortcoming of the various renewal models described in previous sections is that wastage was assumed to be grade specific but not length-of-service specific. In a simple hierarchical system increasing rank will usually go with increasing length of service and hence, to some extent, grade is a proxy for length of service. At best this will only be a crude way of dealing with the fact that propensity to leave depends strongly on length of service, and at worst it will be quite inadequate. In the case of the Markov models of the previous chapter this complication was easily dealt with by defining categories of the model in terms of age, length of service or seniority in addition to grade. Although we can still do this it is no longer reasonable to treat the resulting system as operating according to renewal principles. This is because there is usually no organizational constraint on the number of persons in each age or length-of-service category as the renewal model would require. We therefore require a model in which the overall grade sizes are fixed but in which the distributions of other variables are free to vary within the constraints which the fixed grade sizes impose.

From a mathematical point of view renewal models which incorporate length-of-service dependent wastage rates are awkward to handle in discrete time, but there is no difficulty in writing a computer program to carry out the calculations required by the assumptions. A number of such models are used in practice of which the best known is, perhaps, the renewal version of the KENT model which forms part of the British Civil Service Department's MANPLAN package (Smith (1976)). The following description gives the general structure of such a model. The subdivisions of each grade can be based on any individual attribute but, for brevity, we shall speak here of age.

The two main assumptions concern wastage on the one hand and promotion and recruitment on the other. Let $n_{ij}$ be the initial number in the $j$th age group of the $i$th category and let $w_{ij}$ be the wastage rate for that category. Then we may calculate the expected losses for each category in the usual way as $n_{ij}w_{ij}$ for all $i$ and $j$. This calculation, together with any planned expansion, determines the vacancies entering the system. The second stage is concerned with filling the vacancies by recruitment and promotion. This may be considered in two stages. First we have to decide whether a vacancy is to be filled by recruitment from outside and, if not, from which grade the replacement is to come. This can be done (if past data provides empirical support) by a vacancy transition matrix $[\mathbf{S} \mid \mathbf{s}'_{k+1}]$ as in the previous models. These probabilities could depend on the age of the person being replaced but in the KENT model they are constant, and a grade-specific vacancy transition matrix only is used. By applying these rates to the stocks of vacancies, the total number of people to be moved from all other sources will be determined. At the second stage we have to determine the age distribution of those to be promoted or recruited. In the case of recruits this would normally be done by reference to historical data. The total number of recruits at each level is then divided into age groups to conform to the historical pattern. Promotion can be treated in the same way if historical records of the age distributions of promotees are available. This will determine how many promotees are to be taken from each age group. Once again there will be problems if the number required from any group exceeds the number available and this is more likely to occur if there are many age groups with only a few people in each. The KENT model includes a routine for dealing with this difficulty which involves finding substitutes from neighbouring age groups (see Hopes (1973)). Other promotion rules are possible and one which has some plausibility and is easy to program is the one which selects promotees at random with respect to age. In deterministic terms this means selecting promotees at each age in proportion to the number available. The KENT model includes this option also.

It would be possible to operate such a model by assuming either that vacancies are filled instantaneously or that a vacancy can only move one step per unit time. In either event the main steps at each cycle of the program are as follows:

(a) take out the wastage from each grade;
(b) fill the vacancies in the top grade according to the assumptions of the model;
(c) proceed to the second grade and so on throughout the system until all vacancies are eliminated.

If the system were hierarchical with promotion but no demotion the procedure would start at the top and work down to the bottom. In a more complex network of grades an arbitrary sequence of grades can be adopted.

This could lead to new vacancies being created in grades where others have been filled at an earlier stage. The cycle must then be repeated until all vacancies are eliminated (it is easy to prove that this procedure will terminate with probability 1).

## 5.7 COMPLEMENTS

Renewal theory has been applied outside the manpower field for many years. One of the earliest applications seems to have been in population theory but the main impetus for the development of the theory was its application to problems of industrial replacement where, in the language of this chapter, there was a single grade. An extension to two-grade systems was made by Bartholomew (1963c) and later to multi-grade systems in Bartholomew (1963a). In the latter paper the possible application to manpower planning was briefly indicated. A fuller development of this idea was given in Bartholomew (1967 and 1973a). This work was theoretical in emphasis and used a continuous-time formulation. This made it easy to incorporate the dependence of propensity to leave on length of service or seniority. It also eliminated the possibility of more promotees being required than were available because in continuous time, vacancies occur singly so there will always be at least one person available to fill each one as it occurs. The theory was confined to single hierarchies with one-step promotion and included two promotion rules. According to one, promotees were selected at random from the next lower grade and according to the other the longest-serving candidate was promoted. It proved possible to determine promotion rates and average ages of promotion under various assumptions about loss and promotion.

Although it is common for real organizations to have fixed, or nearly fixed, grade sizes no direct applications of the continuous theory appear to have been made. Instead the usual approach has been to develop computer algorithms for specific models as in the case of the KENT model described in the last section. Further accounts of this will be found in Hopes (1973) (reproduced in Bartholomew (1977d)) and Smith (1976); the model is also now available commercially.

In view of the very widespread use of Markov models in many applied fields it seemed desirable to develop the theory of discrete-time multi-grade systems for manpower modelling. A key idea which facilitated this was provided by the work of White (1970) on vacancy chains. Although he was not concerned with manpower planning in the sense we are using the term, the recognition that vacancies could be modelled by a Markov chain lies behind all of the multi-grade models discussed here. His ideas have been applied by Stewman (1975) to the study of mobility in a state police force as noted earlier. The model in which vacancies can only move one step at a time was first put forward in Bartholomew (1976b).

Although there is certainly some empirical support for the assumptions of the renewal models they need to be tested wherever possible in new applications. Statistically speaking, this involves nothing new and so we have not repeated what was said in the last chapter on the same theme.

It is, perhaps, surprising that the application of renewal models has lagged behind that of Markov models. This may have something to do with the fact that, in the steady state, it is difficult to distinguish between the two models from stock and flow data alone. We have seen that the stocks in a Markov model approach a steady state, and similarly the flows in a renewal model approach a steady state. Thus if one is judging the applicability of the Markov model by the constancy of its flow rates this state of affairs occurs in a renewal model in equilibrium as well as in a Markov model. Therefore, provided that the system is in a steady state either model will perform equally well as far as expected values are concerned. It is in modelling the transitory behaviour of systems that the distinction becomes crucial. This equivalence in equilibrium doubtless goes some way to explain the apparent robustness of Markov chain models in practice.

Much discussion of the appropriateness of the various models turns on how accurately they describe the real-life situation. This is entirely reasonable if one views them as forecasting tools. However, there is a great deal more to manpower planning than the passive extrapolation of current trends. Usually we wish to explore a range of policy options and this may involve relaxing constraints which currently exist or varying the rules under which the system operates. For this purpose one needs a battery of flexible models which allows adequate room to experiment. Markov and renewal models between them have proved in practice that they can cover most of the requirements.

## 5.8 EXERCISES AND SOLUTIONS

**Exercise 5.1**

Estimate the discrete functions described in section 5.2 using the formulae

$$f_i = (G_i - G_{i+1}), \quad w_i = (G_i - G_{i+1})/G_i$$

for the following survivor function:

| $i$ | 0 | 1 | 2 | 3 | 4 | 5 | 6 | 7 |
|---|---|---|---|---|---|---|---|---|
| $G_i$ | 1.00 | 0.95 | 0.67 | 0.59 | 0.56 | 0.54 | 0.52 | 0.51 |

(a) Hence calculate the replacement rate $h_j$ ($j = 1, 2, \ldots, 7$) for the first 7 years assuming the total size of the system remains constant and all the initial population has zero length of service.

(b) Calculate the replacement rate $h_j (j = 1, 2, 3, 4)$ if the population remains constant in size but if the initial population has the length-of-service distribution

| $i$ | 0 | 1 | 2 | 3 | 4 *or more* |
|---|---|---|---|---|---|
| $S_i$ | 0.25 | 0.25 | 0.25 | 0.25 | 0 |

(c) Calculate the required recruitment numbers and the expected leaving rates assuming that initially everyone has zero length of service, that the initial total size is $N(0) = 100$, and that the system expands by 20 each year for the first 3 years and then remains at constant size thereafter.

**Exercise 5.2**

The following is the matrix **S** for the instantaneous movement of vacancies in an organization.

$$\begin{array}{cc} & \textit{destination} \\ & \begin{array}{cccc} 1 & 2 & 3 & \textit{out} \end{array} \\ \begin{array}{r} 1 \\ \textit{origin}\ 2 \\ 3 \end{array} & \left[\begin{array}{ccc|c} 0.70 & 0 & 0 & 0.30 \\ 0.10 & 0.85 & 0 & 0.05 \\ 0 & 0.05 & 0.85 & 0.10 \end{array}\right] \end{array}$$

Calculate the table of expected flows during a year if the grade sizes and wastage rates are

| $i$ | 1 | 2 | 3 |
|---|---|---|---|
| $n_i$ | 500 | 300 | 100 |
| $w_i$ | 0.20 | 0.05 | 0.05 |

**Exercise 5.3**

For Example 5.6 suppose that recruitment is only into the lowest grade, and that 20 per cent of all vacancies are filled every 5 weeks. Set up the Markov-chain type recursive equation for vacancies with an accounting period of 5 weeks, and project the number of vacancies in each grade if all posts are filled, initially.

**Exercise 5.4**

For the system illustrated in Figure 5.1 the numbers in brackets denote the percentage of the vacancies filled by the flow to which they are attached.

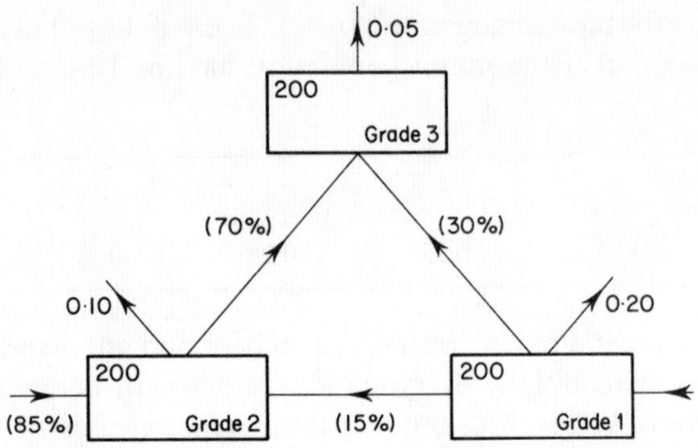


Figure 5.1 Depicting the system of Exercise 5.4

(a) Set up the (1-year) vacancy model and project the number of vacancies starting from 20 in each grade.

(b) As for (a) with vacancies filled every 2 months

(c) As for (b) with only 50 per cent of vacancies filled every 2 months.

(d) As for (c) with the total size of the system growing at 1 per cent every 2 months.

Calculate in each case the total expected promotions from grade 2 to grade 3 in the first year and in equilibrium.

**Exercise 5.5**

(a) Set up the vacancy model for the system depicted in Figure 5.2, where the numbers in brackets are the percentages of vacancies filled by each flow.

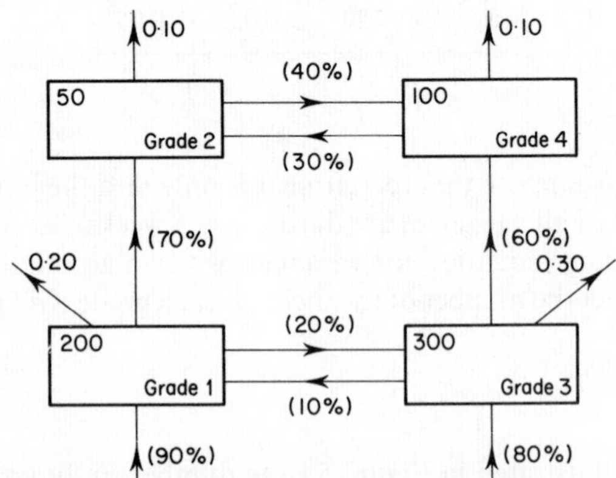


Figure 5.2 Depicting the system of Exercise 5.5

Project the number of vacancies starting from time zero when the system is at constant size.

(b) As for (a) when 50 per cent of vacancies are filled after 4 months.

**Solution to Exercise 5.1**

The estimated values for $f_i$ and $w_i$ and the ratio $S_i/G_i$ are as shown in Table 5.3.

***Table 5.3***

| | $G_i$ | $f_i$ | $w_i$ | $S_i$ | $S_i/G_i$ |
|---|---|---|---|---|---|
| 0 | 1.00 | 0.05 | 0.05 | 0.25 | 0.25 |
| 1 | 0.95 | 0.28 | 0.30 | 0.25 | 0.26 |
| 2 | 0.67 | 0.08 | 0.12 | 0.25 | 0.37 |
| 3 | 0.59 | 0.03 | 0.05 | 0.25 | 0.42 |
| 4 | 0.56 | 0.02 | 0.05 | 0 | 0 |
| 5 | 0.54 | 0.02 | 0.03 | 0 | 0 |
| 6 | 0.52 | 0.01 | 0.03 | 0 | 0 |

(a) The replacement rates for constant total size and zero initial length of service are

$$
\begin{aligned}
h_1 &= f_0 && = 0.05 && = 0.05\\
h_2 &= f_1 + h_1 f_0 && = 0.28 + 0.00 && = 0.28\\
h_3 &= f_2 + (h_1 f_1 + h_2 f_0) && = 0.08 + 0.03 && = 0.11\\
h_4 &= f_3 + (h_1 f_2 + h_2 f_1 + h_3 f_0) && = 0.03 + 0.09 && = 0.12\\
h_5 &= f_4 + (h_1 f_3 + \ldots + h_4 f_0) && = 0.02 + 0.06 && = 0.08\\
h_6 &= f_5 + (h_1 f_4 + \ldots + h_5 f_0) && = 0.02 + 0.06 && = 0.08\\
h_7 &= f_6 + (h_1 f_5 + \ldots + h_6 f_0) && = 0.01 + 0.05 && = 0.06
\end{aligned}
$$

etc.

The replacement rate is low for the first year because in this system propensity to leave reaches its peak in the second year of service.

(b) If the initial length-of-service distribution is given by the column headed $S_i$ above then for constant total size the replacement rates will be

$$
\begin{aligned}
h_1 &= \Sigma\, f_i\, S_i/G_i && = 0.13 && = 0.13\\
h_2 &= \Sigma\, f_{i+1}\, S_i/G_i + h_1 f_0 && = 0.11 + 0.00 && = 0.11\\
h_3 &= \Sigma\, f_{i+2}\, S_i/G_i + (h_1 f_1 + h_2 f_0) && = 0.04 + 0.03 && = 0.07\\
h_4 &= \Sigma\, f_{i+3}\, S_i/G_i + (h_1 f_2 + \ldots + h_3 f_0) && = 0.02 + 0.09 && = 0.11
\end{aligned}
$$

etc.

(c) The increments in total size are $M(1) = M(2) = M(3) = 20$ and $M(4) = 0 = M(5) = M(6) = \ldots$. Since the initial population all have zero length of service the required recruitment numbers and the replacement rates are

$$
\begin{aligned}
R_1 &= 100h_1 + M(1) &&= 25, R_1/N(0) = 0.25 \\
R_2 &= 100h_2 + M(1)h_1 + M(2) &&= 49, R_2/N(1) = 0.41 \\
R_3 &= 100h_3 + (M(1)h_2 + M(2)h_1) + M(3) &&= 38, R_3/N(2) = 0.27 \\
R_4 &= 100h_4 + (M(1)h_3 + M(2)h_2 + M(3)h_1) &&= 21, R_4/N(3) = 0.13 \\
R_5 &= 100h_5 + (M(1)h_4 + M(2)h_3 + M(3)h_2) &&= 18, R_5/N(4) = 0.11 \\
R_6 &= 100h_6 + (M(1)h_5 + M(2)h_4 + M(3)h_3) &&= 14, R_6/N(5) = 0.09 \\
R_7 &= 100h_7 + (M(1)h_6 + M(2)h_5 + M(3)h_4) &&= 12, R_7/N(6) = 0.08
\end{aligned}
$$

etc.

**Solution to Exercise 5.2**

The fundamental matrix is

$$
\mathbf{D} = (\mathbf{I} - \mathbf{S})^{-1} = \begin{bmatrix} 3.33 & 0 & 0 \\ 2.22 & 6.67 & 0 \\ 0.74 & 2.22 & 6.67 \end{bmatrix}
$$

The table of expected movements for people is then calculated from equation (5.12), as in Example 5.5. The matrix of flows during a year (i.e. those flows caused by one round of leaving) is then as shown in Table 5.4.

***Table 5.4***

| | *To grade* | | | |
|---|---|---|---|---|
| | 1 | 2 | 3 | *Outside (= recruits)* |
| *From grade* 1 | 259 | 11 | 0 | 100 |
| 2 | 0 | 94 | 2 | 15 |
| 3 | 0 | 0 | 28 | 5 |
| *Outside (= leavers)* | 111 | 6 | 3 | |

Thus each round of leaving implies the following.

*For the top grade:*
outflow: 5 leavers;
inflow: 3 recruits, and 2 promotions from the middle grade.
'within' flow: 28 job changes or internal promotions.

*For the middle grade:*
outflow: 15 leavers, and 2 promotions to the top grade;

inflow: 6 recruits, and 11 promotions from the bottom grade;
'within' flow: 94 job changes or internal promotions.

*For the lowest grade:*
outflow: 100 leavers, and 11 promotions to the middle grade;
inflow: 111 recruits;
'within' flow: 259 job changes or internal promotions.
As we would expect, outflow = inflow for each grade.

**Solution to Exercise 5.3**

Taking the wastage transition rates for the new 5-week accounting period as 1/10 of the yearly rates we have

$$\mathbf{v}(T) = \mathbf{v}(T-1)\begin{bmatrix} 0.80 & 0 & 0 \\ 0.20 & 0.80 & 0 \\ 0 & 0.20 & 0.80 \end{bmatrix} + [\mathbf{n} - \mathbf{v}(T-1)]\begin{bmatrix} 0.02 & 0 & 0 \\ 0 & 0.005 & 0 \\ 0 & 0 & 0.01 \end{bmatrix}$$

$$= \mathbf{v}(T-1)\begin{bmatrix} 0.78 & 0 & 0 \\ 0.20 & 0.795 & 0 \\ 0 & 0.20 & 0.79 \end{bmatrix} + [6.00 \quad 1.00 \quad 1.00]$$

since $\mathbf{n} = [300 \quad 200 \quad 100]$. The program gives the result shown in Printout 5.6.

```
    K =    ?3
    N =    ?0,0,0
    P =    ?.78,0,0,     .2,.795.0,     0,.2,.79
    R =    ?6,1,1,    1,0,0
  T,% =    ?5,NO
  <*>    ?RUN

    T           1        2        3          TOTAL        R

    0           0        0        0          0( 10%)
    1           6        1        1          8(***%)      8
    2          11        2        2         15(***%)      8
    3          15        3        2         20(***%)      8
    4          18        4        3         25(***%)      8
    5          21        5        3         29(***%)      8
  <*>    ?T=,   10
   10          29        7        4         41(***%)      8
  <*>    ?T=,     20
   20          35        9        5         49(***%)      8
  <*>    ?T=,     30
   30          36        9        5         50(***%)      8
  <*>    ?
```

Printout 5.6 Showing the solution to Exercise 5.3

**Solution to Exercise 5.4**

(a) The equation of the model is

$$\mathbf{v}(T) = \mathbf{v}(T-1)\begin{bmatrix} 0 & 0 & 0 \\ 0.15 & 0 & 0 \\ 0.30 & 0.70 & 0 \end{bmatrix} + [\mathbf{n} - \mathbf{v}(T-1)]\begin{bmatrix} 0.20 & 0 & 0 \\ 0 & 0.10 & 0 \\ 0 & 0 & 0.05 \end{bmatrix}$$

$$= \mathbf{v}(T-1)\begin{bmatrix} -0.20 & 0 & 0 \\ 0.15 & -0.10 & 0 \\ 0.30 & 0.70 & -0.05 \end{bmatrix} + [40 \quad 20 \quad 10]$$

which gives the results shown in Printout 5.7.

```
      K =     ?3
      N =     ?20,20,20
      P =     ?-.2,0,0,        .15,-.1,0,      .3,.7,-.05
      R =     ?40,20,10,       1,0.0
    T,% =     ?5,NO
    <*>     ?RUN

      T            1       2       3          TOTAL       R

      0           20      20      20        60(100%)
      1           45      32       9        86(143%)     70
      2           39      23      10        71(119%)     70
      3           39      24      10        73(121%)     70
      4           39      24      10        73(121%)     70
      5           39      24      10        73(121%)     70
    <*>     ?
```

Printout 5.7 Showing the solution to Exercise 5.4a

This implies, for example, $20 \times 0.7 = 14$ promotions from grade 2 to grade 3 during the first year, and $10 \times 0.7 = 7$ promotions in the steady state.

(b) If the lag in filling vacancies is 2 months instead of 1 year, then changing the accounting period to 2 months and dividing the yearly transition rates by 6:

$$\mathbf{v}(T) = \mathbf{v}(T-1)\begin{bmatrix} 0 & 0 & 0 \\ 0.15 & 0 & 0 \\ 0.30 & 0.70 & 0 \end{bmatrix} + [\mathbf{n} - \mathbf{v}(T-1)]\begin{bmatrix} 0.033 & 0 & 0 \\ 0 & 0.017 & 0 \\ 0 & 0 & 0.008 \end{bmatrix}$$

$$= \mathbf{v}(T-1)\begin{bmatrix} -0.033 & 0 & 0 \\ 0.15 & -0.017 & 0 \\ 0.30 & 0.70 & -0.008 \end{bmatrix} + [6.6 \quad 3.3 \quad 1.65]$$

The result is shown in Printout 5.8.

This shows that there are $0.7(20 + 1 + 2 + 2 + 2 + 2) = 20$ promotions from grade 2 to grade 3 in the first year, and $0.7(6 \times 2) = 8$ promotions in the steady state.

```
    K =    ? 3
    N =    ? 20,20,20
    P =    ? -.033,0,0,    .15,-.017,0,    .3,.7,-.008
    R =    ? 6.6,3.3,1.65,     1,0,0
  T,% =    ? 5,NO
  <*>    ?  RUN

   T         1       2       3          TOTAL       R

   0        20      20      20        60(100%)
   1        15      17       1        33( 56%)     12
   2         9       4       2        15( 25%)     12
   3         7       4       2        13( 22%)     12
   4         8       4       2        14( 23%)     12
   5         7       4       2        14( 23%)     12
  <*>    ?
```

Printout 5.8 Showing the solution to Exercise 5.4b

(c) If, additionally, only 50 per cent of vacancies are filled every 2 months,

$$\mathbf{v}(T) = \mathbf{v}(T-1)\begin{bmatrix} 0.50 & 0 & 0 \\ 0.075 & 0.50 & 0 \\ 0.15 & 0.35 & 0.50 \end{bmatrix}$$

$$+ [\mathbf{n} - \mathbf{v}(T-1)]\begin{bmatrix} 0.033 & 0 & 0 \\ 0 & 0.017 & 0 \\ 0 & 0 & 0.008 \end{bmatrix}$$

$$= \mathbf{v}(T-1)\begin{bmatrix} 0.467 & 0 & 0 \\ 0.075 & 0.483 & 0 \\ 0.150 & 0.350 & 0.492 \end{bmatrix} + [6.60 \quad 3.40 \quad 1.60]$$

The result is shown in Printout 5.9.

```
    K =    ?3
    N =    ?20,20,20
    P =    ?.467,0,0,    .075,.483,0,    .15,.35,.492
    R =    ?6.6,3.4,1.6,    1,0,0
  T,% =    ?5,NO
  <*>    ?RUN

   T         1       2       3          TOTAL       R

   0        20      20      20        60(100%)
   1        20      20      11        52( 87%)     12
   2        19      17       7        44( 73%)     12
   3        18      14       5        37( 62%)     12
   4        17      12       4        33( 55%)     12
   5        16      11       4        30( 50%)     12
  <*>    ?T=,  6
   6        15      10       3        29( 48%)     12
  <*>    ?T=, 12
  12        15       9       3        26( 44%)     12
  <*>    ?
```

Printout 5.9 Showing the solution to Exercise 5.4c

From this we conclude that there will be $0.7(20 + 11 + 7 + 5 + 4 + 4)\,0.5 = 18$ promotions from grade 2 to grade 3 in the first year, and $0.7\,(6 \times 3)\,0.5 = 6$ promotions in the steady state.

(d) If the system also grows by 1 per cent every 2 months then the previous equation requires that $\mathbf{n}$ is replaced by $\mathbf{n}(T-1)$ and that an extra term $0.01\,\mathbf{n}(T-1)$ is added. The recruitment term in the model therefore becomes

$$\mathbf{n}(T-1)\begin{bmatrix} 0.043 & 0 & 0 \\ 0 & 0.027 & 0 \\ 0 & 0 & 0.018 \end{bmatrix} = (1.01)^{T-1}[8.60 \quad 5.40 \quad 3.60]$$

and following on from the previous projection we have the results shown in Printout 5.10.

```
<*>     ?N=N(0)
<*>     ?R=,8.6,5.4,3.6,   1,*.1.01
<*>     ?T%=,   6,NO
<*>     ?RUN

  T          1       2       3          TOTAL        R

  0         20      20      20        60(100%)
  1         22      22      13        58( 97%)       1
  2         23      21      10        54( 90%)       1
  3         23      19       9        50( 84%)       1
  4         22      18       8        48( 80%)       1
  5         22      17       8        47( 78%)       1
  6         22      17       8        46( 76%)       1
<*>     ?
```

Printout 5.10 Showing the solution to Exercise 5.4d

Printout 5.10 shows that there will be $0.7(20 + 13 + 10 + 9 + 8 + 8)0.5 = 24$ promotions from grade 2 to grade 3 during the first year, and $0.7(6 \times 8)\,0.5 = 17$ in the stationary state.

The promotions from grade 2 to grade 3 are summarized in Table 5.5. Note particularly the increase in promotion due to the expansion.

***Table 5.5***

| *Policy for filling vacancies* | *First year* | *Equilibrium* |
|---|---|---|
| all vacancies filled in 1 year | 14 | 7 |
| all vacancies filled in 2 months | 20 | 8 |
| 50% of vacancies filled in 2 months | 18 | 6 |
| 50% of vacancies filled in 2 months and an expansion of 1% per 2 months | 24 | 17 |

**Solution to Exercise 5.5**

(a)

$$\mathbf{v}(T) = \mathbf{v}(T-1)\begin{bmatrix} 0 & 0 & 0.1 & 0 \\ 0.7 & 0 & 0 & 0.3 \\ 0.2 & 0 & 0 & 0 \\ 0 & 0.4 & 0.6 & 0 \end{bmatrix} + [\mathbf{n} - \mathbf{v}(T-1)]\begin{bmatrix} 0.20 & 0 & 0 & 0 \\ 0 & 0.10 & 0 & 0 \\ 0 & 0 & 0.30 & 0 \\ 0 & 0 & 0 & 0.10 \end{bmatrix}$$

$$= \mathbf{v}(T-1)\begin{bmatrix} -0.2 & 0 & 0.1 & 0 \\ 0.7 & -0.1 & 0 & 0.3 \\ 0.2 & 0 & -0.3 & 0 \\ 0 & 0.4 & 0.6 & -0.1 \end{bmatrix} + [40 \quad 5 \quad 90 \quad 10]$$

```
   K =      ? 4
   N =      ? 0,0,0,0
   P =      ? -.2,0,.1,0,     .7,-.1,0,.3,     .2,0,-.3,0,     0,.4,.6,-.1
   R =      ? 40,5,90,10,      1,0,0
 T,% =      ? 5,NO
 <*>     ?    RUN

   T          1       2       3       4           TOTAL        R

   0          0       0       0       0           0( 10%)
   1         40       5      90      10         145(***%)     145
   2         53       8      73      11         146(***%)     145
   3         50       8      80      11         149(***%)     145
   4         52       9      78      11         150(***%)     145
   5         51       9      79      11         150(***%)     145
 <*>     ?T=,10
  10         51       9      78      11         150(***%)     145
 <*>     ?
```

Printout 5.11 Showing the solution to Exercise 5.5a

(b)

$$\mathbf{v}(T) = \mathbf{v}(T-1)\begin{bmatrix} 0.50 & 0 & 0.05 & 0 \\ 0.35 & 0.50 & 0 & 0.15 \\ 0.10 & 0 & 0.50 & 0 \\ 0 & 0.20 & 0.30 & 0.50 \end{bmatrix} +$$

$$[\mathbf{n} - \mathbf{v}(T-1)]\begin{bmatrix} 0.067 & 0 & 0 & 0 \\ 0 & 0.033 & 0 & 0 \\ 0 & 0 & 0.100 & 0 \\ 0 & 0 & 0 & 0.033 \end{bmatrix}$$

```
  K =     ? 4
  N =     ? 0,0,0,0
  P =     ? .433,0,.05,0,    .35,.467,0,.15,    .1,0,.4,0,    0,.2,.3,.467
  R =     ? 13.3,1.67,30,3.33,    1,0,0
T,% =     ? 5,NO
<*>     ?  RUN

  T          1       2       3       4          TOTAL        R

  0          0       0       0       0          0( 10%)
  1         13       2      30       3         48(***%)     48
  2         23       3      44       5         75(***%)     48
  3         29       4      50       6         89(***%)     48
  4         32       5      53       7         97(***%)     48
  5         34       5      55       7        102(***%)     48
<*>     ?T=,10
 10         37       6      57       8        108(***%)     48
<*>     ?T=,20
 20         37       6      57       8        108(***%)     48
<*>     ?
```

Printout 5.12 Showing the solution to Exercise 5.5b

CHAPTER 6

# Career Patterns

## 6.1 INTRODUCTION

In Chapters 4 and 5 we described and illustrated models for predicting stocks and flows in a graded manpower system. By concentrating on the aggregate properties, we paid little attention to the career paths of individuals through the system. Thus, although we were able to make statements about promotions *rates* the analysis gave no direct information on, for example, the chance that any given individual would ultimately be promoted. In this chapter, therefore, we turn to the study of quantities which relate to career prospects for individuals within an organization. There are two kinds of measures which arise in discussions of career patterns. The first are *probabilities* of events such as 'someone will be promoted from his present grade before leaving' or 'a new entrant will ultimately reach the top of the organization.' The second kind have to do with *how long* it takes for such events to occur, for example, the expected time before promotion or the average time spent in a grade. Taken together these two kinds of measures provide a fairly clear picture of career patterns.

This chapter contains three distinct approaches. In the first we shall describe an empirical method of analysing stock data which enables us to identify different career streams. Making certain assumptions we can then often go on to deduce how such things as the average ages on promotion are likely to change under varying conditions of recruitment and expansion. The second approach makes use of elementary probability arguments to relate promotion rates to certain average durations and probabilities of interest in hierarchical systems. The third draws on the theory of absorbing Markov chains to extend and unify these approaches.

In the final section of the chapter we shall consider how far the essentially Markovian approaches of the chapter can be applied to systems operating on renewal principles.

## 6.2 PROMOTION PATTERNS FROM GRADE–AGE STOCK DATA

One way of representing the relationship between age (or length of service) and grade is to construct a diagram like that given as Figure 6.1 for a three-grade system.

The solid curve is the frequency distribution of age for the whole system.

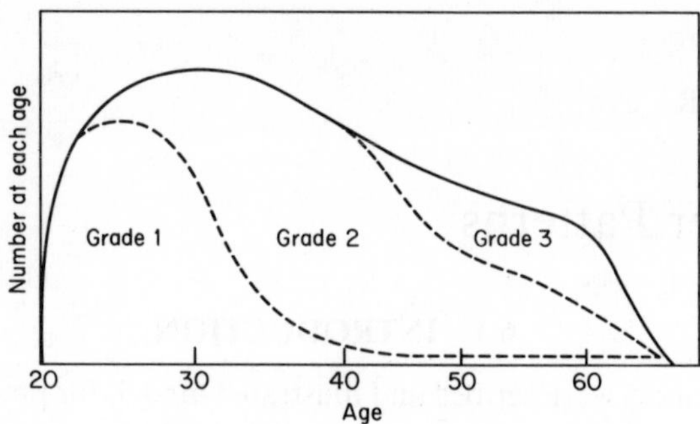


Figure 6.1 A typical grade–age distribution for a three-grade hierarchical system

The dotted curves represent the age distributions within each grade. Thus for any given age the dotted curves divide the total frequency at that age into three parts according to grade. Inspection of such a diagram is quite informative. For example, we can see that few people are left in grade 1 by age 45, and no one reaches grade 3 before age 40. However, it is not easy to see what proportion of people reach each grade, or what their ages are on promotion. A clearer picture can be obtained if we plot the proportion in each grade at each age as in Figure 6.2. For convenience we shall refer to these plots of grade proportions by age as CAMERA diagrams, a name they acquired in early applications. (The origin of this name is obscure but is probably related partly to their connection to the CAMEL model and the development of this work at Cambridge by Morgan and coworkers, and partly to these diagrams being a snapshot of the grade–age distribution.) This method of presenting the data was suggested by Morgan and coworkers (1973). The left-hand series of crosses in Figure 6.2 plots the actual proportions in grade 1

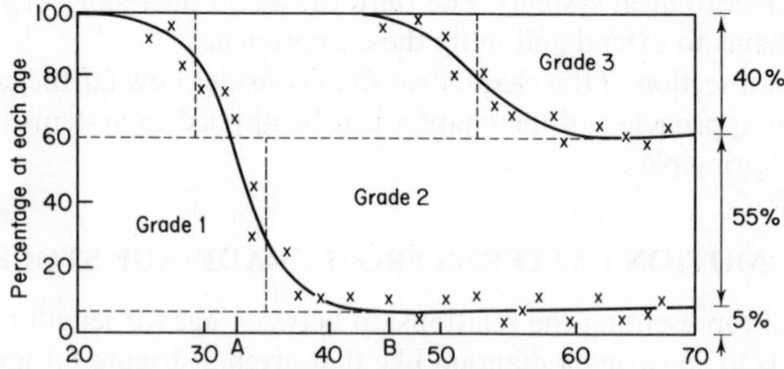


Figure 6.2 A typical CAMERA diagram for systems such as that of Figure 6.1

and the curve drawn through them is an attempt to smooth them and so obtain a clearer picture of the underlying pattern. The right-hand series gives the proportions in grade 2 and below. The horizontal dotted lines are asymptotic to the curves and can be thought of as dividing up the whole group into three career streams defined by the grade eventually reached. To make this interpretation we have to make an assumption of stationarity. This amounts to supposing that the current picture revealed by the diagram would also apply to the experience of a cohort recruited now. In particular we have to assume that because 40 per cent of the current stock at higher ages are in grade 3, then 40 per cent of entrants to grade 1 aged 20 who survive to retirement ultimately reach grade 3 (retirement occurs between 60 and 70 years in this example). This is a very strong assumption which requires stable conditions to exist and to continue in this case for a period of 50 years. However, we do not regard these analyses as predictive but as descriptive of the present position. In other words we are saying that to regard the present state of our system as if it described the experience of a cohort is a useful way of expressing the career experience of present members of the system. This description of the current state of the system is analogous to the census analysis of wastage described in Chapter 2.

We can now interpret the diagram as follows. The horizontal lines divide those who stay in the system until retirement into three bands. The top one, comprising 40 per cent of the whole, consists of those who eventually reach grade 3. The second band of 55 per cent contains those who eventually retire in grade 2, and the bottom band consists of the 5 per cent who never achieve promotion. We call these bands 'career-streams' and note again that they are defined by the grade in which a person retires. The vertical dotted lines are rough estimates of average ages of promotion. If we look at the middle stream we see that the first people to be promoted move at the point marked A which corresponds to age 32. By age 45 (marked B) almost everyone in this stream has been promoted. The average age of promotion is thus somewhere between 32 and 45 and the dotted line is an estimate of the average obtained by eye (it would be rather easier to make an accurate estimate of the median but great precision is rarely required). Those in the top stream have two promotions, on average, at ages 29 and 53.

Thus we may summarize the situation as follows. An entrant who stays until retirement can expect to reach grade 3 with probability 0.40 and grade 2 with probability 0.55. Those who reach grade 3 can expect their first promotion at age 29 and their second at 53. Those who eventually reach grade 2 can expect promotion at age 35. These are the career prospects which could be offered (and expected) if the current grade–age structure were to be maintained.

It is important to remember that the streams cannot be identified except retrospectively. However, a person could get some idea of whether he was likely to reach the top by observing whether his first promotion came nearer to 29 than to 35. Although we have carried out the foregoing discussions in terms of age it could equally well have been expressed in terms of length of service and in some applications it would be more realistic to do so.

*Example 6.1.* Table 6.1a below is a grade–age table of stocks for a four-grade system in which ages are recorded in 5-year age groups. The calculations necessary for the CAMERA diagram are shown in Table 6.1b, which contains the cumulative percentages for each grade within each age band. The CAMERA diagram itself is shown in Figure 6.3, and the career pattern estimated from it in Table 6.2.

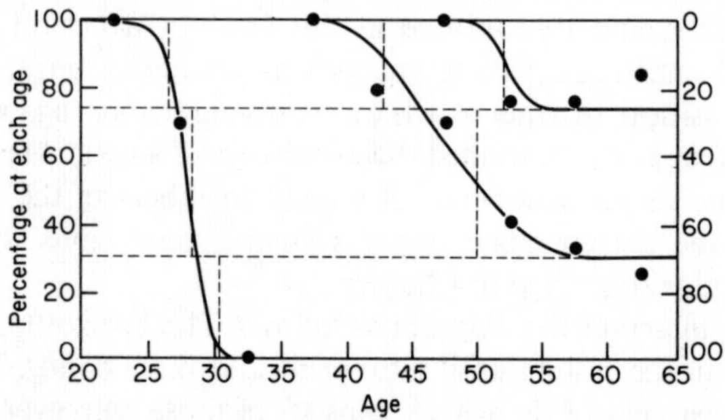


Figure 6.3 The CAMERA diagram for Example 6.1 (see Table 6.1b)

***Table 6.1a Grade–age stock numbers (5-year age bands) for Example 6.1***

| | 20–25 | –30 | –35 | –40 | –45 | –50 | –55 | –60 | –65 | *Total* |
|---|---|---|---|---|---|---|---|---|---|---|
| Director | 0 | 0 | 0 | 0 | 0 | 0 | 9 | 9 | 2 | 20 |
| Chief | 0 | 0 | 0 | 0 | 3 | 8 | 13 | 17 | 7 | 48 |
| Main | 0 | 13 | 27 | 33 | 12 | 18 | 15 | 12 | 3 | 133 |
| Assistant | 37 | 31 | 0 | 0 | 0 | 0 | 0 | 0 | 0 | 68 |
| *Total* | 37 | 44 | 27 | 33 | 15 | 26 | 37 | 38 | 12 | 269 |

***Table 6.1b CAMERA diagram analysis of the data in Table 6.1a (see Example 6.1)***

| | 20–25 | –30 | –35 | –40 | –45 | –50 | –55 | –60 | –65 |
|---|---|---|---|---|---|---|---|---|---|
| Director | — | — | — | — | — | — | 100 | 100 | 100 |
| Chief | — | — | — | — | 100 | 100 | 76 | 76 | 83 |
| Main | — | 100 | 100 | 100 | 80 | 69 | 41 | 32 | 25 |
| Assistant | 100 | 70 | 0 | 0 | 0 | 0 | 0 | 0 | 0 |

***Table 6.2 The promotion prospects or career pattern for Example 6.1 as estimated from Figure 6.3***

| | *Percentage reaching each grade but no higher* | *Approximate ages at each promotion* | | |
|---|---|---|---|---|
| | | A → M | M → C | C → D |
| Director | 25% | 27 | 43 | 52 |
| Chief | 45% | 28 | 50 | — |
| Main | 30% | 31 | — | — |

Using age groups and plotting their values at their mid-points, as in Figure 6.3, has the effect of smoothing the data and thus allowing curves to be fitted more easily. This is a convenient method and does give a useful guide, but it should be used with care as it may blur the promotion picture, particularly if promotion takes place over a narrow age range. In their papers, Morgan and coworkers (1974), and Keenay and coworkers (1977b) suggest using a 7-year moving average for smoothing, and they also fit logistic curves to the points. The reader who wishes to make use of such refinements should consult the original papers.

## 6.3 THE EFFECT OF AGE BULGES ON PROMOTION PROSPECTS

We have already noted that the interpretation we have been placing on the grade–age diagrams involves assumptions of stationarity. It is therefore desirable to go a step further and ask how the patterns revealed by our analysis would change if the age structure were to change—for example as a result of increased recruitment. Morgan and coworkers (1974, and Keenay and coworkers (1977a, 1977b) develop their methods of analysis to investigate these points. We shall not go into the technicalities, but instead, make some qualitative deductions from a pictorial analysis of the situation.

Figure 6.4 is designed to show how the average age of promotion will change if an age 'bulge' passes through the system and if the proportion eventually promoted remains constant. For simplicity we assume:

(a) that there are just two grades which maintain the same relative sizes;
(b) that wastage is age- but not grade-specific;
(c) that recruitment takes place at the lower ages only, say between 25 and 35;
(d) that at any point in time, promotions take place at the same age although this age may change over time;
(e) that the proportion eventually promoted is fixed. In particular we assume that approximately half the people at each age are eligible or suitable for eventual promotion. These are represented by those above the broken line.

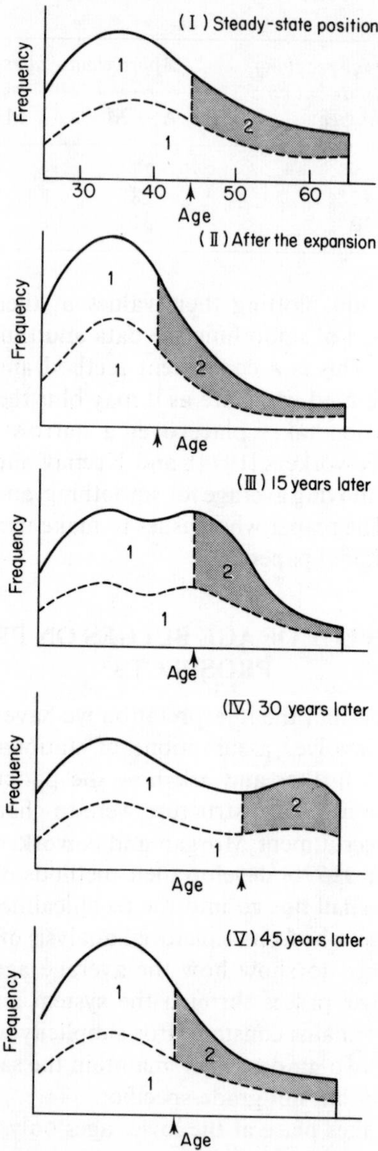


Figure 6.4 Illustrating the effect of an age bulge on age on promotion if the proportion eventually promoted remains constant (see text for assumptions and description)

The successive parts of Figure 6.4 show what will happen as the system goes through a period of expansion, after which it remains constant in size. The first part, (I), shows the initial steady-state age distribution with the shaded area marked '2' representing those people in grade 2. The age of promotion is 45, and the proportion ultimately promoted (=0.5) can be read off from the extreme right-hand side of the diagram.

As a result of the expansion a bulge appears towards the left-hand side. In order to maintain the relative size of grade 2 and not to increase the proportion promoted at each age above the promotion point, the age of promotion must decrease as shown in (II). Fifteen years later, in (III), the peak of the bulge has moved to the right and the promotion age has risen as the bulge provides an increasing number of eligible promotees. After thirty years, in (IV), the promotion age has increased still further because the bulge has now filled up grade 2, and it will not decline again until these people retire. When this happens recruitment will have to rise, thus taking us back to situation (II), although the bulge is likely to be somewhat smaller because of the erosion due to wastage and the variation in the retirement ages over time.

With this analysis in mind we can return to Example 6.1 and make some deductions about the likely present situation and its implications for the possible course of events in the future of that system.

*Example 6.1 (continued)* The analysis which follows must be treated with caution because of the simplifying assumptions which lie behind the foregoing discussion. Nevertheless, we believe that the broad qualitative conclusions which we shall draw are robust. Figure 6.5 shows the age distribution for the data used in this example. Although we have not given enough information to deduce the steady-state distribution a typical form is shown by the dotted curve. If we accept this we see that there is an age bulge in the 50–60 region which suggests that the system may be in the state represented by (IV) in Figure 6.4. Thus the current ages of promotion from M to C and from C to D may be higher than 'normal' and a promotion blockage may

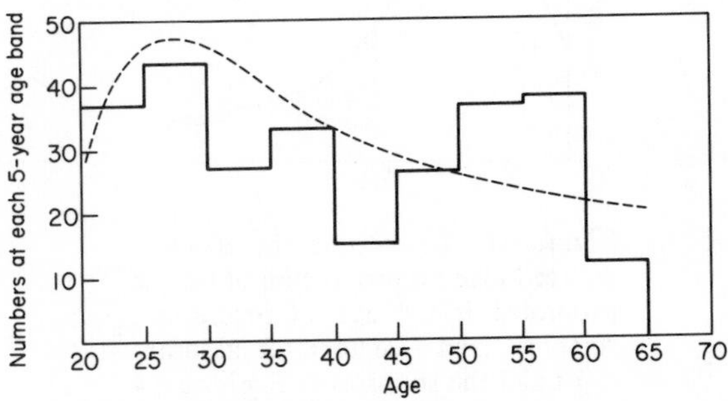


Figure 6.5 The age distribution of Example 6.1

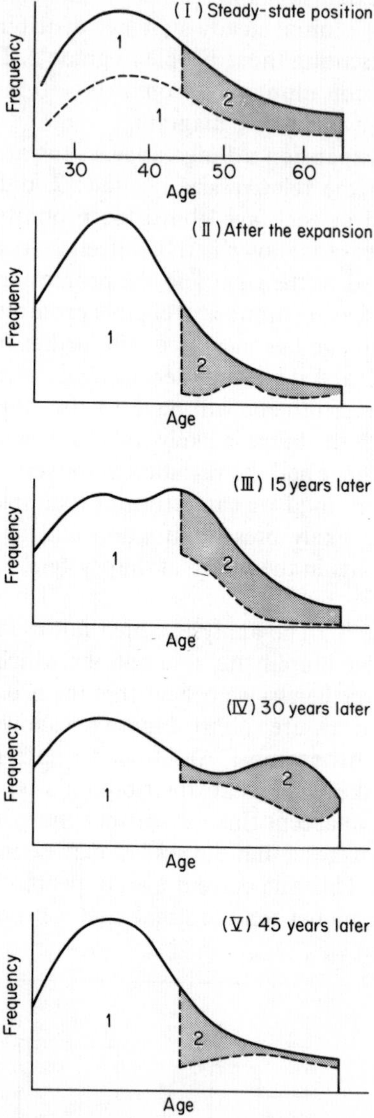


Figure 6.6 Illustrating the effect of an age bulge on proportion of people promoted if the age of promotion remains fixed (see text for assumptions and the solution to Exercise 6.4 for a description)

exist at both of these points. Promotion from A to M on the other hand could be at an earlier age than normal because of the relative scarcity of younger people. ◀

The situation shown in Figure 6.4 was based on assumption (e), that the proportion eventually promoted (in this case 0.5) remained fixed, and that it was only the age of promotion that varied as the age bulge worked its way through the system. If we assume instead that the promotion age remains constant and that the proportion promoted at that age varies in response to the number of promotees required at any time, then we have the situation depicted in Figure 6.6. The reader is invited to provide a commentary in Exercise 6.4.

In practice there are likely to be pressures to maintain both the eventual proportions promoted (as in Figure 6.4) and the age at promotion (as in Figure 6.6). However, if the relative grade sizes are to remain fixed these cannot both be maintained simultaneously unless the age distribution remains stable. Thus some compromise between these two situations must be adopted. Empirical studies indicate that the age on promotion is often more sensitive than the proportion promoted.

## 6.4 THE RELATIONSHIP BETWEEN PROMOTION RATES AND CAREER PATTERNS: A SIMPLE HIERARCHY

The promotion rate from a grade is a measure of the chance of being promoted at a particular time but its value gives little direct insight into the career patterns of those subject to such rates. In this section, therefore, we shall use elementary probability arguments to explore the relationship between the two. Let us begin by calculating some quantities relevant to the length of stay in one particular grade.

Between durations $T$ and $T + 1$ an individual can be promoted, can leave, or can remain in the same grade. Let us assume that the probabilities of these events do not depend on $T$ and are denoted by the transition probabilities $p$, $w$, and $s = 1 - p - w$, respectively. Assume that there is a maximum length of service, $T = k$, in the grade after which the individual retires. The following probabilities are then of interest from a career point of view.

$$P_r = \Pr\{\text{individual is promoted between } T = r \text{ and } T = r + 1\}$$

$$P_u = \Pr\{\text{individual is ultimately promoted}\}$$

If successive transitions are independent then clearly

$$P_r = s^r p \qquad (r = 0, 1, 2, \ldots, k - 1) \tag{6.1}$$

Hence the probability of eventual promotion is

$$P_u = \sum_{r=0}^{k-1} P_r = p(1 - s^k)/(1 - s) \tag{6.2}$$

There would be no difficulty about extending this argument to the case where $p$, $w$, and $s$ depend on $T$—see Forbes (1973). This probability will give the expected proportion of entrants to the grade who will exit by promotion and is of direct interest to the individual and management alike. However, there is a second probability which may be more relevant to the individual in assessing his own prospects within the organization. This is the probability of eventual promotion *if he does not leave*. Let us define, therefore,

$$P_u^* = \Pr\{\text{individual is ultimately promoted} \mid \text{he does not leave before retirement}\}$$

Between $T$ and $T+1$ there are now only two alternatives open to the individual—he is either promoted or he stays. Let the probabilities of these events be $p^*$ and $s^* = 1 - p^*$ respectively. The chance of eventual promotion is thus given by

$$P_u^* = 1 - (s^*)^k \tag{6.3}$$

Note that as $k \to \infty$, $P_u^* \to 1$ as it must since promotion is then the only means of exit.

In order to calculate $P_u^*$ we require a value for $p^*$ which is not necessarily the same as $p$. This value for $p^*$ will depend on what assumption we make about the ordering of the promotion and wastage events. Consider the following three possibilities:

(a) If promotion occurs before wastage, or takes precedence over it (e.g. if people wait to see if they are promoted in any year before deciding whether or not to leave) then

$$p^* = p$$

since the event on which we are conditioning does not take place until after promotion.

(b) If wastage occurs before promotion or takes precedence over it (e.g. if promotions are only made from among those known to be staying through that time interval) then

$$p^* = p/(1 - w).$$

(c) If the promotion and wastage events occur together and if they are independent we require the methods of the theory of competing risk (see for example Chiang (1968)) which gives

$$p^* = 1 - s^{p/(p+w)}.$$

The value of $P_u^*$ will therefore depend on which of these assumptions we adopt.

Different expected durations before promotion for those promoted can be derived corresponding to each of the probabilities $P_u$ and $P_u^*$. These will be denoted by $T_p$ and $T_p^*$ respectively. We consider first how to calculate these expected durations, and then consider their interpretation. The probability that a person is promoted at time $T = r$ given that they are ultimately promoted is $P_r/P_u$, and assuming that on average a person who is promoted in $(T, T + 1)$ is promoted half-way through this interval we have

$$T_p = \sum_{r=0}^{k-1} (r + \tfrac{1}{2})P_r/P_u = (1 - s)^{-1} - \tfrac{1}{2} - ks^k/(1 - s^k) \tag{6.4}$$

A similar calculation can be made for the case where we condition on no leaving before retirement. The expected time to promotion will then be

$$T_p^* = (1 - s^*)^{-1} - \tfrac{1}{2} - k(s^*)^k\{1 - (s^*)^k\}^{-1} \tag{6.5}$$

where, of course, $s^* = 1 - p^*$.

We can make the following connection with Section 6.2 where we saw that the CAMERA diagram depicts the proportion of those at each grade. Since the cohort interpretation of this career pattern is in terms of what would happen to those surviving, the estimated stream sizes and average ages will refer to non-leavers. The CAMERA diagram parameters will therefore be most closely related to $P_u^*$ and $T_p^*$.

*Example 6.2* For the officer system of one of the British women's services already discussed in Example 4.6 we have

$$\text{for grade 1: } p = \frac{101}{997} = 0.101, \qquad w = \frac{171}{997} = 0.170$$

$$\text{for grade 2: } p = \frac{29}{635} = 0.046, \qquad w = \frac{79}{635} = 0.124$$

Substituting these estimates into (6.2) and (6.4) and taking $k = 7$ for grade 1 and $k = 15$ for grade 2 (the actual values) we find

$$\text{for grade 1: } P_u = 0.332, \qquad T_p = 2.23$$

$$\text{for grade 2: } P_u = 0.254, \qquad T_p = 4.43$$

The corresponding calculations when we condition on not leaving give

$$\text{for grade 1: } p^* = 0.111, \qquad P_u^* = 0.562, \qquad T_p^* = 3.03$$

$$\text{for grade 2: } p^* = 0.049, \qquad P_p^* = 0.531, \qquad T_p^* = 6.57$$

where $p^*$ has been calculated using the assumption (c) above (that is, promotion and wastage occur independently).

### The interpretation of $P_u$, $P_u^*$, $T_p$, and $T_p^*$

Example 6.2 shows that within 7 years, out of a group of entrants to grade 1, a proportion $P_u = 0.33$ can expect to be promoted. The remainder will either leave or will attain the maximum length of service and then retire. This chance of eventual promotion applies on average to all entrants. It does not however tell an individual the full extent of his possible promotion chance. This would occur (within the assumptions of our model) if he were not to leave before either promotion or retirement. In other words, he maximizes his chance if he is prepared to wait for promotion if necessary for as long as possible. This chance is given by $P_u^*$, which in this example is 0.56. Note that it is $P_u^*$ and not $P_u$ which corresponds to the value obtained from the CAMERA diagram.

In the same example, the average service of promotees is $T_p = 2.3$ years. This can be thought of as the average experience which promotees will have had in grade 1 on their arrival in grade 2. It could easily be estimated directly in practice by observing the times spent in grade 1 by those promoted to grade 2 and then taking the average. By contrast, $T_p^*$ is a hypothetical quantity which tells an entrant to grade 1 how long he could expect to wait for promotion if he was prepared to wait for the maximum length of time necessary. It may appear paradoxical that $T_p^*$ is greater in Example 6.2 than $T_p$, since by ruling out the leaving option, entrants increase their chance of promotion but apparently also increase the time they have to wait. This result occurs because the chance that an *entrant* will be promoted after a particular time in the grade increases relatively more for the longer durations. In an aggregate sense therefore this produces relatively more long-service promotees, and hence increases the average time to promotion.

To summarize we can make the following interpretations:

$P_u$: the expected proportion of a group of entrants promoted (after $k$ years' service);

$T_p$: the average experience of people being promoted (before $k$ years);

$P_u^*$: the chance of promotion for an average individual who is prepared to wait up to $k$ years;

$T_p^*$: the expected waiting time for this individual if he is promoted.

It is worth noting that in practice there is often little difference between the values $T_p$ and $T_p^*$, particularly when promotion is restricted to some promotion zone.

### Combining $P_u$, $T_p$, $P_u^*$, and $T_p^*$ for adjacent grades

The probabilities and durations for consecutive grades can be combined to give corresponding expressions for promotions between more distant pairs of grades. Thus, using Example 6.2, if $P_u(1)$ is the probability of ultimate promotion from grade 1, and $P_u(2)$ the corresponding probability for grade 2, then the probability of ultimate promotion from grade 1 to grade 3 is

$$P_u(1)P_u(2) = 0.084$$

for the data of Example 6.2. The durations are additive, giving in an obvious notation, that the expected time for promotion from grade 1 to grade 3 is

$$T_p(1) + T_p(2) = 6.74$$

Similar expressions hold when we condition on not leaving. Combining quantities in this way does, however, treat the events as independent. In practice this is unlikely to be true because those promoted early from grade 1 (the 'high fliers') are likely to be promoted early from grade 2. Also, if there is a wide variation in the time of promotion from grade 1 it may not be reasonable to suppose that the maximum length of service, $k$, is the same for all entrants to grade 2.

## 6.5 PROMOTION ZONES

An unrealistic feature of our treatment in the last section is the assumption that an individual's chance of promotion does not depend on how long he has been in the grade. In practice this is rarely the case. The chance of promotion is usually very low initially, then it rises to a peak, after which it may decline. A crude but useful approximation to this situation can be made by introducing the idea of a promotion zone. Let us assume that the length of service in a grade, which we shall refer to as seniority, is divided into three zones with boundaries $a$ and $b$. Below seniority $a$ and above seniority $b$ no one is promoted. The middle interval $(a, b)$ is the promotion zone and we denote by $p_2$ the promotion probability (at each time point) for people in this interval. The wastage and staying probabilities for the three zones will be $w_1, w_2, w_3$ and $s_1, s_2, s_3$, respectively. Under these conditions the indices of promotion prospects are easily found as follows using (6.2) and (6.4)

$$\begin{aligned} P_u &= \Pr\{\text{survives to } a\} \Pr\{\text{ promotion occurs during } (a, b)\} \\ &= s_1^a p_2(1 - s_2^{b-a})(1 - s_2)^{-1} \qquad (6.6) \end{aligned}$$

$$\begin{aligned} T_p &= a + \text{expected time to promotion in } (a, b) \\ &= a + (1 - s_2)^{-1} - \tfrac{1}{2} - (b - a)s_2^{b-a}(1 - s_2^{b-a})^{-1} \qquad (6.7) \end{aligned}$$

The corresponding expressions from (6.3) and (6.5) are

$$P_u^* = 1 - (s_2^*)^{b-a} \qquad (6.8)$$

$$T_p^* = a + (1 - s_2^*)^{-1} - \tfrac{1}{2} - (b - a)(s_2^*)^{b-a}\{1 - (s_2^*)^{b-a}\}^{-1} \qquad (6.9)$$

Note that the zone following the promotion zone influences none of these values since $k$, $w_3$, and $s_3$ do not appear in these equations. (The value for $k$ is needed, however, if $p_2$ has to be estimated indirectly using the approximation given later.)

*Example 6.3* We take the same system as in Example 6.2 but introduce promotion zones as set out in Tables 6.3 and 6.4. (For this example these zones correspond quite well with the actual promotion zones by seniority—see Forbes (1973).) The wastage and promotion rates are based on seniority data from the original study which has not been included here. Note again that the data given in the last row of each table are not required for calculating the career pattern statistics derived above. These data have been included in the table in order to emphasize the breakdown into three zones.

***Table 6.3 Zone-specific transition rates from grade 1 for Example 6.3***

| | | *Transition rates* | | |
|---|---|---|---|---|
| *Time in grade* | $i$ | $p_i$ | $w_i$ | $p_i^*$ |
| before promotion zone | 1 | 0 | 0.129 | 0 |
| promotion zone (3, 7) | 2 | 0.337 | 0.215 | 0.388 |
| after promotion zone (7, 7) | 3 | 0 | 0 | 0 |

Substitution in the formulae above, or use of the program in Appendix B, gives for promotion from grade 1 to grade 2:

average proportion of entrants promoted $(P_u) = 0.39$;
average experience of promotees $(T_p) = 4.1$ years;
chance of promotion for someone who is prepared to wait for up to 7 years $(P_u^*) = 0.86$;
time this person can expect to wait $(T_p^*) = 4.4$ years.

***Table 6.4 Zone-specific transition rates from grade 2 for Example 6.3***

| | | *Transition rates* | | |
|---|---|---|---|---|
| *Time in grade* | $i$ | $p_i$ | $w_i$ | $p_i^*$ |
| before promotion zone (0, 6) | 1 | 0 | 0.131 | 0 |
| promotion zone (6, 12) | 2 | 0.138 | 0.089 | 0.145 |
| after promotion zone (12, 15) | 3 | 0 | 0.095 | 0 |

For promotion from grade 2 to grade 3:

$$P_u = 0.21; \qquad T_p = 8.3 \textit{ years}:$$

$$P_u^* = 0.61; \qquad T_p^* = 8.6 \textit{ years}.$$

Combining the foregoing results to obtain the corresponding values for promotion from grade 1 to grade 3 we find:

$$P_u = 0.08; \qquad T_p = 12.4 \textit{ years};$$

$$P_u^* = 0.52; \qquad T_p^* = 13.0 \textit{ years}.$$

◀

These calculations should be compared with those made before we introduced the promotion zones. (Both sets of values are summarized in rows 1 and 2 of Table 6.5.) Although there is some change in the promotion probabilities, the major change is in the waiting times. These calculations emphasize the desirability of incorporating promotion zones into the model if it is known that the chance of promotion depends strongly on seniority.

## An approximation when only grade-specific rates are available

A practical difficulty which often arises is that although we can specify a promotion zone we may not have the data to estimate directly the zone-specific rate. If the grade-specific rates only are available the following procedure may be useful.

(a) Estimate the promotion rate $p_2$ using the approximation

$$p_2 = p(1 - s^k)/(s^a - s^b) \tag{6.10}$$

where $p$ and $s$ are the grade-specific rates and $k$ is the maximum possible length of service in the grade. This formula is based on the assumption that the seniority distribution within the grade is stationary and approximates to that implied by the grade-specific rates (further details can be found in Forbes (1977)).

(b) Assume that the wastage rates for each zone are equal and estimate them by the grade-specific rate, i.e. put $w_1 = w_2 = w_3 = w$.

Note that these are rather strong assumptions. They should be checked wherever possible, and the value of $p_2$ together with the other results should always be inspected carefully, bearing in mind the assumptions. (The above procedure can be improved if we have some knowledge of the zone-specific rates.)

*Example 6.4* Using this approximation in our previous example, equation (6.10) gives the following:

for grade 1: $p_2 = 0.324$ compared with the direct estimate of 0.337

for grade 2: $p_2 = 0.196$ compared with the direct estimate of 0.138

The approximation is thus good in the first case but rather poorer in the second. However, the real test is how well the approximation reproduces the promotion indices. This can be judged from row 3 of Table 6.5. ◀

***Table 6.5 Promotion indices calculated on various assumptions for the data on one of the women's services***

| | *Promotion from grade* 1 | | | | *Promotion from grade* 2 | | | |
|---|---|---|---|---|---|---|---|---|
| | $P_u$ | $T_p$ | $P_u^*$ | $T_p^*$ | $P_u$ | $T_p$ | $P_u^*$ | $T_p^*$ |
| 1. Ignoring promotion zones | 0.33 | 2.3 yrs | 0.56 | 3.0 yrs | 0.25 | 4.4 yrs | 0.53 | 6.6 yrs |
| 2. Using promotion zones and zone-specific data | 0.39 | 4.1 yrs | 0.86 | 4.4 yrs | 0.21 | 8.3 yrs | 0.61 | 8.6 yrs |
| 3. Using promotion zones with $p_2$ estimated from grade-specific data | 0.35 | 4.2 yrs | 0.83 | 4.5 yrs | 0.25 | 8.0 yrs | 0.76 | 8.3 yrs |

It is clear from Table 6.5 that the main differences, at least for our example, are between the method which ignores the zones and the others. No serious errors result in this system from using the approximation in the third row.

## 6.6 PROMOTION PATTERNS USING MARKOV CHAIN THEORY

The passage of an individual through a manpower system operating according to the assumptions of the Markov model can be studied by means of the theory of absorbing Markov chains. This approach is more general than that of the last section in that the system does not have to be hierarchical, but it is more restrictive in that we have to take $a = 0$ and $b = k = \infty$. In other words, promotion is assumed to be possible immediately on entering a grade, and continues indefinitely with no upper limit on the length of stay in each grade. This latter limitation will not be serious if the average stay in a grade is relatively short compared with the normal career length.

The theory of absorbing Markov chains is available in most standard texts and one of the best references is still Kemeny and Snell (1960). An account of the application of the theory to manpower planning will be found in Bartholomew (1973a, p. 65 ff). Here we shall merely quote theoretical results as required. In the terminology of the theory, the grades of the system comprise the transient states of the chain and 'left' constitutes a single absorbing state. To illustrate the theory we take the parameters as estimated for the women's officer system.

Thus we have

$$\mathbf{P} = \begin{bmatrix} 0.729 & 0.101 & 0 \\ 0 & 0.830 & 0.046 \\ 0 & 0 & 0.901 \end{bmatrix}, \qquad \mathbf{w}' = \begin{bmatrix} 0.170 \\ 0.124 \\ 0.099 \end{bmatrix}$$

First we compute the so-called fundamental matrix $\mathbf{D} = (\mathbf{I} - \mathbf{P})^{-1}$ which for the example is

$$\mathbf{D} = \begin{bmatrix} 3.69 & 2.19 & 1.02 \\ 0 & 5.89 & 2.73 \\ 0 & 0 & 10.10 \end{bmatrix} \qquad \begin{matrix} \textit{Row totals} \\ 6.90 \\ 8.62 \\ 10.10 \end{matrix}$$

The element in the $i$th row and $j$th column of $\mathbf{D}$ is the expected length of time that an entrant to $i$ will ultimately spend in $j$. The row totals are therefore the expected times spent in the system before leaving and after entry to the grade corresponding to the row. These are useful indices in their own right and they show that the higher one moves up this particular system, the longer one is likely to stay. The diagonal elements are of particular interest since they are similar to the quantities $T_p$ calculated on the assumption that $a = 0$ and $b = k = \infty$. In order to bring them into line with the earlier calculations we need to subtract $\frac{1}{2}$ from the diagonal element since previously we assumed that individuals leave half-way through an interval rather than at the end. The off-diagonal elements in $\mathbf{D}$ are of little interest in themselves since they are not conditional on reaching grade $j$. Thus their values are small because they contain contributions of zero time from all those people who do not reach grade $j$. In fact the $(j, j)$th diagonal element gives the (conditional) time in this grade for grade $i$ entrants.

The probabilities that entrants to $i$ will ultimately spend time in $j$ are obtained from $\mathbf{D}$ by dividing each column by its diagonal element. In our example this gives the matrix

$$\mathbf{\Pi} = \begin{bmatrix} 1 & 0.373 & 0.101 \\ 0 & 1 & 0.271 \\ 0 & 0 & 1 \end{bmatrix}$$

Thus an entrant at the bottom has a 10 per cent chance of reaching the top. (Note that (0.373) (0.271) = 0.101 as it should for a simple hierarchy because of the independence of successive moves implied by the Markov assumption.) The probabilities given here are, of course, the same as the $P_u$'s under the assumptions $a = 0$ and $b = k = \infty$.

Similar results can be obtained if we condition on the event of not leaving.

Thus if the new transition matrix, $\mathbf{P}^*$ say, is obtained by dividing each row of $\mathbf{P}$ by its sum

$$\mathbf{P}^* = \begin{bmatrix} 0.878 & 0.122 & 0 \\ 0 & 0.947 & 0.053 \\ 0 & 0 & 1.000 \end{bmatrix}$$

This method of calculating the conditional probabilities corresponds to the use of assumption (b) to derive $p^*$ (see Section 6.4). Inversion of $(\mathbf{I} - \mathbf{P}^*)$ gives

$$\mathbf{D}^* = \begin{bmatrix} 8.20 & 18.87 & \infty \\ 0 & 18.87 & \infty \\ 0 & 0 & \infty \end{bmatrix}$$

The column elements are identical because if there is no leaving everyone who enters any grade must survive to enter subsequent grades. As before, the conditional waiting times turn out to be longer than the unconditional ones.

The infinite values arise because there is now no movement out of grade 3. In other words it is an absorbing state. Numerically these values can cause obvious inversion problems which can be avoided by reducing $\mathbf{P}^*$ by omitting rows and columns corresponding to these absorbing states. This will not affect times spent in other states since by definition no one returns from an absorbing state. In the present example the reduced form is

$$\mathbf{P}^* = \begin{bmatrix} 0.878 & 0.122 \\ 0 & 0.947 \end{bmatrix} \quad \text{giving } \mathbf{D}^* = \begin{bmatrix} 8.20 & 18.87 \\ 0 & 18.87 \end{bmatrix}$$

which contains all the unknown values required in the full size $\mathbf{D}^*$. Exercise 6.9 and its solution illustrate how these ideas can be extended to an absorbing set of grades.

There is no point in calculating $\mathbf{\Pi}^*$ to give the probabilities of attaining different levels since all the non-zero elements will obviously be 1. This follows because, with $k = \infty$, everyone will ultimately reach any level. This highlights the fact that it is only because $k$ will be finite in practice, and possibly quite small, that the probabilities $P_u^*$ have any practical interest.

## 6.7 RELEVANCE TO THE RENEWAL MODEL

Throughout the greater part of this chapter we have made the Markovian assumption that loss and promotion probabilities were constant and the same for all people. It is natural to ask whether similar analyses can be made for systems which operate according to renewal principles. The theory in this case is rather more difficult though some further discussion of the point is given in Section 6.8. However, for many purposes the Markovian analysis will be sufficient. We have already used the fundamental matrix in Chapter 5 to study the flow of vacancies which were there assumed to conform to a Markov chain. A vacancy chain is not, of course, the same thing as a career path, though the probability that a given individual will be selected to fill a vacancy is a relevant indicator of the opportunities which a system offers. A more substantial reason for claiming that the Markovian analysis is useful for renewal systems is based on equilibrium considerations. We have seen in Chapter 5 that when the various renewal models have reached equilibrium the flow rates are, on average, constant. Under such conditions the renewal system will be, in average terms, indistinguishable from a Markov system. In that sense, therefore, the results derived in this chapter can be applied to renewal systems provided that they are at or near their steady state.

## 6.8 COMPLEMENTS

The analysis of career patterns by the methods of Section 6.2 was developed by Morgan and the most up-to-date account is contained in Keenay and coworkers (1977b). They have formalized the type of argument we used in Section 6.3. This method leads to charts from which one can read off the effect of changes in the structure of the system on the average ages of promotion. A brief account and an illustration of such a chart is given in Hopes (1973). Computer programs are available to carry out these calculations—see Forbes (1976).

The discussion of the relationship between promotion rates and career patterns in Sections 6.4 and 6.5 is based upon Forbes (1973). There is no difficulty in principle in obtaining output on career patterns for any simulation model of a manpower system which keeps track of individuals. For models in which a system is simulated in aggregate terms it is usually necessary to do some theoretical work in order to obtain career patterns. This has been done for the KENT model of the Civil Service Department's MANPLAN system and the theory is given in Jones (1978). In a paper on the relationship between personal attributes and job performance, Wise (1975) used promotion probabilities as indices of performance and he developed maximum likelihood methods of estimating these from the kind of data available to him. In particular he considers the probability

$$\Pr\{\text{individual is at level } k \text{ after } t \text{ years} \mid \text{he has not left}\}$$

The calculation of conditional probabilities and durations such as $P_u^*$ and $T_p^*$ is really an application of the theory of competing risks. This is usually presented in continuous time—see, for example, Chiang (1968). The connection may be more apparent if we cast Example 6.2 (grade 2 → 3) in a medical context. Suppose that for someone now aged 60 there are only two possible causes of death, namely, heart-attack with probability 0.046 each year, and cancer with probability 0.214. Then according to our calculations there is a chance of $P_u = 0.25$ of dying of a heart-attack before age 75. Similarly, by interchanging the rates for 'promotion' and 'wastage', and recalculating $P_u$, we find that the chance of dying of a cancer is 0.68 and thus the chance of surviving to age 75 is 0.07. Furthermore, the expected age of death is $60 + T_p = 64.4$. (Interchanging 'promotion' and 'wastage' does not alter $T_p$.) Suppose next that we are interested in how these risks would change if a complete cure for cancer were to be discovered. The chance of dying from a heart-attack before 75 would now be $P_u^* = 0.53$ at an expected age of $60 + T_p^* = 66.6$. There is now a greater chance of dying of heart-attack (since some who would have otherwise died of cancer will now survive to succumb to a heart-attack) but the overall chance of dying before age 75 is reduced from 0.93 ($= 0.25 + 0.68$) to 0.53.

There appears to be very little theoretical work on promotion zones though they often correspond, at least roughly, to what happens in practice. In real life the chance of promotion is more likely to be a smoothly changing function of seniority than the step function assumed in our analysis. Even so we regard our analysis as a move in the right direction and our crude model does appear to give a good approximation at least for the data tested. The point is developed further in Forbes (1973).

We noted in Section 6.6 that the use of Markov chain theory implied that there was no upper limit on the length of time that could be spent in a grade. In theory this is a shortcoming of the model when applied to any human system. Gani (1963) discussed this in relation to his application of the model to a university system. He showed that the effect was not likely to be serious, by showing that the probability of staying an 'unreasonably' long time was negligible. This point, however, should not be overlooked in practice.

In Section 6.6 we defined one absorbing state to cover all leavers. There is no difficulty about subdividing this category according to reason for leaving. We can then use the theory to calculate the probabilities of leaving for each particular reason and the expected lengths of service conditional upon each reason for leaving. As well as calculating average times to absorption, it is also possible to find the variances and higher moments of such quantities. An example, showing how to find the variances of various sojourn times for the Markov chain model is given in Bartholomew (1973a, p. 70).

The theory of renewal systems in continuous time provides methods for determining such quantities as the average length of service at promotion under a variety of promotion rules. A full discussion will be found in Bartholomew (1973a, Chapter 8) but the following argument shows that it is sometimes possible to obtain approximate results by very simple methods. Suppose that we have a two-grade hierarchy with $n_1$ places in the lower grade and $n_2$ in the higher. When a vacancy occurs in the higher grade it is filled by selecting the longest-serving member of the lower grade. This promotion rule implies that everyone in the higher grade must have served for at least as long as anyone in the lower grade. Hence an individual is promoted at the time when his length-of-service ranking moves from $n_1$ to $n_1 + 1$. Approximately, this will be at length of service $x$ where $x$ is the solution of

$$\int_0^x S(y)\,dy = n_1/(n_1 + n_2)$$

where $S(y)$ is the length-of-service distribution of serving members. In general, $S(y)$ will also depend on the age of the system so that the length of service on promotion will change over time until the structure attains its equilibrium.

## 6.9 EXERCISES AND SOLUTIONS

### Exercise 6.1

Organization A has a six-grade simple hierarchy with promotion only into the grade immediately above. Entrants are taken into the bottom two grades over a wide age range 20–45. From Table 6.6 draw the histogram of the age distribution, and comment on its shape and its likely influence on the career pattern. Plot the CAMERA diagram and estimate by eye average ages of promotion to each grade. Estimate also the age-stream career prospects.

***Table 6.6 Age distribution for Exercise 6.1***

| | *Ages* | | | | | | | | |
|---|---|---|---|---|---|---|---|---|---|
| *Grade* | 20–25 | –30 | –35 | –40 | –45 | –50 | –55 | –60 | –65 |
| 6 | 0 | 0 | 0 | 0 | 5 | 21 | 20 | 21 | 7 |
| 5 | 0 | 0 | 0 | 2 | 22 | 46 | 53 | 37 | 12 |
| 4 | 0 | 0 | 3 | 46 | 143 | 121 | 120 | 83 | 35 |
| 3 | 0 | 1 | 76 | 368 | 406 | 236 | 219 | 111 | 39 |
| 2 | 3 | 202 | 329 | 142 | 41 | 14 | 2 | 7 | 13 |
| 1 | 82 | 117 | 17 | 1 | 2 | 0 | 0 | 0 | 0 |
| *All* | 85 | 320 | 425 | 559 | 619 | 438 | 414 | 259 | 106 |

### Exercise 6.2

Tables 6.7 and 6.8 show the stock by age and grade for university teachers in 1961 and 1974. For both sets of data draw the age-distribution histograms and plot the CAMERA diagrams. Estimate the career prospects in 1961 and

***Table 6.7 University teachers 1961 (non-medical) (Reproduced by permission of the Association of University Teachers)***

| Grade | Ages -25 | -30 | -35 | -40 | -45 | -50 | -55 | -60 | -65 |
|---|---|---|---|---|---|---|---|---|---|
| Professor | 0 | 0 | 30 | 122 | 142 | 179 | 189 | 116 | 78 |
| Reader & Senior Lecturer | 0 | 4 | 97 | 311 | 251 | 223 | 164 | 136 | 74 |
| Lecturer | 390 | 1223 | 1239 | 969 | 420 | 230 | 109 | 71 | 29 |
| *Total* | 390 | 1227 | 1366 | 1392 | 813 | 632 | 462 | 323 | 181 |

† Stocks over 65 have been excluded. These numbered 25 in 1961.

***Table 6.8 University teachers 1974 (non-clinical) (Reproduced by permission of the Association of University Teachers)***

| Grade | Ages -25 | -30 | -35 | -40 | -45 | -50 | -55 | -60 | -65 |
|---|---|---|---|---|---|---|---|---|---|
| Professor | 0 | 0 | 45 | 205 | 450 | 755 | 725 | 490 | 410 |
| Reader & Senior Lecturer | 0 | 10 | 245 | 1050 | 1355 | 1325 | 1025 | 615 | 365 |
| Lecturer | 280 | 3940 | 5420 | 3960 | 2045 | 1290 | 780 | 445 | 245 |
| *Total* | 280 | 3950 | 5710 | 5215 | 3850 | 3370 | 2530 | 1550 | 1020 |

† Stocks over 65 have been excluded. These numbered 195 in 1974.

***Table 6.9 Age distribution for Exercise 6.3***

| Grades | Ages <20 | 20– | 25– | 30– | 35– | 40– | 45– | 50– | 55– | 60– | 65–70 | *All* |
|---|---|---|---|---|---|---|---|---|---|---|---|---|
| 6 | 0 | 0 | 0 | 0 | 0 | 0 | 3 | 12 | 9 | 5 | 0 | 29 |
| 5 | 0 | 0 | 0 | 0 | 0 | 22 | 35 | 33 | 31 | 10 | 0 | 131 |
| 4 | 0 | 0 | 0 | 12 | 75 | 109 | 158 | 157 | 48 | 21 | 1 | 581 |
| 3 | 0 | 0 | 10 | 159 | 121 | 165 | 162 | 168 | 42 | 14 | 5 | 846 |
| 2 | 0 | 57 | 178 | 196 | 223 | 611 | 585 | 306 | 80 | 21 | 20 | 2277 |
| 1 | 109 | 817 | 660 | 396 | 755 | 1106 | 1039 | 608 | 242 | 41 | 8 | 5781 |
| *All* | 109 | 874 | 848 | 763 | 1174 | 2031 | 1982 | 1284 | 452 | 112 | 34 | 6945 |

1974, and comment on their differences and on any possible effects on the career patterns due to the age distributions.

**Exercise 6.3**

For the data of Organization B shown in Table 6.9 draw the histogram of the numbers in each age group, plot the CAMERA diagram and estimate the career prospects. Comment on these results given that recruitment is mainly in age range 20–30.

**Exercise 6.4**

Provide a commentary for the situations illustrated in Figure 6.6 making the following assumptions:

(a), (b), and (c) as at the beginning of Section 6.3;
(d) promotions always and only occur at the same age (in this case 45);
(e) the proportion of people promoted at this age (and hence the proportion eventually promoted) is *not* fixed but varies in response to the number required at any time.

**Exercise 6.5**

Using the data of Example 6.2 and the methods of Section 6.4, calculate the career pattern comparable with that given by the Markov chain method of Section 6.6.

**Exercise 6.6**

For Example 6.3, check the sensitivity of the career pattern to possible changes in the wastage rate by calculating the career pattern with the wastage rates increased and decreased by a factor of $\frac{1}{3}$.

**Exercise 6.7**

For a simple three-grade hierarchical organization, anticipated future promotion and wastage rates are given in Table 6.10 (the lowest grade is grade 1). The recent past has seen considerable growth in the system, while in the future a slight contraction is expected. These situations are reflected in the recent high promotion rates and the lower promotion rates anticipated in the future. In order to see what these imply in terms of career prospects, calculate and compare the career patterns for each situation. Assume that

***Table 6.10 Transition rates for Exercise 6.7***

| | *From grade 1* | | *From grade 2* | |
|---|---|---|---|---|
| | *promotion* | *wastage* | *promotion* | *wastage* |
| Recent transition rates | 10% | 20% | 5% | 10% |
| Anticipated future transition rates | 5% | 20% | 4% | 10% |

the promotion zones and retirement points are as follows:

| | *promotion zone* | | *retirement* |
|---|---|---|---|
| | *a* | *b* | *k* |
| in grade 1 | 4 | 10 | 45 |
| in grade 2 | 7 | 20 | 35 |

**Exercise 6.8**

For the five-grade hierarchical system of Example 4.2 which has transition matrix

$$\mathbf{P} = \begin{bmatrix} 0.65 & 0.20 & 0 & 0 & 0 \\ 0 & 0.70 & 0.15 & 0 & 0 \\ 0 & 0 & 0.75 & 0.15 & 0 \\ 0 & 0 & 0 & 0.85 & 0.10 \\ 0 & 0 & 0 & 0 & 0.95 \end{bmatrix}$$

calculate the career patterns:

(a) using the methods of Section 6.4, i.e. assuming no promotion zones or retirement points;

(b) using the Markov-chain theory (see Section 6.6);

(c) using the methods of Section 6.5, the approximate formula for $p_2$, and the following promotion zones and retirement points:

| *From grade* | *promotion zone* *a* | *b* | *retirement* *c* |
|---|---|---|---|
| 1 | 3 | 15 | 45 |
| 2 | 4 | 15 | 40 |
| 3 | 5 | 20 | 35 |
| 4 | 8 | 15 | 30 |

Compare the results for (a), (b). and (c).

**Exercise 6.9**

Use the Markov chain theory to calculate the career pattern for the following non-hierarchical system with transition rates as shown in the Figure 6.7. The wastage rates are shown in brackets.

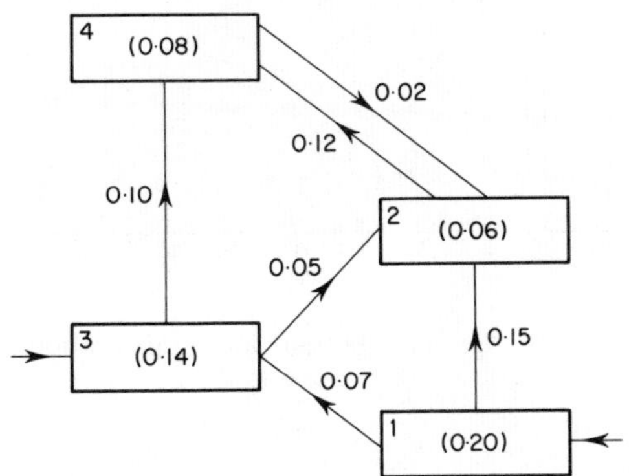


Figure 6.7 The system for Exercise 6.9

**Exercise 6.10**

For a three-grade hierarchical system suppose that the bottom two grades are divided into zones as in the model of Section 6.5, and that this gives the transition matrix below. Use the Markov-chain theory to determine, as far as possible, the career pattern of the organization.

$$\left[\begin{array}{ccc:ccc:c}
0.70 & 0.10 & 0 & 0 & 0 & 0 & 0 \\
0 & 0.78 & 0.02 & 0.10 & 0 & 0 & 0 \\
0 & 0 & 0.95 & 0 & 0 & 0 & 0 \\
\hdashline
0 & 0 & 0 & 0.88 & 0.07 & 0 & 0 \\
0 & 0 & 0 & 0 & 0.83 & 0 & 0.10 \\
0 & 0 & 0 & 0 & 0 & 0.95 & 0 \\
\hdashline
0 & 0 & 0 & 0 & 0 & 0 & 0.90
\end{array}\right]$$

**Solution to Exercise 6.1**

The histogram of the age distribution is shown in Figure 6.8.

Without information on the age-specific wastage rates, inflow levels, and expansion, it is difficult to assess how far the age distribution differs from the

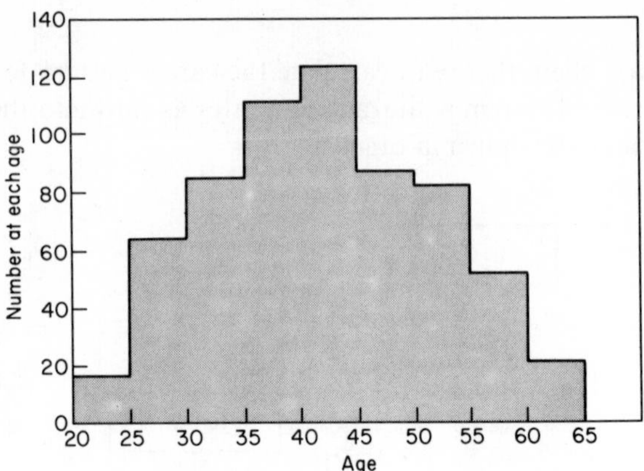


Figure 6.8 Age distribution for Organization A (Exercise 6.1)

stationary one and hence how much it is likely to affect the observed career pattern. The age distribution which is shown in Figure 6.8 has no obvious peaks or troughs. The main peak is somewhat later than usual but is correctly placed in relation to the broad age range of entrants (20–45). For the distribution after age 45, suppose we make the following assumptions and then consider their implications in order to assess their plausibility.

Assume firstly that the wastage at all ages between 45 and 65 is constant with rate $\lambda$, say, and secondly that past expansion has been at a constant rate $\alpha$. With these assumptions, and since there is no recruitment between 45 and 65, the stable age distribution will be exponential. Thus with approximately 120 at age 45 and 20 at age 65 the data give

$$e^{-(65-45)(\lambda+\alpha)} = \frac{20}{120}, \quad \text{and hence } (\lambda + \alpha) = 0.09$$

Now the wastage rates in this organization are probably of the order of 3–5 per cent between these ages, which would in turn imply a long-term expansion of approximately 4–6 per cent. Although this may be rather high for long-term growth, it is not sufficiently unlikely to reject our assumptions as unrealistic.

With the limited information available we can only conclude that the age distribution could be stable and that the observed career pattern is therefore not likely to be too distorted by age bulges. However, from the relationship we have noted in this chapter between promotion and expansion, it is likely that the observed career pattern will depend on the continuation of the long-term expansion.

For the CAMERA diagram we first calculate for each age band the percentages in each grade and below as shown in Table 6.11 and then plot the values as in Figure 6.9.

***Table 6.11***

| | *Ages* | | | | | | | | |
|---|---|---|---|---|---|---|---|---|---|
| *Grade* | 20–25 | –30 | –35 | –40 | –45 | –50 | –55 | –60 | –65 |
| 6 | — | — | — | — | 100 | 100 | 100 | 100 | 100 |
| 5 | — | — | — | 100 | 99 | 95 | 95 | 92 | 93 |
| 4 | — | — | 100 | 100 | 96 | 85 | 82 | 78 | 82 |
| 3 | — | 100 | 99 | 91 | 73 | 57 | 53 | 46 | 49 |
| 2 | 100 | 100 | 81 | 26 | 7 | 3 | 0 | 3 | 12 |
| 1 | 96 | 37 | 4 | 0 | 0 | 0 | 0 | 0 | 0 |

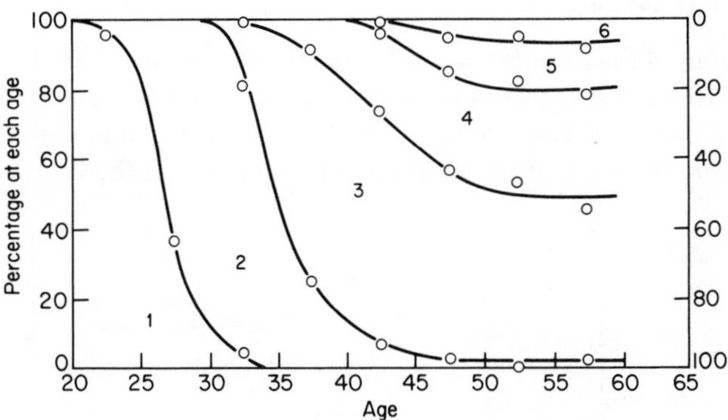


Figure 6.9 CAMERA diagram for Organization A

The age range of 60–65 has been ignored in drawing the curves in Figure 6.9. In practice there are often different retirement rates in each grade, with higher grades usually having earlier retirement, so that the curves often move upwards over the retirement range. This appears to be happening in this system at least for the curve corresponding to promotion from grade 2 to grade 3. Apart from this the curves have exactly the shapes we expect. By inspection the average ages of promotion are approximately:

| | *Promotions* | | | | |
|---|---|---|---|---|---|
| | 1 → 2 | 2 → 3 | 3 → 4 | 4 → 5 | 5 → 6 |
| Averages | 27 | 36 | 42 | 46 | 48 |

and the career prospects are shown in Table 6.12.

***Table 6.12***

| Grade | Stream size, i.e. percentage reaching each grade but no higher | Approximate ages at promotion within each stream | | | | |
|---|---|---|---|---|---|---|
| | | 1 → 2 | 2 → 3 | 3 → 4 | 4 → 5 | 5 → 6 |
| 6 | 7 | 23 | 31 | 35 | 42 | 46 |
| 5 | 13 | 24 | 32 | 38 | 47 | |
| 4 | 31 | 26 | 34 | 45 | | |
| 3 | 47 | 28 | 38 | | | |
| 2 | 2 | 33 | | | | |

Note that people eventually reaching grade 6 may be spending relatively short times in grades 3 and 5.

**Solution to Exercise 6.2**

The histogram of the numbers at each age for 1961 and 1974 are both shown in Figure 6.10 and have been drawn on different scales to facilitate comparison of their overall shapes. A factor of 4 was used for this scaling: this is approximately the ratio of the total size of the system at the two dates.

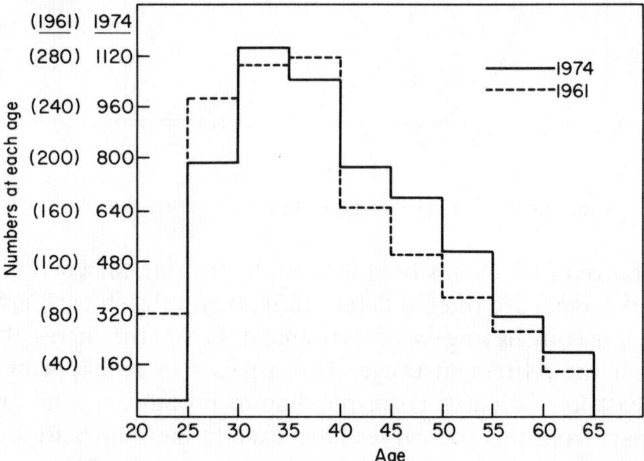


Figure 6.10 Histogram of the age distribution for university teachers 1961, 1974

There are no obvious bulges or troughs in either age distribution. People enter this system at all ages, although the majority are in the age range 20–35. The peak is therefore appropriately placed. It is interesting to note that there is an apparent ageing of the population With entrants mainly at younger ages, we would usually expect the opposite during an expansion such as this system has experienced. This ageing could therefore imply either a change in the wastage pattern or recruitment at the higher ages. Note that people below the minimum of the lecturer scale (e.g. demonstrators) were excluded from the 1974 data, but included in 1961. However, this incompatibility probably involves only the lowest age band.

Ignoring this difference between the two distributions and assuming they are nearly stationary, we can employ similar arguments to those used in the solution to Exercise 6.1 to calculate the implied constant (expansion + wastage) rate for ages over 35 years. This turns out to be approximately 7 per cent for the 1974 distribution, giving a decrease from 1040 to 200 over an age range of 25 years. This does not agree well with the known average expansion rate of 11 per cent per annum at these ages. This discrepancy may be due to recruitment at these higher ages, since any such recruitment will have the effect of reducing the calculated value for the (expansion + wastage) rate.

The percentages at each age in each grade and below are given in Table 6.13.

***Table 6.13***

| | | *Ages* | | | | | | | | |
|---|---|---|---|---|---|---|---|---|---|---|
| *Year* | *Grade* | –25 | –30 | –35 | –40 | –45 | –50 | –55 | –60 | –65 |
| 1961 | Reader/SL | — | 100 | 98 | 92 | 83 | 72 | 59 | 64 | 57 |
| | Lecturer | 100 | 100 | 91 | 70 | 52 | 36 | 24 | 22 | 16 |
| 1974 | Reader/SL | — | 100 | 99 | 96 | 88 | 78 | 71 | 68 | 60 |
| | Lecturer | 100 | 100 | 95 | 76 | 53 | 38 | 31 | 29 | 24 |

These values are plotted in the CAMERA diagram of Figure 6.11.

The promotion curves in Figure 6.11 imply that proportionally fewer people are reaching Professor and Reader/Senior Lecturer in 1974 than in 1961. Also, in 1974 the age on promotion is 1 or 2 years later to Reader/Senior Lecturer, and 3 or 4 years later to Professor. Estimating the career prospects by eye gives the results shown in Table 6.14.

In practice this system, like many others, is not homogeneous. In terms of careers it is composed of areas such as Arts, Science, Technology, Medicine, and disciplines such as History, Physics, etc. How fine a breakdown it is advisable to attempt will depend on the aim of the analysis and the features of the system. In practice the breakdown will usually be dictated by the data available.

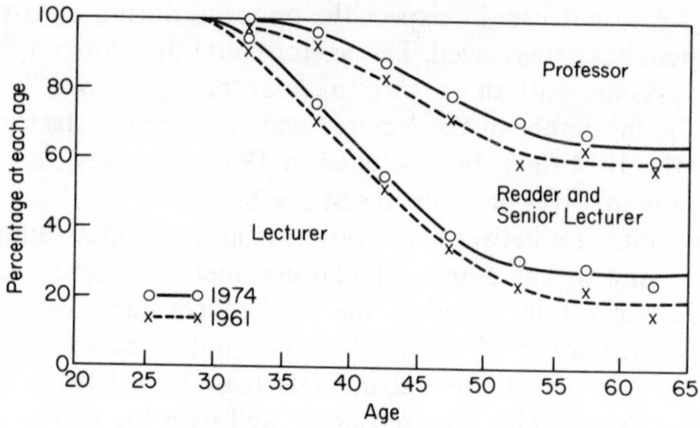


Figure 6.11 CAMERA diagram for university teachers 1961, 1974 (Exercise 6.2)

***Table 6.14***

| *Year* | *Grade* | *Stream size, i.e. percentage reaching each grade but no higher* | *Approximate ages at promotion within each stream* L → RSL | RSL → P |
|---|---|---|---|---|
| 1961 | Professor | 40 | 35 | 43 |
| | Reader/SL | 40 | 47 | — |
| | Lecturer | 20 | — | — |
| 1974 | Professor | 36 | 35 | 46 |
| | Reader/SL | 36 | 45 | — |
| | Lecturer | 28 | — | — |

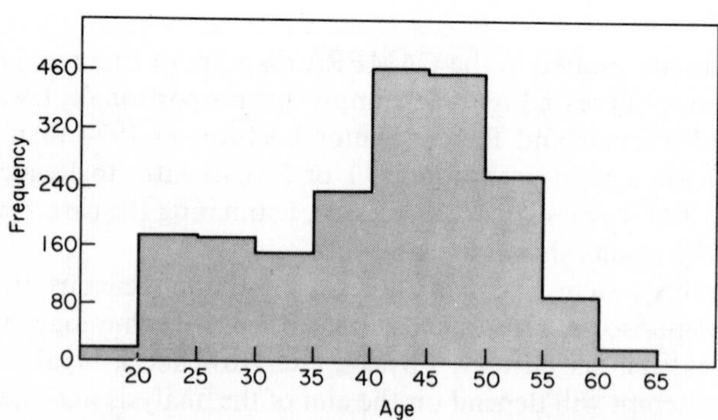


Figure 6.12 Histogram for the age distribution for Organization B

**Solution to Exercise 6.3**

If the age of the majority of entrants to Organization B is less than 30, Figure 6.12 suggests that there is a considerable age bulge between 40 and 50. We can expect this to have a marked effect on the career pattern of the organization. Table 6.15 shows the calculated percentages in each grade and below.

*Table 6.15*

| | Ages | | | | | | | | | | |
|---|---|---|---|---|---|---|---|---|---|---|---|
| *Grades* | <20 | 20– | 25– | 30– | 35– | 40– | 45– | 50– | 55– | 60– | 65–70 |
| 5 | — | — | — | — | — | 100 | 100 | 99 | 98 | 96 | 100 |
| 4 | — | — | — | 100 | 100 | 99 | 98 | 96 | 91 | 87 | 100 |
| 3 | — | — | 100 | 98 | 94 | 93 | 90 | 84 | 81 | 68 | 97 |
| 2 | — | 100 | 99 | 78 | 83 | 85 | 82 | 71 | 71 | 55 | 82 |
| 1 | 100 | 93 | 78 | 52 | 64 | 55 | 52 | 47 | 54 | 37 | 24 |

These values are plotted in Figure 6.13. Since the top three promotion curves would be very close in this CAMERA diagram, the top two promotions have been omitted for the sake of clarity.

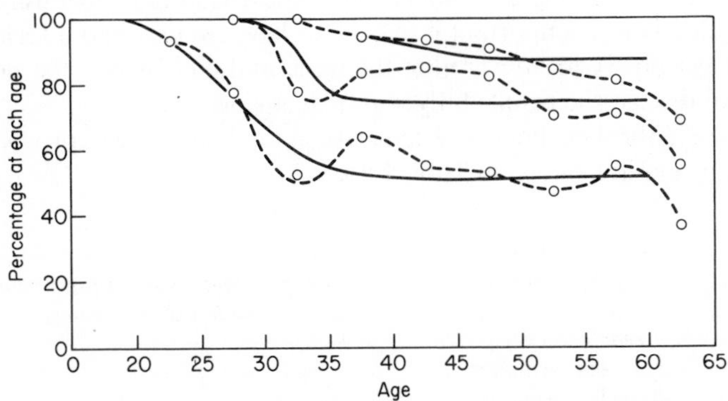


Figure 6.13 CAMERA diagram for Organization B

As anticipated from the age bulge, the points in this CAMERA diagram do not follow the smooth curves we have found in the previous examples. The unbroken curves have the reverse S-shape usually observed and have been fitted by eye. The points in the 60–65 age range have been ignored in fitting the unbroken line, because of possible differential retirement effects between the grades.

Although the diagram does give information on the proportions that have been promoted in the past on the assumption that promotion ages have remained approximately constant (cf. the assumption of Figure 6.6), we cannot deduce the complementary information about promotion ages if proportions promoted had remained constant (cf. the assumption of Figure 6.4). With the former assumption we can broadly distinguish the type of changes in the promotion pattern that we would expect as an age bulge passes through the system (cf. Figures 6.4 and 6.6). Thus it can be seen that except for the 30–35 age group the percentage reaching each grade was higher in the past, presumably due to a higher number of vacancies relative to the number available at the promotion age. Those in front of the bulge appear to have experienced better promotion opportunities than later entrants. The 30–35 group appears to be experiencing relatively favourable promotion chances although these might normally not be expected to improve until vacancies increase when those in the bulge start retiring. If, however, as we are assuming, these promotions ages are relatively fixed then the very low stocks at these ages have more than outweighed the reduced number of vacancies. Two other possibilities are that the system has started to expand, or that people in the 30–35 age group have been promoted with a view to developing them to avoid the succession problems which would otherwise occur in the higher grades when the bulge does retire. A third possibility is a change in the mix of entrants: perhaps relatively more able people are now being recruited. Probably a combination of these factors is operating. This illustrates clearly how important it may be to have background information about the system when interpreting the data, and emphasizes the need to consult and discuss the results fully with management.

Using the unbroken lines, the tentative underlying career prospects as estimated by eye are shown in Table 6.16.

***Table 6.16***

| *Grade* | *Stream size, i.e. percentage reaching each grade but no higher* | *Approximate ages at promotion within each stream* | | |
|---|---|---|---|---|
| | | 1 → 2 | 2 → 3 | 3 → 4 + |
| 4 + 5 + 6 | 13 | 22 | 31 | 40 |
| 3 | 11 | 26 | 34 | — |
| 2 | 24 | 32 | — | |
| 1 | 52 | — | | |

**Solution to Exercise 6.4**

Part (II) of Figure 6.6 shows that in order to maintain the relative size of grade 2 during the expansion, the proportion of people being promoted has been increased. Fifteen years later, in part (III), when the peak of the bulge has moved to the right, the proportion of people being promoted has dropped, as the bulge provides increasing numbers of people at the promotion age. After thirty years, in part (IV), the proportion promoted has dropped still further because the bulge is now filling grade 2 and blocking promotion chances. When these people start to retire and the numbers required for promotion increases, the proportion promoted will increase again, as illustrated in part (V). Recruitment too will increase and the bulge and this whole process may repeat itself, although probably in a damped-down form.

## Solution to Exercise 6.5

The comparable career patterns require: $a = 0$, $b = 999$, and:

for grade 1 → 2: $p_2 = 0.101$, $w_1 = 0$, $w_2 = 0.170$

for grade 2 → 3: $p_2 = 0.046$, $w_1 = 0$, $w_2 = 0.124$

together with the use of assumption (b) of Section 6.4 in calculating the conditional promotion probability. Using these values gives the career pattern shown in Table 6.17. The equivalent Markov chain results are taken directly from the example in section 6.6

***Table 6.17***

| | *Grade 1 → 2* | | | | *Grade 2 → 3* | | | |
|---|---|---|---|---|---|---|---|---|
| | $P_u$ | $T_p$ | $P_u^*$ | $T_p^*$ | $P_u$ | $T_p$ | $P_u^*$ | $T_p^*$ |
| Methods of Section 6.4 | 0.373 | 3.19 | 1.000 | 7.72 | 0.271 | 5.38 | 1.000 | 18.54 |
| Markov chain method | 0.373 | 3.69 | 1.000 | 8.20 | 0.271 | 5.89 | 1.000 | 18.87 |

The two methods are therefore almost identical except for a difference of the order of 0.5 in the expected times. This is due to the Markov method assuming that people leave at the end of each interval and so contribute a full year to the duration. The method of Section 6.4 on the other hand assumes leavers leave on average at the mid-point of the interval and therefore contribute only 0.5 of their final year.

## Solution to Exercise 6.6

The appropriate zone-specific wastage rates based on those in Example 6.3 are set out in Table 6.18. Using these and the promotion rates and zones given in Example 6.3 we obtain the following promotion indices where the

*Table 6.18*

| | From grade 1 | | From grade 2 | |
|---|---|---|---|---|
| | $w_1$ | $w_2$ | $w_1$ | $w_2$ |
| Wastage $\frac{1}{3}$ down | 0.086 | 0.143 | 0.087 | 0.059 |
| Wastage as at present | 0.129 | 0.215 | 0.131 | 0.089 |
| Wastage $\frac{1}{3}$ up | 0.172 | 0.287 | 0.175 | 0.119 |

values relate respectively to wastage down by $\frac{1}{3}$/present wastage/wastage up by $\frac{1}{3}$:

| | $P_u$ | $T_p$ | $P_u^*$ | $T_p^*$ |
|---|---|---|---|---|
| $1 \to 2$ | 0.50/0.39/0.30 | 4.3/4.1/4.0 | 0.84/0.86/0.88 | 4.5/4.4/4.4 |
| $2 \to 3$ | 0.30/0.21/0.14 | 8.4/8.3/8.2 | 0.60/0.61/0.62 | 8.6/8.6/8.5 |

Before summarizing the effect of changes in the wastage let us be clear what assumptions lie behind these results. We are assuming that the zone-specific promotion rates remain constant so that the chance of promotion in this sense is unaltered. The main point therefore is that we are assuming a *push* situation. (The *numbers* promoted will depend on the numbers in the promotion zone. If wastage goes up these zone numbers will go down, and vice versa. In the context of a renewal system, however, we can expect the opposite effect: if wastage in the grade above increases, then so does promotion in order to fill the increased number of vacancies.) One way of interpreting the exercise is in terms of the errors in the calculated career pattern if the promotion rates were correct but the wastage rates were incorrectly estimated or specified.

The pattern of these results for this example implies that as wastage increases the proportion of entrants eventually promoted decreases. This is as we would expect since fewer survive to, and through, the promotion zone. By contrast, the chance of promotion for an individual who does not leave increases slightly with wastage. This is caused by a small increase in the calculated value of the conditional promotion rate $p^*$, and is a consequence of using assumption (c) of Section 6.4. The average experience of promotees drops as wastage increases, presumably because the distribution of people within the promotion zone shifts towards the lower durations, while the waiting time of the non-leaving individual decreases, again as a consequence of the small increase in $p^*$. Perhaps the most important result of this sensitivity analysis is that the magnitude of the changes in the level of wastage appears to have little effect except on the proportion of entrants promoted.

## Solution to Exercise 6.7

The career patterns implied by both the recent and the anticipated future (grade-specific) promotion rates ($p$) are given in the Table 6.19.

*Table 6.19*

| | *Recent value/anticipated future value* | | | | | |
|---|---|---|---|---|---|---|
| *Grade* | $p$ | $p_2$ | $P_u$ | $T_p$ | $P_u^*$ | $T_p^*$ |
| $1 \rightarrow 2$ | 0.10/0.05 | 0.47/0.19 | 0.29/0.19 | 5.0/5.7 | 0.99/0.77 | 5.3/6.3 |
| $2 \rightarrow 3$ | 0.05/0.04 | 0.18/0.13 | 0.30/0.26 | 9.9/10.4 | 0.93/0.86 | 10.9/11.5 |
| $1 \rightarrow 3$ | – | – | 0.09/0.05 | 14.9/16.1 | 0.92/0.66 | 16.2/17.8 |

Consider grade 1; note first that a grade-specific promotion rate of 10 per cent corresponds to a very much higher zone-specific rate of 47 per cent. The expected alteration in the zone rate is also much more dramatic than that of the grade rate with the latter changing from 10 to 5 per cent, while the former drops from 47 to 19 per cent. Recently 29 per cent of entrants could have expected to be promoted (see $P_u$) while in future this will be only 19 per cent. Similarly $P_u^*$ implies that an individual, who is prepared not to leave, recently had an almost certain chance of promotion (i.e. 99%), but that this will now be reduced to 77 per cent. The average experience of promotees ($T_p$) is not likely to change so much: we can expect an increase from 5.0 to 5.7 years. Correspondingly, the waiting time for the individual who does not leave ($T_p^*$) only increases from 5.3 to 6.3 years. The results for grade 2 are similar but less dramatic.

The overall feature of these results is that the proportion promoted and the chance of promotion appear to be more severely changed than the experience of promotees and the waiting time to promotion. This may be a consequence of the promotion zones and retirement points used. In an organization such as this these are unlikely to be clearly defined and in practice it would be advisable to check the sensitivity of the results by repeating the calculations with a range of values for $a$, $b$, and $k$. It may also be a consequence of the fact that the model assumes a rather sharp change in the promotion rate at the boundaries of the promotion zone.

## Solution to Exercise 6.8

(a) The career patterns using no promotion zone or retirement point constraints are shown in Table 6.20. As expected from the model, the times to promotion are small with these assumptions because promotion is assumed to be possible immediately on entry to the grade. The chance of promotion

*Table 6.20*

| | $P_u$ | $T_p$ | $P_u^*$ | $T_p^*$ |
|---|---|---|---|---|
| 1 → 2 | 0.57 | 2.4 | 1.00 | 3.8 |
| 2 → 3 | 0.50 | 2.8 | 1.00 | 5.2 |
| 3 → 4 | 0.60 | 3.5 | 1.00 | 5.5 |
| 4 → 5 | 0.67 | 6.2 | 1.00 | 9.0 |

for an individual who does not leave is also probably unrealistically high because no upper limit is assumed on time in the grade. The $T_p$, $T_p^*$, and $P_u^*$ quantities are therefore of doubtful value. From experience with other examples the proportion of entrants promoted, $P_u$, is probably more accurate.

(b) The career patterns derived from the Markov chain theory are similar to those of (a) above except that the expected times are 0.5 of a year greater. (This difference is due to the assumptions of the two methods: see solution to Exercise 6.5.)

*Table 6.21*

| | $p$ | $p_2$ | $P_u$ | $T_p$ | $P_u^*$ | $T_p^*$ |
|---|---|---|---|---|---|---|
| 1 → 2 | 0.20 | 0.73 | 0.51 | 3.6 | 1.00 | 3.7 |
| 2 → 3 | 0.15 | 0.64 | 0.42 | 4.8 | 1.00 | 4.9 |
| 3 → 4 | 0.15 | 0.64 | 0.51 | 5.8 | 1.00 | 6.0 |
| 4 → 5 | 0.10 | 0.54 | 0.61 | 9.2 | 1.00 | 9.3 |

(c) Note in Table 6.21 how the zone-specific promotion rate is much higher than the grade rate. As anticipated $P_u$ (the proportion of entrants promoted) is of the same order as those calculated in (a). For this model $T_p$ (the average experience of promotees) and $T_p^*$ (the waiting time for non-leavers) are very similar: we have observed this effect in other examples. Despite the promotion zones the proportion of non-leavers promoted is still 100 per cent. This last result, in particular, implies that the (grade-specific) promotion rates in this example are probably unacceptable. (The steady state promotion rates quoted in Exercise 4.2 for grade 1, grade 2, etc., are 9, 7.5, 5 and 2.5 per cent.)

**Solution to Exercise 6.9**

The Markov chain method assumes particular importance for non-hierarchical situations such as this example, since the other methods described for calculating career patterns can only be used for hierarchical systems. The transition matrix is

$$\mathbf{P} = \begin{bmatrix} 0.58 & 0.15 & 0.07 & 0 \\ 0 & 0.82 & 0 & 0.12 \\ 0 & 0.05 & 0.71 & 0.10 \\ 0 & 0.02 & 0 & 0.90 \end{bmatrix}$$

giving

$$\mathbf{D} = \begin{bmatrix} 2.38 & 2.55 & 0.57 & 3.63 \\ 0 & 6.41 & 0 & 7.69 \\ 0 & 1.55 & 3.45 & 5.31 \\ 0 & 1.28 & 0 & 11.54 \end{bmatrix} \qquad \mathbf{\Pi} = \begin{bmatrix} 1.00 & 0.40 & 0.17 & 0.31 \\ 0 & 1.00 & 0 & 0.67 \\ 0 & 0.24 & 1.00 & 0.46 \\ 0 & 0.20 & 0 & 1.00 \end{bmatrix}$$

As explained in Section 6.6, the elements $D_{ij}$ in the matrix **D** give the expected times spent in grade $j$ for an entrant to grade $i$, and also the off-diagonal elements are of little relevance because they result from treating people who do not reach a grade as spending zero time there. That is, these expected times are not conditional on reaching the grade. The matrix **Π** gives the proportion of entrants to the row grade who reach the column grade. Thus, for example, 31 per cent of grade 1 entrants and 46 per cent of grade 3 entrants reach the top grade. Furthermore, 20 per cent of people who reach the top grade are later transferrred to grade 2.

Conditioning on not leaving, the transition matrix becomes

$$\mathbf{P}^* = \begin{bmatrix} 0.725 & 0.188 & 0.088 & 0 \\ 0 & 0.872 & 0 & 0.128 \\ 0 & 0.058 & 0.826 & 0.116 \\ 0 & 0.022 & 0 & 0.978 \end{bmatrix}$$

If we re-arrange the grades in the order 1, 3, 2, 4, the matrix has the form

$$\mathbf{P}^* = \left[\begin{array}{cc|cc} 0.725 & 0.088 & 0.188 & 0 \\ 0 & 0.826 & 0.058 & 0.116 \\ \hline 0 & 0 & 0.872 & 0.128 \\ 0 & 0 & 0.022 & 0.978 \end{array}\right]$$

indicating that grades 2 and 4 form a closed set. That is, once entered, the set can never be left. As shown in Section 6.6 we can find the expected lengths of stay by finding the fundamental matrix of the top left-hand corner sub-matrix of $\mathbf{P}^*$ from which we find

$$\mathbf{D}^* = \begin{bmatrix} 3.64 & 1.84 \\ 0 & 5.75 \end{bmatrix}$$

**Solution to Exercise 6.10**

The matrices obtained are:

$$\mathbf{D} = \left[\begin{array}{ccc|ccc|c} 3.33 & 1.52 & 0.61 & 1.26 & 0.52 & 0.52 & 0.52 \\ 0 & 4.55 & 1.82 & 3.79 & 1.56 & 1.56 & 1.56 \\ 0 & 0 & 20.00 & 0 & 0 & 0 & 0 \\ \hline 0 & 0 & 0 & 8.33 & 3.43 & 3.43 & 3.43 \\ 0 & 0 & 0 & 0 & 5.88 & 5.88 & 5.88 \\ 0 & 0 & 0 & 0 & 0 & 20.00 & 0 \\ \hline 0 & 0 & 0 & 0 & 0 & 0 & 10.00 \end{array}\right]$$

$$\mathbf{\Pi} = \left[\begin{array}{ccc|ccc|c} 1.00 & 0.33 & 0.03 & 0.15 & 0.09 & 0.03 & 0.05 \\ 0 & 1.00 & 0.09 & 0.45 & 0.27 & 0.08 & 0.16 \\ 0 & 0 & 1.00 & 0 & 0 & 0 & 0 \\ \hline 0 & 0 & 0 & 1.00 & 0.58 & 0.17 & 0.34 \\ 0 & 0 & 0 & 0 & 1.00 & 0.29 & 0.59 \\ 0 & 0 & 0 & 0 & 0 & 1.00 & 0 \\ \hline 0 & 0 & 0 & 0 & 0 & 0 & 1.00 \end{array}\right]$$

$$\mathbf{D}^* = \left[\begin{array}{ccc|ccc|c} 8.00 & 7.52 & 3.31 & 11.28 & 5.45 & 5.56 & 5.56 \\ 0.00 & 7.52 & 3.31 & 11.28 & 5.45 & 5.56 & 5.56 \\ 0 & 0 & 20.00 & 0 & 0 & 0 & 0 \\ \hline 0 & 0 & 0 & 13.51 & 6.54 & 6.67 & 6.67 \\ 0 & 0 & 0 & 0 & 6.54 & 6.67 & 6.67 \\ 0 & 0 & 0 & 0 & 0 & 20.00 & 0 \\ \hline 0 & 0 & 0 & 0 & 0 & 0 & 10.00 \end{array}\right]$$

$$\Pi^* = \left[\begin{array}{ccc|ccc|c} 1.00 & 1.00 & 0.17 & 0.83 & 0.83 & 0.28 & 0.56 \\ 0.00 & 1.00 & 0.17 & 0.83 & 0.83 & 0.28 & 0.56 \\ 0 & 0 & 1.00 & 0 & 0 & 0 & 0 \\ \hline 0 & 0 & 0 & 1.00 & 1.00 & 0.33 & 0.67 \\ 0 & 0 & 0 & 0 & 1.00 & 0.33 & 0.67 \\ 0 & 0 & 0 & 0 & 0 & 1.00 & 0 \\ \hline 0 & 0 & 0 & 0 & 0 & 0 & 1.00 \end{array}\right]$$

Suppose the lower limits to the promotion zones are $a_1$ and $a_2$, then from the above matrices we obtain the results shown in Table 6.22.

***Table 6.22***

| | $1 \rightarrow 2$ | $3 \rightarrow 4$ |
|---|---|---|
| Proportion of entrants promoted | 0.15 | 0.34 |
| Average experience of promotees | $a_1 + 4.55$ | $a_2 + 5.88$ |
| Chance of promotion for a person who does not leave | 0.83 | 0.67 |
| Time before promotion for this person | $a_1 + 7.52$ | $a_2 + 6.54$ |

CHAPTER 7

# Stationarity and Control

## 7.1 INTRODUCTION

The models described in Chapters 4 and 5 were concerned with the changing pattern of stocks and flows as time passes. However, for planning purposes, the ideal situation would often be one in which such quantities did not change but, instead, maintained the same values through time. Of course, this is not the same as saying that there is no movement but simply that the net effect of all movements is one of 'no change'. We use the term 'stationarity' to describe such a set of circumstances. This idea plays an important role in manpower planning, where the objectives can often be stated in terms of achieving a stationary state in which the principal variables have stable and acceptable values.

Since the stocks and flows are interrelated it is clearly not possible to arbitrarily assign desired values to all the variables involved, though this is often mistakenly supposed to be the case by policy makers. For example, fixing some quantities like promotion rates and grade sizes will inevitably constrain the options open for recruitment. This being so, it is necessary to explore the nature of these constraints in order to map out what is possible. If one is aiming for a stationary state then it is obviously sensible to make this analysis under assumptions of stationarity.

The setting up of a stationary state as a desirable goal in manpower planning leads on naturally to the idea of control. This is concerned with how to choose values for those variables (e.g. promotion rates or recruitment parameters) which are under the manager's control in such a way that the aims of the organization are achieved. Indeed, this is an essential function of management. Our earlier emphasis on forecasting must not be allowed to distract attention from the importance of control. Forecasting in this context is seldom an end in itself but, rather, a first step towards control. By telling us what is likely to happen under a range of options it helps us to choose between them in the light of their consequences.

Approaching the problem from the forecasting direction involves an element of trial and error. A control theory on the other hand starts with the goal and works backwards to determine what strategies will lead to that goal.

A certain amount of confusion surrounds the terminology of the subject. The terms 'in equilibrium', 'steady state', and 'stable state' are sometimes used in much the same sense as we have written of a stationary state. In the demographic literature a distinction is drawn between a stable and a stationary population. Although we shall use the two terms as synonyms, the distinction which lies behind the demographic usage is an important one. Thus a system which is expanding at a constant rate cannot be stationary in an absolute sense since the stock numbers will continue to grow indefinitely. For many purposes, however, it is the relative sizes of the stocks that matter and it is perfectly possible for these to remain in constant proportion while the system, as a whole, grows. Elsewhere we have used the term quasi-stationary to describe this case, but here we prefer to include it under the general heading of stationarity.

We have already met the idea of stationarity when discussing the limiting behaviour of Markov and renewal models. In the former case, for example, we saw that if the parameters remained fixed the expected values of the stocks would eventually converge to limiting values. These limits also represented a stationary state because, once attained, they would be maintained. Likewise, we saw that the same was true of renewal models where the flows approached a steady state. In an average sense, therefore the Markov and renewal models are indistinguishable when the equilibrium has been reached (but only in an average sense because quantities which are fixed in one model will vary about a fixed average in the other). For planning purposes, however, it is the average values which are most important and so we can ignore the distinction between 'push' and 'pull' flows for the purposes of a stationary analysis and use whichever model is most convenient. Usually this will be the Markov model but in Sections 7.4 and 7.6 we shall adopt a renewal point of view.

The chapter falls into two parts. In the first (Sections 7.2 to 7.5) we shall show how to use the theory of the Markov chain model to make explicit the relationships between such things as promotion and expansion rates. The steady-state equations may be regarded as defining the constraints which the parameters must satisfy. As such, they determine the degree of freedom open to a manager who seeks to make changes in the system. For this reason it is useful to investigate how changes in one set of parameters require consequential changes in others if the steady state is to be maintained. In the second part of the chapter (Section 7.6) we shall examine, in detail, the relationship between age on promotion and the proportion of people promoted. This is most easily done using a continuous-time formulation which is closely related to the work of Chapter 6 on career prospects. Where appropriate, the results of this chapter will be expressed in the same notation to emphasize the continuity.

## 7.2 THE STATIONARY MARKOV MODEL

For a system of fixed size we saw in equation (4.7) that the expected structure satisfies

$$\bar{\mathbf{n}}(T) = \bar{\mathbf{n}}(T-1)\{\mathbf{P} + \mathbf{w}'\mathbf{r}\} \tag{7.1}$$

A steady-state structure $\mathbf{n}$ therefore satisfies

$$\mathbf{n} = \mathbf{n}\{\mathbf{P} + \mathbf{w}'\mathbf{r}\} \tag{7.2}$$

and these equations therefore determine the constraints which the parameters of the model must satisfy if $\mathbf{n}$ is to be maintained. Since the rate of expansion is often a critical factor in manpower planning work, we shall generalize the problem by incorporating an expansion rate and deal with the quasi-stationary behaviour of the stock proportions. To do this let $M(T)$ denote the change in size between $(T - 1)$ and $T$. If $\alpha$ is the rate of expansion, we have

$$M(T) = \bar{\mathbf{n}}(T - 1)\mathbf{1}'\alpha \tag{7.3}$$

where $\mathbf{1}$ is now a vector of 1's. Note that $\alpha$ will be negative if the system is contracting. Under the above conditions we may write

$$\bar{\mathbf{n}}(T)\mathbf{1}' = (1 + \alpha)\bar{\mathbf{n}}(T - 1)\mathbf{1}' \tag{7.4}$$

If we now introduce the vector of proportions

$$\mathbf{q}(T) = \bar{\mathbf{n}}(T)\{\bar{\mathbf{n}}(T)\mathbf{1}'\}^{-1} \tag{7.5}$$

and substitute (7.5) and (7.4) into (4.7) then

$$(1 + \alpha)\mathbf{q}(T) = \mathbf{q}(T - 1)\{\mathbf{P} + (\mathbf{w}' + \mathbf{1}'\alpha)\mathbf{r}\} \tag{7.6}$$

which has a stationary structure $\mathbf{q}$ satisfying

$$(1 + \alpha)\mathbf{q} = \mathbf{q}\{\mathbf{P} + (\mathbf{w}' + \mathbf{1}'\alpha)\mathbf{r}\} \tag{7.7}$$

which is equivalent to (7.2) when $\alpha = 0$. This equation will form the starting-point for our investigations.

## Some special cases

To illustrate the use which can be made of (7.7) let us consider a simple hierarchical system with three grades, promotion being into the next higher grade only. The notation is the same as in Chapter 4 except that we abbreviate the promotion rate for grade 1, which was formerly $p_{12}$, as $p_1$, and for grade 2, which was $p_{23}$, as $p_2$. In this case (7.7) is equivalent to

$$\left.\begin{aligned} &q_2p_2 - q_3(w_3 + \alpha) + (l + \alpha)r_3 = 0 \\ &q_1p_1 - q_2(w_2 + \alpha) - q_3(w_3 + \alpha) + (l + \alpha)(r_1 + r_2) = 0 \end{aligned}\right\} \tag{7.8}$$

where $l = q_1w_1 + q_2w_2 + q_3w_3$ is the overall loss rate. An alternative direct derivation of (7.8) can be obtained by equating the inflow and the outflow for each grade. There are two noteworthy things which are apparent from these equations (and, also, from (7.7)). Firstly, the wastage rates and the expansion rate are always linked together in the form $w_i + \alpha$ for $i = 1, 2, 3$.

This implies the similarity of effect of wastage and expansion. Thus the steady state can be maintained if a change in the expansion rate is exactly balanced by a change of the same magnitude but in the opposite direction, in the wastage rates. Secondly, the promotion rates are linear functions of the expansion and wastage rates. This implies that any change in either requires a linear change in the promotion rates if the structure is to be maintained.

*Example 7.1* To illustrate the use of the equations consider the system with parameter values as given in Figure 7.1.

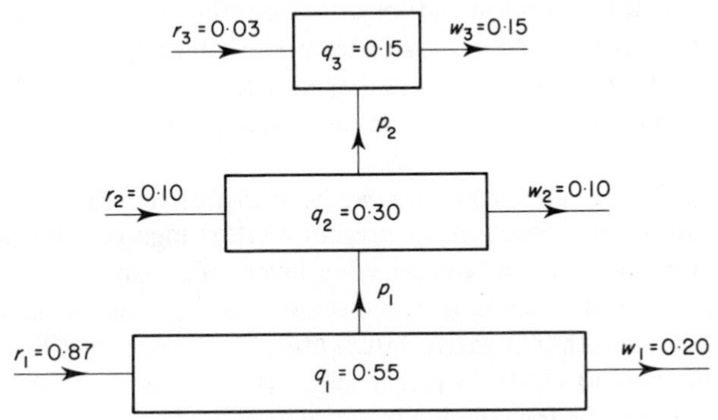


Figure 7.1 Example of a three-grade system

If the system is expanding at a rate $\alpha$ then it follows from (7.8) that

$$p_1 = 0.057 + 0.6\alpha$$

$$p_2 = 0.059 + 0.4\alpha.$$

For a constant size system the promotion rates would thus be 0.057 and 0.059 respectively. With 3 per cent expansion they would rise to 0.075 and 0.071, and at 10 per cent expansion, to 0.117 and 0.099. Thus we see that promotion rates corresponding to an expansion rate of 10 per cent are roughly twice those that would obtain in the constant size situation. In the case of contraction the rates would be similarly reduced, but note that a 10 per cent contraction would make $p_1$ negative which indicates that such a rate of contraction is incompatible with this steady state. ◀

*Example 7.2* Suppose now that the wastage **w** and the structure **q** are the same as in Example 7.1 but that the promotion rates are fixed with $p_1 = 0.11$, and $p_2 = 0.09$. We can then use (7.8) to explore the relationship between **r** and $\alpha$ in the steady state by writing the equations in the form

$$\left.\begin{aligned}(l+\alpha)r_3 &= n_3w_3 - n_2p_2 + n_3\alpha \\ (l+\alpha)r_2 &= n_2w_2 - n_1p_1 + n_2p_2 + n_2\alpha\end{aligned}\right\} \tag{7.9}$$

Notice that the $r_2$ and $r_3$ are not linear functions of $\alpha$. If $\alpha = 0.03$ (3 per cent expansion) then

$$\mathbf{r} = (0.971, 0.029, 0.000)$$

If, however, $\alpha = 0$ we find

$$\mathbf{r} = (1.077, -0.049, -0.028)$$

The negative elements in $\mathbf{r}$ show that there is no allocation of recruits which by itself will maintain that structure under conditions of zero growth. The problem of delineating the circumstances under which it is possible to maintain structures without violating the basic restraints of the model is an important one to which we shall return in the next section. ◀

A negative element appearing in $\mathbf{r}$ can be interpreted in terms of an outflow rather than an inflow, which in the present context means redundancy. Thus the last example leads us to wonder what levels of expansion or contraction are possible without causing any redundancies. For the parameter values used above it is clear that negative values must first appear in $\mathbf{r}$, for some value of $\alpha$ in the interval (0, 0.03). In fact it may easily be verified that $\alpha = 0.03$ is the smallest value of $\alpha$ for which $\mathbf{r}$ is non-negative.

Instead of the promotion rates being fixed, it might be possible for them to vary within certain limits. We could then ask what values of $\alpha$ are consistent with the promotion rates satisfying these conditions.

*Example 7.3* Suppose that $p_2$ is constrained to lie in the interval (0.04, 0.11), then since $p_2 = 0.059 + 0.4\alpha$,

$$0.04 \leqslant 0.059 + 0.4\alpha \leqslant 0.11$$

so that

$$\frac{0.04 - 0.059}{0.40} \leqslant \alpha \leqslant \frac{0.11 - 0.059}{0.40}$$

or

$$-0.0475 \leqslant \alpha \leqslant 0.128$$

For any $\alpha$ within this interval the corresponding value of $p_2$ will satisfy the required condition. Similarly, if $p_1$ lies in the interval (0.05, 0.15) then

$$-0.012 \leqslant \alpha \leqslant 0.155$$

If both restrictions are to be satisfied then

$$-0.012 \leqslant \alpha \leqslant 0.128$$ ◀

**Changes in the values of the parameters**

In a planning exercise we are often interested in the consequential effect of a change in one group of parameters on another if the stationary state is to be maintained. Because of their simple form it is easy to express (7.7) in terms of first differences of the parameters. Suppose, for example, that $\mathbf{P}$ is changed to $\mathbf{P} + \Delta\mathbf{P}$ where $\Delta\mathbf{P}\mathbf{1}' = \mathbf{0}$, and $\mathbf{r}$ to $\mathbf{r} + \Delta\mathbf{r}$ where $\Delta\mathbf{r}\mathbf{1}' = 0$, then it follows from (7.7) that

$$\mathbf{q}\Delta\mathbf{P} = -(l + \alpha)\Delta\mathbf{r} \tag{7.10}$$

*Example 7.4* If we return to the case of a simple three-grade hierarchy we find that (7.10) gives

$$\left.\begin{aligned} q_1\Delta p_1 &= (\alpha + l)\Delta r_1 = -(\alpha + l)(\Delta r_2 + \Delta r_3) \\ q_2\Delta p_2 &= -(\alpha + l)\Delta r_3 \end{aligned}\right\} \tag{7.11}$$

Thus, for example, an increase in the relative number recruited at levels 2 and 3 leads to a decrease in the promotion rate from level 1. Because of the linearity of the equations the effect of a given percentage change in $\mathbf{r}$ (or $\mathbf{P}$) produces the same change in $\mathbf{P}$ (or $\mathbf{r}$) whatever the initial level of the parameters. Suppose that $\Delta r_3 = 0.02$ (corresponding to an increase in $r_3$ from 3 to 5 per cent in Example 7.1) and $\Delta r_2 = 0.05$. If $l$ and $\alpha$ are as before, then

$\Delta p_2 = -0.013$, giving $p_2 = 0.058$: a decrease of about 20 per cent;

$\Delta p_1 = -0.025$, giving $p_1 = 0.050$: a decrease of about 33 per cent.

Furthermore, whatever the values of $p_1$ and $p_2$ they will always change by the amounts $\Delta p_1$ and $\Delta p_2$ if the above changes take place in $\mathbf{r}$. ◀

## 7.3 GENERAL TREATMENT OF THE PROBLEM OF MAINTAINING A STRUCTURE

Examples 7.1 and 7.2 were special cases of a more general problem concerned with maintaining a given grade structure. This is a particular aspect of the control of a Markov system which has been discussed in Bartholomew (1973a, 1975b, 1976a, 1977c). The problem of maintaining a structure $\mathbf{q}$ is essentially that of finding values of $\mathbf{P}$, $\mathbf{w}$, and $\mathbf{r}$ for which (7.7) holds for the given $\mathbf{q}$ and $\alpha$. In practice it is rarely possible or desirable to vary all the parameters over their full range. It is convenient to consider each group of parameters—concerned with promotion, wastage, and recruitment—separately. In practice it is rarely feasible to exert any direct control over $\mathbf{w}$ since this largely depends on individual choice. We shall therefore assume that $\mathbf{w}$ is not subject to control.

Promotion rates are usually subject to direct control, but even so there are often organizational and other constraints which prevent them from being varied too much from their current values. In any event, changes in the promotion rates cause changes in career prospects which it may be wise to minimize. Of course, if the grades are defined in terms of age or length of service, promotion corresponds to ageing and obviously cannot be controlled This leaves $\mathbf{r}$, the recruitment vector, as the set of parameters through which control can most easily be exercised. Even then there are often practical difficulties about recruiting relatively large numbers of people at the higher levels. We shall use the terms *recruitment control* and *promotion control* respectively to describe the problems of choosing $\mathbf{r}$ and $\mathbf{P}$ to maintain a desired structure.

Under recruitment control, with rate of expansion $\alpha$, the structure $\mathbf{q}$ will be maintained if an $\mathbf{r}$ can be found to satisfy

$$(1+\alpha)\mathbf{q} = \mathbf{qP} + (\mathbf{qw}' + \alpha)\mathbf{r} \tag{7.12}$$

If such an $\mathbf{r}$ exists it is obviously given by

$$\mathbf{r} = \{\mathbf{q}(\mathbf{I}-\mathbf{P}) + \alpha\mathbf{q}\}(\mathbf{qw}' + \alpha)^{-1} \tag{7.13}$$

The elements of this vector always sum to 1, but they will only be non-negative if

$$(1+\alpha)\mathbf{q} \geqslant \mathbf{qP} \tag{7.14}$$

Thus for given $\alpha$ and $\mathbf{P}$ (7.14) provides an easy arithmetical check on whether a particular structure $\mathbf{q}$ is maintainable.

*Example 7.5* Suppose that

$$\mathbf{P} = \begin{bmatrix} 0.71 & 0.14 & 0 & 0 \\ 0 & 0.80 & 0.05 & 0 \\ 0 & 0 & 0.85 & 0.05 \\ 0 & 0 & 0 & 0.85 \end{bmatrix}$$

and we wish to know whether $\mathbf{q} = (0.40, 0.25, 0.25, 0.10)$ is maintainable when the total size of the system is held constant. We find

$$\mathbf{qP} = (0.2840, 0.2560, 0.2250, 0.0975)$$

The elements of $\mathbf{q}$ all exceed those of $\mathbf{qP}$ except in the case of the second element, where $0.25 < 0.2560$. Hence since (7.14) is not satisfied this structure is therefore not maintainable. However, if the rate of expansion were $\alpha$ it would be maintainable if

$$(1+\alpha) \times 0.25 \geqslant 0.2560$$

that is, if $\alpha > 0.024$. ◀

When control has to be exercised by promotion, $\mathbf{w}$ and $\mathbf{r}$ are fixed and the problem is then to find a **P** satisfying (7.7). In the case of recruitment control the unknown was a vector in an equation which had a unique solution. With promotion control the unknown in (7.7) is the matrix **P** and, in general, there will be infinitely many solutions. From a planning point of view this is an advantage because it allows a measure of choice and so provides the opportunity to take other factors into account. To find whether there is an admissible solution (that is a **P** with non-negative elements and row sums not exceeding 1) we rewrite the equation in the form

$$\mathbf{qP} = (1 + \alpha)\mathbf{q} - \mathbf{q}(\mathbf{w}' + \mathbf{1}'\alpha)\mathbf{r} = (1 + \alpha)\mathbf{q} - (l + \alpha)\mathbf{r} \tag{7.15}$$

Since, for any admissible **P**, the vector **qP** must have non-negative elements it is obviously necessary that every element on the right-hand side of (7.15) be non-negative. This will be so if

$$(1 + \alpha)\mathbf{q} > \mathbf{q}\{\mathbf{w}'\mathbf{r} + \alpha\mathbf{1}'\mathbf{r}\} = (l + \alpha)\mathbf{r} \tag{7.16}$$

*Example 7.6* Suppose that $k = 4$, $\mathbf{w} = (0.15, 0.15, 0.10, 0.15)$, and $\mathbf{r} = (0.70, 0.30, 0, 0)$ and that we wish to know whether the structure $\mathbf{q} = (0.25, 0.25, 0.25, 0.25)$ can be maintained. For this structure $l = \mathbf{qw}' = 0.1375$ and (7.16) holds if

$$\mathbf{q} \geqslant \left(\frac{0.1375 + \alpha}{1 + \alpha}\right)(0.70, 0.30, 0, 0)$$

This will be so if:

(a) $0.25 \geqslant \left(\frac{0.1375 + \alpha}{1 + \alpha}\right)(0.70)$;

and

(b) $0.25 \geqslant \left(\frac{0.1375 + \alpha}{1 + \alpha}\right)(0.30)$

Inequality (a) is true whenever $\alpha < 0.3418$ and (b) always holds with $\alpha$ in that range. ◀

The calculations of Example 7.6 do not of themselves show that the **q** considered there is maintainable, because the argument leading to (7.16) only establishes its necessity. It is, however, also sufficient as the following argument shows. The condition (7.16) says that a structure **q** can only be maintained if the number of recruits coming into each grade does not exceed the desired number in that grade. Provided that this is true it will be possible to re-allocate individuals within the system in such a way that the structure is precisely maintained. This implies that an admissible **P** exists. For example, it may easily be verified that if $\alpha = 0$,

$$\mathbf{P} = \begin{bmatrix} 0.570 & 0.105 & 0.070 & 0.105 \\ 0.045 & 0.730 & 0.030 & 0.045 \\ 0 & 0 & 0.900 & 0 \\ 0 & 0 & 0 & 0.850 \end{bmatrix}$$

will maintain the structure $\mathbf{q} = (\frac{1}{4}, \frac{1}{4}, \frac{1}{4}, \frac{1}{4})$ with $\mathbf{w}$ and $\mathbf{r}$ as in Example 7.6. There are, of course other transition matrices which would achieve the same end (see Example 7.7).

In practice it is rather unusual to have the degree of flexibility implied by the analysis carried out above. In a simple hierarchy of grades, for example, demotion is often nonexistent and promotion occurs only by one step at a time. In that case all of the elements of $\mathbf{P}$ are zero apart from those on the main and super-diagonals. This means that there are only $(k - 1)$ independent parameters in $\mathbf{P}$ and hence there exists a unique solution to (7.15). It is given by

$$p_{i,i+1} = \frac{1}{q_i}\left\{(l + \alpha)\sum_{j=1}^{i} r_i - \sum_{j=1}^{i} q_j(w_j + \alpha)\right\} \quad (i = 1, 2, \ldots, k - 1) \qquad (7.17)$$

which is a generalization of (7.8). If, in addition, $0 \leqslant p_{i,i+1} \leqslant 1 - w_i$ $(i = 1, 2, \ldots, k - 1)$ then the structure $\mathbf{q}$ is maintainable and the promotion probabilities are given by (7.17). Note again that the wastage rates and the expansion rate are linked together in the form $(w_j + \alpha)$.

*Example 7.7* If we take the data of Example 7.6 but now require a solution for $\mathbf{P}$ to have the super-diagonal form we find from (7.17) that

$$p_{12} = 0.3475 + 2.55\alpha$$
$$p_{23} = 0.4750 + 3.50\alpha$$
$$p_{34} = 0.4500 + 3.25\alpha$$

With $\alpha = 0$ this gives a very different $\mathbf{P}$ to that previously obtained for this problem but it produces the same net result. As $\alpha$ increases a point will be reached at which one of the promotion rates $p_{i,i+1}$ will exceed $1 - w_i$. The structure will then cease to be maintainable. This will happen first in the case of $p_{23}$ when $\alpha = 0.107$; thus, provided that the expansion rate does not exceed 10.7 per cent a suitable strategy can be found. In practice, of course, the promotion rates required to maintain the structure might be unacceptably high. ◀

Promotion rates near zero or $1 - w_i$ are likely to be unrealistic in practice. Hence, more generally, one might wish to solve (7.15) subject to constraints of the form

$$a_{ij} \leqslant p_{ij} \leqslant b_{ij}$$

where in some cases the two bounds might be equal and, possibly, zero. The mathematical problem which this formulation poses is the same as finding a feasible solution to a linear program. Algorithms are available for this purpose but we shall not develop the subject here.

## 7.4 THE STATIONARY STATE FROM THE POINT OF VIEW OF A RENEWAL MODEL

We observed at the beginning of this chapter that in equilibrium the Markov and renewal models were equivalent, in an average sense. The stationary state of a system can therefore be studied from either point of view and so far we have found it more convenient to adopt the standpoint of the Markov model. However, when a system operates according to renewal principles with fixed grade sizes it is more natural, for example, to specify recruitment in terms of the proportion of vacancies to be filled by this means rather than by means of the proportion of recruits to be allocated to that grade. We shall illustrate how this may be done for the three-grade hierarchy with promotion into the next higher grade only.

In the Markov formulation the number recruited into the top grade is $N(l + \alpha)r_3$, where $N$ is the total size of the system. In a renewal model this would be expressed in terms of the proportion of vacancies filled from outside. The expected number of vacancies arising in grade 3 is $n_3(w_3 + \alpha)$ and if a proportion $(1 - s_3)$ are filled from outside the recruitment flow is $n_3(w_3 + \alpha)(1 - s_3)$. Equating this to the corresponding expression for the Markov model we have

$$N(l + \alpha)r_3 = n_3(w_3 + \alpha)(1 - s_3)$$

or

$$r_3 = q_3(w_3 + \alpha)(1 - s_3)/(l + \alpha) \tag{7.18}$$

This equation establishes the relationship between $r_3$ and $s_3$. Similarly, for the second grade

$$N(l + \alpha)r_2 = \{n_3(w_3 + \alpha)s_3 + n_2(w_2 + \alpha)\}(1 - s_2)$$

and so

$$r_2 = \{q_3(w_3 + \alpha)s_3 + q_2(w_2 + \alpha)\}(1 - s_2)/(l + \alpha) \tag{7.19}$$

Using these relationships. (7.8) may be written

$$\left.\begin{aligned} p_2 &= \{q_3(w_3 + \alpha)\}s_3/q_2 \\ p_1 &= \{q_2(w_2 + \alpha) + q_3(w_3 + \alpha)s_3\}s_2/q_1 \end{aligned}\right\} \tag{7.20}$$

*Example 7.8* These equations may be used to illustrate the effect on promotion chances of recruitment into higher grades. The values in Table 7.1 relate to a system with structure $\mathbf{q} = (0.55, 0.30, 0.15)$ and wastage $\mathbf{w} = (0.15, 0.05, 0.07)$.

***Table 7.1 The effect on promotion chances of recruitment at higher levels***

| *Proportion of vacancies in each grade filled by recruitment* | | | *Promotion rate from grade 1* | | *Promotion rate from grade 2* | |
|---|---|---|---|---|---|---|
| $1 - s_1$ | $1 - s_2$ | $1 - s_3$ | $\alpha = 0.00$ | $\alpha = 0.03$ | $\alpha = 0.00$ | $\alpha = 0.03$ |
| 1 | 0 | 0 | 0.046 | 0.071 | 0.035 | 0.050 |
| 1 | 0.3 | 0.2 | 0.030 | 0.046 | 0.028 | 0.040 |
| 1 | 0.5 | 0.4 | 0.019 | 0.030 | 0.021 | 0.030 |

Reading down the columns of promotion probabilities, we see how promotion chances decline as the amount of recruitment at higher levels increases. Even in the case of the modest rise from the first to the second row in which 30 per cent of vacancies in grade 2 and 20 per cent of vacancies in grade 3 are filled by recruitment, the promotion probability from grade 1 drops from 4.6 to 3 per cent. The position is similar for promotion from grade 2. The effects of expansion, $\alpha$, are also very clear. Roughly speaking, the probabilities when growth is 3 per cent are half as large again as in the no-growth situation. ◀

As pointed out in the last chapter, grade-specific promotion rates are not always the most informative measures for presenting information about career prospects. It is possible to go further and express the results of Table 7.1 in terms of the quantities $P_u$, $P_u^*$, $T_p$ and $T_p^*$. This is set as Exercise 7.4 at the end of the chapter.

## 7.5 THE PROBLEM OF ATTAINING A DESIRED GRADE STRUCTURE

The whole of the argument thus far has been in terms of averages. For example, when we showed above that the recruitment vector $\mathbf{r}$ given by (7.13) would maintain the structure $\mathbf{q}$, this was to be interpreted as a statement about the average numbers in the grades. In reality, the random variation in the numbers leaving and being promoted means that the structure cannot be maintained exactly. Thus at any given time the current structure $\mathbf{n}$ may differ from the desired structure $\mathbf{n}^*$, say. The problem then becomes one of choosing a policy to attain $\mathbf{n}^*$ with $\mathbf{n}$ as a starting-point.

Exactly the same problem arises whenever it is desired to change a structure for any other reason. There has been a considerable amount of research on this topic, some of which is referred to in the Complements section of this chapter (Section 7.7). Here we shall simply describe two kinds of strategy which may be adopted.

### Fixed strategies

The problem of making the transition from $\mathbf{n}$ to $\mathbf{n}^*$ can always be solved by using the strategy which maintains $\mathbf{n}^*$ (if such exists). To see this, consider the case of recruitment control and let $\mathbf{R}$ denote the vector of recruitment numbers. Then the value $\mathbf{R}^*$ of $\mathbf{R}$ which maintains $\mathbf{n}^*$ satisfies

$$\mathbf{n}^* = \mathbf{n}^*\mathbf{P} + \mathbf{R}^*$$

giving

$$\mathbf{R}^* = \mathbf{n}^*(\mathbf{I} - \mathbf{P})$$

It is clear that this sequence will converge to $\mathbf{n}^*$ and in that sense the fixed strategy will attain $\mathbf{n}^*$. The same argument may be used for control by promotion by taking as $\mathbf{P}^*$ any transition matrix which maintains $\mathbf{n}^*$.

Thus, whether we wish to maintain or attain a given structure the fixed strategy, if it exists, will always achieve the purpose. It will not do so precisely but, after a sufficiently long time, the average stock numbers will be close to those required.

### Adaptive strategies

The disadvantage of the fixed strategy is that it takes no account of the extent and nature of the difference between the current structure and the goal. Intuitively, one would wish to concentrate recruitment or promotion into categories where there were particularly large deficiencies and so bring the structure near to the goal more quickly. This idea has led to the investigation of strategies whose object is to bring the structure as near to the goal as possible at each step. In general, these strategies have shown themselves to be superior to the fixed strategies. We shall confine the discussion here to recruitment control. This is particularly important when the 'grades' are defined by age (or length of service) and the object is to control the age structure. In this case 'promotion' corresponds to ageing and is not, of course, subject to control.

The basic idea of an adaptive strategy is to choose the recruitment numbers so that the expected discrepancy between the goal and the structure attained in one step is as small as possible. This requires us to introduce a measure of distance between two structures. There are many possibilities but the two strategies which we shall describe are based on the distance function

$$D = \sum_{i=1}^{k} |n_i^* - \bar{n}_i(1)|^a \qquad a = 1 \text{ or } 2 \tag{7.21}$$

where $\bar{\mathbf{n}}(1)$ is the expected structure after one further step. Since $\bar{\mathbf{n}}(1)$ is a function of the recruitment numbers, the mathematical problem is to minimize $D$ with respect to the elements of the recruitment vector. Let $\mathbf{X}$ denote the vector of numbers in the grades after wastage, promotions, and transfers have taken place but before any recruits are introduced. If the decision on recruitment has to be made before the other flows are known then $\mathbf{X}$ will be equal to $\mathbf{nP}$, the predicted stocks remaining. If the decision can be postponed until after these flows have taken place the elements of $\mathbf{X}$ will be known from observation. The vector of differences $\mathbf{n}^* - \mathbf{X}$ may be referred to as the vector of 'recruitment needs'. A positive element denotes that a grade falls short of its target and a negative value that it contains an excess. We would not expect any sensible strategy to recruit into a grade which is already overstocked.

It is easy to see that $D$ is minimized, whatever the value of $a$, by choosing

$$\mathbf{R} = \max(\mathbf{0}, \mathbf{n}^* - \mathbf{X}) \tag{7.22}$$

where the maximum is determined element by element. In words, this means that we bring each deficient grade up to its target level and leave the others alone. This is very simple to specify but it has one big shortcoming which draws attention to a deficiency in our formulation of the problem. It is clear that if this strategy is implemented, the total stock will be too large on account of the existing excess in each of the grades which did not receive any recruits. Whether or not this is important will depend on the practical context but it seems desirable for many purposes to introduce the restraint that the total size must remain fixed. The minimization problem was solved in this case in Bartholomew (1973a) for $a = 1$ and 2 with the following results.

If $a = 1$ there is more than one allocation of recruits yielding the same minimum value of $D$. Any such allocation has the following properties:

(a) Recruits go only into those grades with a deficiency, i.e. those in which $n_i^* > X_i$.

(b) For such grades the number of recruits should not exceed the deficiency, i.e. $R_i \leqslant n_i^* - X_i$.

If $a = 2$ there is just one allocation which achieves the minimum. It can be shown that the number recruited into grade $i$ has to be

$$R_i = \begin{cases} n_i^* - X_i - b & \text{if } n_i^* - X_i > b \\ 0 & \text{otherwise} \end{cases} \tag{7.23}$$

where $b$ is a constant chosen such that

$$\sum_{i=1}^{k} R_i = \sum_{i=1}^{k} (n_i^* - X_i - b)$$

As before, recruits go only into those grades with a deficit although not all deficient grades will necessarily be allocated recruits. The constant $b$ can be found by trial and error or by using an iterative formula given in Bartholomew (1973a eq. 4.66). Notice that the strategy minimizing $D$ in the case of $a = 2$ also provides a minimum with $a = 1$.

*Example 7.9* Suppose that the desired structure is

$$\mathbf{n}^* = (50, 62, 40, 31, 17) \text{ where } \sum n_i^* = 200$$

and that the observed (or predicted) stocks after wastage and transfer but before recruitment are

$$\mathbf{X} = (45, 63, 45, 20, 11) \text{ where } \sum X_i = 184$$

The problem is to find how many people should be recruited into each grade so that the stocks shall be as near to those given by $\mathbf{n}^*$ as possible.

If we follow the strategy of (7.22) we make each deficient grade up to the target level. That is

$$\begin{aligned}\mathbf{R} &= (50\text{–}45, 0, 0, 31\text{–}20, 17\text{–}11)\\ &= (5, 0, 0, 11, 6)\end{aligned}$$

This yields the new stock vector

$$\mathbf{n}(1) = (50, 63, 45, 31, 17) \quad \text{where} \sum n_i(1) = 206$$

which is as close as possible to $\mathbf{n}^*$ but gives a total size of 206 against the original figure of 200.

Turning next to strategies which keep the total size fixed there will be $\sum_{i=1}^{5} (n_i^* - X_i) = 16$ recruits to be allocated to grades, 1, 4 and 5. The following are two vectors which satisfy the requirement for $a = 1$:

$$\mathbf{R} = (4, 0,0, 8, 4)$$
$$\mathbf{R} = (5, 0, 0, 5, 6)$$

In both cases the sum of the absolute deviations of the resulting stock vector from $\mathbf{n}^*$ is 12.

If we wish to minimize $D$ with $a = 2$ we must find $b$ such that $\sum_{i=1}^{5} R_i = 16$ where

$$R_1 = (50\text{–}45) - b, R_4 = (31\text{–}20) - b, R_5 = (17\text{–}11) - b$$

It is easy to see that this is achieved by taking $b = 2$ in which case

$$\mathbf{R} = (3, 0, 0, 9, 4)$$

This solution also has a sum of absolute deviations of 12 but it achieves the minimum sum of square deviations with a value of 38. These results are set out in Table 7.2. If it happens in the foregoing calculation that the value chosen for $b$ is such as to make one of the $R_i$'s negative then, in accordance with (7.23), that $R_i$ must be put equal to zero and a new $b$ determined, and so on until all the resulting $R_i$'s are positive. ◀

***Table 7.2*** ***Comparison of strategies over one year for attaining the goal*** $\mathbf{n}^* = (50, 62, 40, 31, 17)$ ***by recruitment control starting from*** $X = (45, 63, 45, 20, 11)$

| *Recruitment vector* $\mathbf{R}$ | *Stock vector attained* $\mathbf{n}(1)$ | $\Sigma n_i(1)$ | $\Sigma\lvert n_i^* - n_i(1)\rvert$ | $\Sigma(n_i^* - n_i(1))^2$ |
|---|---|---|---|---|
| 1. (5, 0, 0, 11, 6) | (50, 63, 45, 31, 17) | 206 | 6 | 26 |
| 2. (4, 0, 0, 8, 4) | (49, 63, 45, 28, 15) | 200 | 12 | 40 |
| 3. (5, 0, 0, 5, 6) | (50, 63, 45, 25, 17) | 200 | 12 | 62 |
| 4. (3, 0, 0, 9, 4) | (48, 63, 45, 29, 15) | 200 | 12 | 38 |

## 7.6 A STATIONARY ANALYSIS IN CONTINUOUS TIME

We have already found in the chapters on wastage and renewal models that it is sometimes more convenient to work in continuous time. This is especially true of those aspects of stationary behaviour which are described in terms of lengths of time rather than frequencies. We shall therefore adopt this approach to study the relationship between size, recruitment, wastage, and promotion. The ideas are similar to those touched upon in the Complements of Chapter 6 (Section 6.8) but the assumptions underlying our present analysis will be highly simplified. In spite of the lack of realism which this entails it does enable us to achieve a degree of insight into the interrelationships of variables in a manpower system without going beyond elementary mathematical arguments.

Consider first a system with a single grade. Suppose that all recruits enter at age $A$ at a constant rate of $R$ per unit time. Assume that all have the same survivor function $G(x)$ where $x$ denotes length of service. Suppose that everyone surviving to age $A + K$ retires, so that $G(x) = 0$ for $x > K$. The expected number with length of service in $(x, x + \delta x)$ will be $RG(x)\delta x$, that is, the number of those who entered a time $x$ ago who have survived to the present. (Note that the steady state will be reached at $x = K$ by which time all the original members will have left.) The total expected size is thus

$$N = R \int_0^K G(x)\,\mathrm{d}x$$

It was shown in equation (2.21) that

$$\int_0^K G(x)\,\mathrm{d}x = D_0 = \mu$$

where $\mu$ is the mean length of service, so we may write

$$N = R\mu \tag{7.24}$$

This is essentially the same result as we reached by renewal arguments in Section 5.2, where it would have appeared in the form

$$R = N\mu^{-1} \tag{7.25}$$

In Chapter 5, $N$ was supposed fixed and $R$ was arrived at as the steady-state recruitment level. Here we have taken $R$ as fixed and reached (7.24) as the steady-state formula for $N$. Under those circumstances the two approaches are equivalent in an average sense and the formula can be used to give any one variable in terms of the other two.

In this model $A + K$ is the retirement age and this implies that all entrants have the same maximum length of service. If age varies at entry there will be different maxima for different individuals. However, if most survival times are relatively small compared with the maximum possible length of service there may be little error in taking $K = \infty$.

Wastage is represented in (7.24) by $\mu$, the average completed length of service. It may sometimes be more informative to express that relationship in terms of the overall wastage rate for the system. This is easily done in the steady state since recruitment and wastage will then be in balance, giving

$$w = R/N$$

from which we deduce that $\mu = w^{-1}$, and so (7.24) becomes

$$N = Rw^{-1} \tag{7.26}$$

**The age of promotion and the proportion promoted**

Next we bring promotion into the picture. First we suppose that the promotion under consideration only occurs at one age $A + T$ and that the proportion of those reaching this age who are promoted is $\beta$. We also assume that propensity to leave is unaffected by promotion so that the promoted and the unpromoted have the same survivor function.

Under these assumptions the expected number of people with length of service in $(x, x + \delta x)$ (where $x > T$) *who have also been promoted* is

$$\beta R G(x) \delta x$$

Hence the proportion of people in the system who have been promoted is

$$\begin{aligned} \gamma &= \beta \int_T^K G(x)\,\mathrm{d}x \Big/ \int_0^K G(x)\,\mathrm{d}x \\ &= \beta \mu^{-1} \int_T^K G(x)\,\mathrm{d}x \\ &= \beta w \int_T^K G(x)\,\mathrm{d}x \end{aligned} \tag{7.27}$$

Thus for any given survivor function it is easy to deduce the relationship between the age of promotion ($T$), the chance of promotion ($\beta$) and the proportion in the system above the promotion point ($\gamma$). This relationship would help us to decide whether, for example, to delay promotion in order to increase the proportion ultimately promoted, or vice versa.

The foregoing theory is easily extended to cope with expansion and contraction. Thus if the recruitment rate at time $x$ is $R(x)$ then the expected number of people with length of service in $(x, x + \delta x)$, which is denoted by $S(x)\delta x$, will be given by

$$S(x)\delta x = R(-x)\, G(x) \delta x$$

and hence

$$\gamma = \beta \int_T^K S(x)\,\mathrm{d}x \Big/ \int_0^K S(x)\,\mathrm{d}x \tag{7.28}$$

*Example 7.10* The foregoing formulae will be illustrated using data for the university teaching profession in the United Kingdom. Recently, the grade structure of the profession was as shown in Table 7.3. For the sake of illustration, assume that there is a constant propensity to leave of $m(x) = 0.02$ for $x < K$. (This is an oversimplification because $m(x)$ is known to be relatively high early in the service period but it is roughly constant over the middle part of the career.) On this assumption

$$G(x) = \exp(-0.02x)$$

***Table 7.3 Grade structure of Example 7.10***

| *Rank* | *Proportion* |
|---|---|
| Professor | 0.12 |
| Reader/Senior Lecturer | 0.20 |
| Lecturer | 0.68 |
| *Total* | 1.00 |

Suppose, also, that the number of recruits has been increasing (or decreasing) at a constant rate $\alpha$ so that

$$R(x) = R\exp(\alpha x)$$

then

$$S(x) = R\exp[-(\alpha + 0.02)x]$$

and

$$\gamma = \beta \int_T^K \exp[-(\alpha + 0.02)x]\,\mathrm{d}x \bigg/ \int_0^K \exp[-(\alpha + 0.02)x]\,\mathrm{d}x$$

$$= \beta\{\exp[-(\alpha + 0.02)T] - \exp[-(\alpha + 0.02)K]\}/\{1 - \exp[-(\alpha + 0.02)K]\}$$

Let us take the average age of promotion to Reader/Senior Lecturer as 38 and to Professor as 42 and treat all promotions as though they occurred at those ages. Assume that everyone enters the profession at age 25 and leaves at 65 so that $T = 13$ and 17, and $K = 40$. If we take the proportions ($\gamma$) above each promotion point as given in the table above we can use the formula to calculate $\beta$ as a function of $\alpha$. Some numerical values are given in the Table 7.4.

***Table 7.4 Proportion of non-leavers promoted*** ($\beta$) ***as a function of the growth or contraction rate*** ($\alpha$)

| *Promotion* | $\alpha$ | −3% | 0% | +3% |
|---|---|---|---|---|
| to Professor | | 0.19 | 0.25 | 0.36 |
| to Reader/Senior Lecturer | | 0.45 | 0.55 | 0.72 |

These figures illustrate yet again how relatively small changes in the growth rate often have very substantial implications for the promotion opportunities which the system provides. The formula can also be used to determine how the age on promotion depends on $\alpha$ if $\gamma$ and $\beta$ are fixed, and an example is included in the exercises. ◀

## 7.7 COMPLEMENTS

Stationary models were among the first to be used in manpower planning, drawing heavily on the stable-population theory of demography. An elementary account is given by Rowntree in Smith (1970) and an application in a Civil Service context by Rowntree in Bartholomew and Smith (1971). A good example of an application involving steady-state assumptions in the field of educational planning is provided by Oliver and coworkers (1972) (reproduced in Bartholomew (1977d)). This example is very much in the spirit of Section 7.2 but involves a much more elaborate version of the model.

The general problem of maintaining a structure by recruitment or promotion control discussed in Section 7.2 is treated in greater depth in Bartholomew (1973a, Chapter 4). Here we have concentrated on simple arithmetical tests of whether or not a given structure can be maintained. This is sufficient for many practical purposes but for a full understanding of the situation it is desirable to know something about the class of structures which can be maintained. This leads to the idea of a maintainable set (or region) of structures which can be maintained by recruitment or promotion as the case may be. For recruitment control with simple hierarchical systems, the maintainable set is liable to consist of a rather small set of top-heavy structures. From this we may infer that this form of control is unlikely to be effective for the more typical bottom-heavy structure, especially where recruitment is concentrated at the bottom end.

All of the work described so far is deterministic and it is desirable to know how the strategies determined by the theory would perform in a real-life stochastic environment. Research to this end is described in Bartholomew (1975b and 1977c) where, among other things, it is shown how to calculate the probability that a structure will be maintained at any time. This leads on to the question of how best to get back to the desired structure once chance variation has blown the structure 'off course'. Some theory supported by simulation has provided answers to these questions. Perhaps the most important practical conclusion to emerge is that the average structure actually maintained may not be the one which the deterministic theory predicts. Such conclusions warn against the uncritical acceptance of the results of deterministic analyses.

Bartholomew (1973a and 1977c) also treats the problem of attaining a desired structure discussed here in Section 7.5. The adaptive strategies described are based on the idea of getting as close as possible to the goal in one step. This control problem can be set in a wider context by introducing a longer (possibly much longer) time horizon. The main technique involved is mathematical programming and this is capable of handling large-scale problems with many constraints. This approach is very much akin to the work of Charnes, Cooper, Niehaus, and others. A good introduction to their work is given in Charnes, Cooper, and Niehaus (1972).

The continuous-time theory in Section 7.6 is based upon Morgan (1971) and Forbes, Morgan, and Rowntree (1975). These authors give a fuller treatment which, among other things, allows wastage to be age (or length of service) dependent. Their more detailed analysis requires a computer for the calculations but their results suggest that the crude assumptions made here give quite a good overall picture. As with all stationary models these generalizations cannot encompass the transitory phase of the approach to equilibrium and hence they are of little use in building up a detailed projection of the future; for this we need dynamic models. Despite this obvious drawback, stationary models do give a great deal of insight into the underlying relationships in a manpower system. They typically require much less data than dynamic models and this often means that results can be obtained much more quickly than with other models without recourse to computers. This makes them particularly suited to preliminary analyses where they pave the way for the use of the more sophisticated models outlined in earlier chapters.

## 7.8 EXERCISES AND SOLUTIONS

### Exercise 7.1

The officers' system of one of the British womens' services (see also Example 4.6 and Exercise 4.5) had the values shown in Table 7.5.

***Table 7.5 The system of Exercise 7.1***

| *Grade* $i$ | *Structure* $q_i$ | *Wastage* $w_i$ | *Promotion* $p_i$ | *Recruitment* $r_i$ |
|---|---|---|---|---|
| 4 | 0.045 | 0.098 | — | 0 |
| 3 | 0.116 | 0.100 | 0.033 | 0 |
| 2 | 0.306 | 0.124 | 0.046 | 0 |
| 1 | 0.533 | 0.170 | 0.102 | 1 |

Given these values of $q$, $w$, and $r$, express the promotion rates in terms of an expansion rate $\alpha$. Hence deduce the maximum possible rates of expansion and contraction. In particular determine whether a contraction of 11 per cent is possible using only promotion control. What are the limits to expansion and contraction if each $p_i$ is not to fall below half, or exceed twice, its current value?

Calculate the recruitment distributions necessary for an expansion of 7.2 per cent and a contraction of 2.0 per cent. Check whether these rates of expansion and contraction are possible using the general condition (7.14).

Also use the general condition (7.16) to test whether an 11 per cent contraction is possible under promotion control. Why does this result differ from that previously obtained?

### Exercise 7.2

The present parameters for an age system are shown in Table 7.6. Test whether the present numbers in each class are maintainable by recruitment control, and calculate the age spread of recruits necessary to maintain the present structure at 5 per cent expansion.

*Table 7.6 The system of Exercise 7.2*

| Age class | 'Promotion' rate | Wastage rate, **w** | Recruitment distribution, **r** | Structure **q** |
|---|---|---|---|---|
| –25 | 0.36 | 0.16 | 0.73 | 0.219 |
| –30 | 0.11 | 0.18 | 0.27 | 0.388 |
| –35 | 0.17 | 0.13 | 0 | 0.149 |
| –45 | 0.07 | 0.07 | 0 | 0.178 |
| –55 | — | 0.23 | 0 | 0.066 |

**Exercise 7.3**

For the four-grade non-hierarchical system shown in Figure 4.1 test whether the present grade structure can be maintained under recruitment control and under promotion control assuming in each case that the other parameters of the system are as shown. For the promotion control situation assume the recruitment distribution is as given in Printout 4.4.

**Exercise 7.4**

Table 7.1 shows the effect of different amounts of recruitment on the grade-specific promotion rates. Express these rates in terms of the more relevant zone-specific rates using the approximation which is given by equation (6.10). Assume that for the bottom grade the boundaries of the promotion zone and the retirement point are at 5, 15, and 40 years of service respectively; and that for the second grade the corresponding values are 10, 20, and 35 years' service in the grade. Calculate also the promotion pattern in terms of the parameters $P_u$, $P_u^*$, $T_p$, and $T_p^*$. (Both the approximation and the promotion pattern parameters can be obtained using the program of Appendix B.)

**Exercise 7.5**

Suppose that the present situation for the three-grade hierarchical organization of Example 4.1 is as shown in Table 7.7. Calculate and compare the expected grade sizes over the next three years for the fixed (steady-state) recruitment policy, and for the adaptive recruitment strategy based on minimizing the squared distance function (7.21). (Round your calculated values at each stage to the nearest whole number.)

**Exercise 7.6**

Derive the relationship corresponding to (7.28) between time before promotion ($T$), chance of promotion ($\beta$), proportion promoted ($\gamma$) and rate of

*Table* 7.7

| Grade | Promotion rate | Wastage rate | Present size | Target size |
|---|---|---|---|---|
| 1 | 0.10 | 0.15 | 180 | 180 |
| 2 | 0.05 | 0.10 | 145 | 135 |
| 3 | — | 0.20 | 35 | 45 |
| *Totals* | | | 360 | 360 |

expansion ($\alpha$) when the survivor function has the mixed exponential form

$$G(x) = \theta\,e^{-\lambda_1 x} + (1 - \theta)\,e^{-\lambda_2 x}$$

Using the mixed exponential fitted in Example 3.4 (i.e. $\hat{\theta} = 0.276$, $\hat{\lambda}_1 = 0.118$, $\hat{\lambda}_2 = 0.882$) and assuming $K = \infty$, find how long people can expect to wait before promotion if $\beta = 0.5$, $\gamma = 0.1$, and the system remains at constant size. What happens to $T$ in the two situations in which $\beta = 1/3$, and $\alpha = -0.03$?

**Solution to Exercise 7.1**

For the parameter values given the value of $l$ is 0.14456. Final results are usually rounded to 3 decimal places since the data and the assumptions of the model rarely allow any greater accuracy. For intermediate results, however, it is usually necessary to carry more decimal places to ensure that the final results are accurate to at least 3 places. Using (7.17), or the generalization of (7.8), gives

$$p_1 = 0.1012 + 0.876\alpha$$

$$p_2 = 0.0523 + 0.526\alpha$$

$$p_3 = 0.0380 + 0.388\alpha$$

Thus for a constant size system the promotion rates would be $p_1 = 0.101$, $p_2 = 0.052$, and $p_3 = 0.038$. These are similar to the current values, implying that the system is probably stable: this is confirmed by the fact that the projected grade sizes for the first five years shown in Printout 4.7 of Example 4.6 are almost constant. The maximum limits to expansion and contraction under promotion control are obtained by setting each of the $p_i$ equal to their maximum and minimum possible values. These are $(1 - w_i)$ and zero. Thus for the maximum values we obtain $\alpha = 0.832$, $\alpha = 1.566$, and $\alpha = 2.222$. Since these must all be satisfied the maximum expansion is 83.2 per cent. Similarly, the minimum values given by $p_i = 0$ are $\alpha = -0.116$, $\alpha = -0.099$, and $\alpha = -0.098$. Hence the maximum contraction is 9.8 per cent. Thus an 11 per cent contraction is not possible under promotion control. These limits on $\alpha$ assume rather extreme values for $p_i$. In the more realistic case where each $p_i$ is limited to half or twice its present value we obtain, $(-0.057 \leqslant \alpha \leqslant 0.118)$, $(-0.055 \leqslant \alpha \leqslant 0.076)$, and $(-0.055 \leqslant \alpha \leqslant 0.072)$. Since all these must be satisfied the maximum possible rates of expansion and contraction with these assumptions are 7.2 and 5.5 per cent respectively.

The recruitment distribution corresponding to particular values of the other parameters can be obtained from the obvious generalization of equation (7.9). Thus for $\alpha = 0.072$ we require

$$\mathbf{r} = (0.847, 0.091, 0.045, 0.017)$$

and for $\alpha = -0.020$

$$\mathbf{r} = (1.078, -0.068, -0.008, -0.002)$$

Thus $\mathbf{q}$ can be maintained by means of recruitment control at an expansion of 7.2 per cent, but cannot be maintained at a contraction of 2 per cent. To use the general condition (7.14) to check these recruitment control results, we must first evaluate

$$\mathbf{qP} = (0.3880, 0.3083, 0.1146, 0.0444)$$

Then for $\alpha = 0.072$

$$(1+\alpha)\mathbf{q} = (0.5714, 0.3280, 0.1244, 0.0482) \geqslant \mathbf{qP}$$

implying $\mathbf{q}$ is maintainable. For $\alpha = -0.020$

$$(1+\alpha)\mathbf{q} = (0.5223, 0.2999, 0.1137, 0.0441) \ngeqslant \mathbf{qP}$$

showing that in this case $\mathbf{q}$ is not maintainable. Both these results agree with our previous findings.

For the general promotion control condition (7.16) we have

$$(l+\alpha)\mathbf{r} = (0.0346, 0, 0, 0)$$

when $\alpha = -0.110$ and

$$(1+\alpha)\mathbf{q} = (0.4744, 0.2723, 0.1032, 0.0401)$$

Hence $(1+\alpha)\mathbf{q} \geqslant (l+\alpha)\mathbf{r}$, which implies that an 11 per cent contraction is possible. The reason for this apparent disagreement with our previous result is that the general condition (7.16) assumes that any flow within the system is both possible and controllable. This is not the case in this system, which is a simple hierarchy. Our previous result is therefore the appropriate one.

**Solution to Exercise 7.2**

The first part of this exercise involves using condition (7.14) to test whether the present structure $\mathbf{q}$ is maintainable when $\alpha = 0$. Since

$$\mathbf{qP} = (0.105, 0.354, 0.147, 0.178, 0.063),$$

and

$$\mathbf{q} = (0.219, 0.388, 0.149, 0.178, 0.066)$$

the inequality $\mathbf{q} \geqslant \mathbf{qP}$ just holds since the fourth elements are equal for these vectors. Although this strictly implies that $\mathbf{q}$ is maintainable this is obviously a borderline case. Any small change in any of the parameters could alter the situation.

Recruitment policies by age are given by (7.13), from which with $\alpha = 0.05$ and $(l + \alpha) = \mathbf{qw}' + 0.05 = 0.2019$, we obtain the necessary age spread for recruits as

$$\mathbf{r} = (0.618, 0.263, 0.047, 0.042, 0.030)$$

**Solution to Exercise 7.3**

To investigate the maintainability of non-hierarchical systems we require the general tests of (7.14) and (7.16). Both these need $(1 + \alpha)\mathbf{q}$ which in this case is

$$(1 + 0.02)\mathbf{q} = (0.389, 0.291, 0.097, 0.243)$$

In testing for recruitment control we require

$$\mathbf{qP} = (0.221, 0.301, 0.094, 0.258)$$

Since $(1 + 0.02)\mathbf{q} \ngeqslant \mathbf{qP}$ the system cannot be maintained at 2 per cent expansion. For promotion control we require

$$(l + \alpha)\mathbf{r} = (0.1257 + 0.02)\mathbf{r} = (0.112, 0, 0.034, 0)$$

Since $(1 + 0.02)\mathbf{q} \geqslant (l + \alpha)\mathbf{r}$ the system can be maintained at an expansion of 2 per cent. Note, however, that because (7.16) is a condition for general systems this control may involve using flows within the system that are not marked as possible in Figure 4.1.

**Solution to Exercise 7.4**

Table 7.8 contains the approximate zone-specific rates. These are larger than the corresponding grade rates of Table 7.1, and give a better indication of the promotion chance of people eligible for promotion.

The promotion patterns corresponding to these zone rates are given in Tables 7.9 and 7.10 which show that the proportions eventually promoted are more sensitive to recruitment and expansion than are the times before

*Table 7.8*

| Proportions of vacancies in each grade filled by recruitment | | | Zone-specific promotion rates at various levels of expansion $\alpha$ | | | |
|---|---|---|---|---|---|---|
| | | | From grade 1 | | From grade 2 | |
| $1 - s_1$ | $1 - s_2$ | $1 - s_3$ | $\alpha = 0.00$ | $\alpha = 0.03$ | $\alpha = 0.00$ | $\alpha = 0.03$ |
| 1 | 0 | 0 | 0.145 | 0.241 | 0.133 | 0.199 |
| 1 | 0.3 | 0.2 | 0.090 | 0.145 | 0.102 | 0.152 |
| 1 | 0.5 | 0.4 | 0.056 | 0.090 | 0.076 | 0.110 |

*Table 7.9*

| Proportion of vacancies in each grade filled by recruitment | | | Promotion pattern from grade 2 | | | | | | | |
|---|---|---|---|---|---|---|---|---|---|---|
| | | | $\alpha = 0.00$ | | | | $\alpha = 0.03$ | | | |
| $1-s_1$ | $1-s_2$ | $1-s_3$ | $P_u$ | $T_p$ | $P_u^*$ | $T_p^*$ | $P_u$ | $T_p$ | $P_u^*$ | $T_p^*$ |
| 1 | 0 | 0 | 0.212 | 7.6 | 0.819 | 8.6 | 0.273 | 7.0 | 0.952 | 7.8 |
| 1 | 0.3 | 0.2 | 0.157 | 8.0 | 0.643 | 9.2 | 0.212 | 7.6 | 0.819 | 8.6 |
| 1 | 0.5 | 0.4 | 0.109 | 8.2 | 0.466 | 9.5 | 0.157 | 8.0 | 0.643 | 9.2 |

*Table 7.10*

| Proportion of vacancies in each grade filled by recruitment | | | Promotion pattern from grade 1 | | | | | | | |
|---|---|---|---|---|---|---|---|---|---|---|
| | | | $\alpha = 0.00$ | | | | $\alpha = 0.03$ | | | |
| $1-s_1$ | $1-s_2$ | $1-s_3$ | $P_u$ | $T_p$ | $P_u^*$ | $T_p^*$ | $P_u$ | $T_p$ | $P_u^*$ | $T_p^*$ |
| 1 | 0 | 0 | 0.378 | 13.4 | 0.768 | 13.8 | 0.449 | 12.9 | 0.898 | 13.2 |
| 1 | 0.3 | 0.2 | 0.325 | 13.7 | 0.669 | 14.1 | 0.406 | 13.3 | 0.817 | 13.6 |
| 1 | 0.5 | 0.4 | 0.268 | 13.9 | 0.558 | 14.3 | 0.339 | 13.6 | 0.696 | 14.0 |

promotion. This may be due to the times to promotion being constrained within promotion zones. Notice too that neither of these measures appears to be as sensitive as the (per annum) zone rates or the grade rates.

**Solution to Exercise 7.5**

As shown in Section 7.5 the fixed (steady-state) recruitment policy is given by

$$\mathbf{R}^* = \mathbf{n}^*(\mathbf{I} - \mathbf{P}) = (45, 2, 2)$$

Using this recruitment vector the expected grade sizes over the next three years can then be calculated from the basic equation

$$\mathbf{n}(T+1) = \mathbf{n}(T)\mathbf{P} + \mathbf{R}^* \qquad T = 0, 1, 2$$

These results are summarized in Table 7.11.

***Table 7.11***

| Time $T$ | Recruitment vector $\mathbf{R}^*$ and (total recruits) | | | | Stock vector attained $\mathbf{n}(T)$ and (total stocks) | | | | Distance $\sum\|n_i^* - n_i(T)\|$ | Distance $\sum(n_i^* - n_i(T))^2$ |
|---|---|---|---|---|---|---|---|---|---|---|
| 0 | — | — | — | — | 180 | 145 | 35 | (360) | 20 | 200 |
| 1 | 45 | 2 | 2 | (49) | 180 | 142 | 37 | (359) | 16 | 130 |
| 2 | 45 | 2 | 2 | (49) | 180 | 141 | 39 | (360) | 12 | 72 |
| 3 | 45 | 2 | 2 | (49) | 180 | 140 | 40 | (360) | 10 | 50 |
| | *Target* $\mathbf{n}^*$ = | | | | 180 | 135 | 45 | (360) | | |

The adaptive strategy for recruitment which minimizes the squared distance function is given by (7.23), and is illustrated in Example 7.9. For the first year the expected stocks remaining in each grade after wastage and transfer are

$$\mathbf{X} = \mathbf{n}(0)\mathbf{P} = (135, 141, 35) \qquad (\text{total} = 311)$$

Subtracting this vector from the required target stock vector gives the recruitment needs, i.e.

$$\mathbf{n}^* - \mathbf{X} = (45, -6, 10)$$

The recruitment distribution, whose total must equal 49 in order to keep the size of the system constant, is determined by (7.23). In other words we allocate these 49 people in such a way that no one is recruited into grades with an excess (grade 2 in this case), and the numbers recruited into each of the remaining grades are less than their needs by the same number, $b$. In this case the adjustment is $b = 3$ so that our recruitment vector is

$$\mathbf{R}(1) = (42, 0, 7)$$

The expected stock vector attained after the first year is therefore

$$\mathbf{n}(1) = \mathbf{X} + \mathbf{R}(1) = (177, 141, 42)$$

The recruitment and stock vectors for years 2 and 3, which can be calculated by repeating this procedure, are shown in the Table 7.12.

As we would expect the adaptive strategy performs better than the fixed strategy, and indeed actually reaches the target after three years. The main differences are that the adaptive policy recruits no one into grade 2, and more

*Table 7.12*

| Time $T$ | Recruitment vector $\mathbf{R}(T)$ and (total recruits) | Stock vector attained, $\mathbf{n}(T)$, and (total stocks) | Distance $\sum\lvert n_i^* - n_i(T)\rvert$ | Distance $\sum(n_i^* - n_i(T))^2$ |
|---|---|---|---|---|
| 0 | — — — — | 180 145 35 (360) | 20 | 200 |
| 1 | 42 0 7 (49) | 177 141 42 (360) | 12 | 54 |
| 2 | 45 0 10 (55)† | 178 138 44 (360) | 6 | 14 |
| 3 | 46 0 3 (49) | 180 135 45 (360) | 0 | 0 |
| | *Target,* $\mathbf{n}^*$ = | 180 135 45 (360) | | |

† For this year the adjustment $b = 1\frac{1}{2}$. The solution adopted for rounding this number was to put $b = 1$ for one grade and $b = 2$ for another, thus preserving the correct total number of recruits. We could therefore equally well use the recruitment vector (46, 0, 9). This makes no difference to either of the distances of $\mathbf{n}(2)$, and the target is still achieved at $T = 3$.

people into the top grade, particularly during the second year. This enables the second grade to run down and the top grade to increase as quickly as possible.

**Solution to Exercise 7.6**

With a mixed exponential CLS distribution and an expansion rate of $\alpha$ the stationary length of service distribution will be

$$S(x) = R(-x)\,G(x) = R\,\mathrm{e}^{-\alpha x}\,G(x)$$
$$= R\{\theta\,\mathrm{e}^{-(\lambda_1+\alpha)x} + (1-\theta)\,\mathrm{e}^{-(\lambda_2+\alpha)x}\}$$

Thus writing $\lambda' = (\lambda + \alpha)$, $a = \theta/\lambda_1'$, and $b = (1-\theta)/\lambda_2'$

$$\int_T^K S(x)\,\mathrm{d}x = R\{a(\mathrm{e}^{-\lambda_1' T} - \mathrm{e}^{-\lambda_1' K}) + b(\mathrm{e}^{-\lambda_2' T} - \mathrm{e}^{-\lambda_2' K})\},$$

and the required relationship based on (7.28) is

$$\gamma = \beta\left[\frac{a(\mathrm{e}^{-\lambda_1' T} - \mathrm{e}^{-\lambda_1' K}) + b(\mathrm{e}^{-\lambda_2' T} - \mathrm{e}^{-\lambda_2' K})}{a(1 - \mathrm{e}^{-\lambda_1' K}) + b(1 - \mathrm{e}^{-\lambda_2' K})}\right]$$

Note that when $K$ is large enough to assume $\mathrm{e}^{-\lambda_i' K} \approx 0$ this relationship simplifies to the following

$$\gamma = \beta(a\,\mathrm{e}^{-\lambda_1' T} + b\,\mathrm{e}^{-\lambda_2' T})/(a + b)$$

When $\beta = 0.5$, $\gamma = 0.1$, and $\alpha = 0$ we have

$$0.1 = 0.5(2.39\,\mathrm{e}^{-0.118T} + 0.821\,\mathrm{e}^{-0.882T})/3.160$$

This form can be further simplified by assuming $T$ is sufficiently large for $e^{-0.882T} \approx 0$. With this assumption we obtain

$$e^{-0.118T} = 0.270 \text{ giving } T = 11.1 \text{ years.}$$

(Checking the approximation:

$$0.821\, e^{-0.882 \times 11.1} = 0.00005 \approx 0 \text{ as assumed.})$$

When $\beta$ changes from $\frac{1}{2}$ to $\frac{1}{3}$ the same procedure and assumptions give

$$e^{-0.118T} = 0.405 \text{ implying } T = 7.7 \text{ years.}$$

(Checking: $0.821\, e^{-0.882 \times 7.7} = 0.00092 \approx 0$.)

When $\beta = 0.5$ as before but $\alpha = -0.03$, $a = 3.136$ and $b = 0.850$ so that using the same approximation

$$e^{-0.088T} = 0.254, \text{ giving } T = 15.6 \text{ years.}$$

(Checking: $0.850\, e^{-0.852T} \approx 0$.)

The first results shows that in this system with 10 per cent of people in a particular grade and above, and with 50 per cent of non-leavers eventually being promoted, entrants can expect to wait 11.1 years for promotion. The second implies that if the proportion of non-leavers being promoted was reduced to 33 per cent then the time before promotion would decrease to 7.7 years. Thus in this case fewer people would be promoted but they would spend longer in this grade and above. The final result shows that a contraction increases the time before promotion. In particular, a 3 per cent contraction increases the waiting time from 11.1 years to 15.6 years.

CHAPTER 8

# Statistical Techniques for Demand Forecasting

## 8.1 INTRODUCTION

Forecasting the demand for manpower is usually the most difficult part of balancing the manpower equation. There is no set of 'demand models' to set alongside the supply models which have formed the main subject of the book so far. This is hardly surprising when one considers the diversity of factors which enter into the typical demand forecasting situation. The factors which govern the availability of jobs depend on such things as the demand for a product or a service and the state of technology and method of working. In manufacturing industry, for example, account must be taken of changes in technology and organization, levels of taxation and earnings, and the performance of competitors at home and overseas. In local government, on the other hand, it will be the character of the area, the political hue of the party in power, government legislation, and the level of public spending which will largely influence staffing levels. It is clear even from these brief examples that any attempt at demand forecasting must be intimately related to the special features of the application in view. It is equally clear that it is not a purely statistical problem—indeed statistical considerations often play only a minor role. Nevertheless, enough experience has accumulated in recent years to identify some useful statistical techniques which form the subject of the present chapter.

The difference between supply and demand forecasting can, perhaps, best be appreciated by observing that supply models are concerned essentially with what is common to organizations—namely people—whereas demand forecasting deals with those aspects which distinguish one firm of industry from another—namely jobs. Thus, whereas we can expect to develop a methodology of supply forecasting which is widely applicable, the same cannot be said for demand forecasting. This fact is reflected in the structure and contents of the present chapter. Our object here will be to identify statistical methods which can sometimes be used to advantage in demand forecasting. For the most part these will be standard methods which are adequately expounded in existing textbooks so there will be no need to explain and illustrate them here. Our emphasis will be on identifying situations in which standard techniques are applicable, illustrating their use, as far as possible, by reference to published examples. In this chapter there are no exercises and there is no Complements section. The material which might have gone into the latter is included in the body of the chapter. Indeed, the whole chapter might be regarded as an extended Complements section linking a body of statistical theory to problems arising in manpower demand forecasting.

There is one point of terminology which often causes some confusion and this concerns the precise meaning of the word 'demand'. Some writers, especially those concerned with manpower planning at the national level, use the words 'need', 'requirement', and 'demand' in distinct senses. *Need* refers to the number of people required to provide some ideal level of service or output, *requirement* to the number technically necessary to produce a given level of output, and *demand* to the number of places which can be provided at current wage rates. It is not always easy to make this distinction and the use of these terms as just defined is far from universal. In this chapter we shall not, therefore, attempt to maintain distinctions between terms which in ordinary English are almost synonymous. However, it is important to be clear about what is meant by *demand* in any particular context and we shall try to follow this advice throughout the chapter. In general, demand will refer to the number of jobs which will become available which is roughly consistent with the use of 'demand' and 'requirement' described above.

## 8.2 DEMAND FORECASTING STRATEGIES

Before discussing individual methods we shall make a general review of two possible approaches to the problem of forecasting manpower demand.

The first is to base the forecast on past manning levels. At its simplest this means collecting data on the numbers employed at a sequence of times in the past and using some techniques of time series analysis to extrapolate the series into the future. This method has many shortcomings, on which we shall elaborate later, but a fundamental weakness is that it equates the attained levels of manning with required levels. Thus if, for example, there had been a chronic shortage of recruits in the past the attained levels would have failed to reach the levels really required. Hence any forecasts based on actual manning would tend to underestimate the number of jobs. An obvious way of avoiding the difficulty is to work with the time series of vacancies but, in practice, this is less likely to be available.

A second and usually more satisfactory approach is to forecast the workload and then convert this into manning levels. In some cases this may be a relatively straightforward matter. For example, in forecasting the demand for primary school teachers this approach would require us to forecast the number of children of primary age and then to convert this into a teacher requirement by applying a teacher/pupil ratio. In other cases the relationship between the work to be done and the size of the workforce required may be more complex. For example, a government department may handle flows of work of many different kinds and there may be a greater or lesser degree of substitutability of people between jobs. This approach calls on two branches of statistical theory; time series methods are needed to forecast the workload series, and correlation techniques are required to translate these forecasts into manpower demands.

This dichotomy of methods is not exhaustive and neither do all approaches fall neatly into one or other category. Sometimes the connection between workload and manning involves lagged relationships as, for example, when an increase in workload creates a recruiting and training demand which only manifests itself in increased numbers at work at a later date. In such a case the workload series is said to 'lead' the manning levels. If such a lagged relationship can be established, the current values of the workload series can be used to make short-term forecasts of the leading series. A similar situation arises in the development stage of a new technological process where the manning requirements are known before the process goes into full production.

The search for leading indicators is an important aspect of demand forecasting and the leading series need not be a direct measure of predicted workload. Often, for example, it will be related to planned expenditure as in some branches of the public service where the potential demand ('need') is almost unlimited. The statistical problem will then be that of converting forecasts of spending into the number of jobs which can be supported by that level of expenditure.

## 8.3 FORECASTING MANPOWER TRENDS

In essence the first approach involves looking for a relationship between the historical manpower series and time, and then extrapolating the trend into the future. A simple example is given by Rowntree and Stewart in Smith (1976) and quoted in modified form in Bartholomew, Hopes, and Smith (1976). They plotted the numbers in the Principal grade of the Civil Service for each year from 1963 to 1973 and noted that there had been growth at a constant rate of about 6 per cent per annum. Accordingly they fitted an exponential growth curve by least squares giving

$$S_t = \exp\{7.670 + 0.0163t\} \tag{8.1}$$

where $S_t$ is the number of Principals in year $t$ measured from 1963. Table 8.1 gives the observed and fitted numbers in each year together with some forecast values based on equation (8.1). The fit was very close between 1963 and 1973 and this provided some justification for believing that the relationship would continue into the short-term future.

*Table 8.1* ***Actual and fitted numbers of Civil Service Principals assuming growth according to equation (8.1)***

| *Year* | 1963 | 1964 | 1965 | 1966 | 1967 | 1698 | 1969 | 1970 | 1971 |
|---|---|---|---|---|---|---|---|---|---|
| Actual | 2245 | 2329 | 2374 | 2479 | 2682 | 2943 | 3038 | 3251 | 3451 |
| Fitted | 2142 | 2278 | 2423 | 2576 | 2740 | 2914 | 3098 | 3295 | 3504 |

| *Year* | 1972 | 1973 | 1974 | 1975 | 1980 | 1985 | 1990 | 2000 |
|---|---|---|---|---|---|---|---|---|
| Actual | 3732 | 4105 | 4279 | 4468 | — | — | — | — |
| Fitted | 3726 | 3963 | 4214 | 4481 | 6094 | 8288 | 11,272 | 20,848 |

This simple example also illustrates the hazards of a purely statistical approach to forecasting uninformed by practical knowledge of the situation to which the figures related. Equation (8.1) implies, as we have noted, an annual growth rate of about 6 per cent per annum. If this were to continue for more than a few years the number of Principals would be impossibly high and this, in turn, would imply much larger numbers of supporting staff. Any forecasts using equations like (8.1) need to be supplemented by non-statistical information about how long the system can sustain growth at the present rate. The extrapolation of simple trends should always be approached with caution and never used uncritically for long-term forecasting. In fact, the cut-backs in public expenditure which followed this period have arrested the rate of increase so that the forecasts given in the table would have to be modified.

Apart from exponential and linear trends, the pattern of variation in time most likely to be encountered is a cyclical one. This may involve a regular cycle arising from seasonal variations in the demand for a product or a more irregular one associated, for example, with a trade cycle. The former can be dealt with by including trigonometric terms in the function to be fitted to the series. In organizations which do not practice 'hiring and firing' policies to meet seasonal and other fluctuations, it will usually be the average workload which will determine the manning levels. In such cases it will usually be desirable to eliminate the seasonal variations before attempting to identify longer-term trends. This may be done most simply by working with annual figures, but methods exist for 'de-seasonalizing' series. These and other methods will be found in standard texts on time series analysis such as Kendall (1976).

The second main approach to the extrapolation of time series depends on establishing a relationship between successive members of a series. A forecasting equation then expresses future values as a function of past values rather than as a function of time. One such method is based on exponentially weighted averages deriving from the work of Holt (1957), Winters (1960), and Brown (1963). In its simplest form this method forecasts the next member of a series on the basis of the current value and the value forecast for that value. Thus let $\hat{y}_t(1)$ denote the forecast for time $t + 1$ made at time $t$ and $y_t$ be the actual value at $t$. Then the forecasting equation is

$$\begin{aligned} \hat{y}_t(1) &= \hat{y}_{t-1}(1) + (1 - \lambda)\{y_t - \hat{y}_{t-1}(1)\}, \quad |\lambda| < 1 \\ &= (1 - \lambda)y_t + \lambda\hat{y}_{t-1}(1) \end{aligned} \tag{8.2a}$$

The first version above shows that each forecast is updated by some fraction of the error of the previous forecast; the second shows it to be a weighted average of the current value and its forecast made at the previous point in time. Yet another representation of (8.2a) is

$$\hat{y}_t(1) = (1 - \lambda) \sum_{i=1}^{\infty} y_{t-i}\lambda^i \tag{8.2b}$$

where the upper limit of $\infty$ is used to denote the entire length of series. In this version $\hat{y}_t(1)$ is seen to be a weighted average of all previous values with weights decreasing exponentially. This accords with the commonsense view that the most recent values should be given the most weight. The main problem in practice is to choose a suitable value for $\lambda$. The smaller $\lambda$, the greater is the weight given to the most recent values. Cox (1961) has shown that if $\{y_t\}$ is a Markov series then the optimum value of $\lambda$ as judged by the expected mean-square error of prediction is not very critical. In practice it is desirable to proceed empirically by examining the performance of different values of $\lambda$ had they been used on the historical series. Cox (1961) also proposed a modified predictor which was further generalized by Blight (1968).

The method can be extended to deal with the case when a trend or seasonal component is expected to be present. Young and Vassiliou (1974) used such a method in a manpower context based on the equation.

$$\hat{y}_t(1) = \hat{y}_{t-1}(1) + (1 - \lambda)\{y_t - \hat{y}_{t-1}(1)\} + r_{t-1} \tag{8.3a}$$

where $r_{t-1}$ represents a trend term which is computed recursively from

$$r_t = r_{t-1} + \mu\{y_t - \hat{y}_{t-1}(1)\} \qquad 0 < \mu < 1 \tag{8.3b}$$

The forecast value for time $t + h$, denoted by $\hat{y}_t(h)$, is then

$$\hat{y}_t(h) = \hat{y}_t(1) + (h - 1)r_t \tag{8.3c}$$

In their application, Young and Vassiliou took $\lambda = 0.8$ and $\mu = 0.02$ and obtained satisfactory results when using the method on the historical data available. Applications of these methods have been made in sales forecasting (see, for example, Harrison (1965, 1967) and Ward (1963)) but are relatively untried as a means of projecting manpower series. However, in so far as manpower demand in manufacturing and sales organizations depends on sales, these applications are relevant to the subject matter of Section 8.4 which is concerned with predicting workload series.

Autoregressive models are similar in that they lead to prediction equations in which the forecast value is expressed as a linear function of previous values. For example, a second-order model would assume the relationship

$$y_t + \alpha_1 y_{t-1} + \alpha_2 y_{t-2} = e_t \tag{8.4}$$

where $e_t$ is a sequence of independent and identically distributed random variables. The model may also be thought of as the linear regression of $y_t$ on $y_{t-1}$ and $y_{t-2}$, and this is more natural for prediction purposes. If such a model is fitted to a historical series the regression equation can be used to predict further terms in the series.

The culmination of this kind of approach is to be found in the autoregressive moving-average models of Box and Jenkins (1970) which have found extensive applications in time series forecasting. One of the main practical difficulties encountered in applying these methods to manpower forecasting is that the series available are usually too short. However, an application due to Cameron and Nash (1974) applies the technique to some series consisting of 43 members. In their work each series represented a workload rather than a manning level and so the example properly belongs to the following section. We mention it here because the method is equally applicable to the prediction of manning series.

As personnel data bases become widespread, longer series will become available and it should then be possible to apply time series methods to greater effect. However, in practice there will often be other data which are relevant apart from past manning levels. In particular, information about other series which are correlated with manning levels as well as information on workloads as functions of market conditions and changing technology will often be available. In the following sections we shall describe some techniques which are capable of utilizing such information to improve forecasts.

## 8.4 FORECASTING WORKLOADS

In some cases, the projection of workloads and of manning levels is, from a statistical point of view, essentially the same problem. If 'work' consists of a single type of activity then the total output is a measure of the amount of work done and future levels can be predicted from a historical record by the time series methods discussed in the last section. Similarly, if output is geared to sales then a sales forecast is also a prediction of the output required; an example of Box–Jenkins methods applied in this field is given by Chatfield and Prothero (1973). In areas such as the provision of education and health care the workload is often, approximately at least, a simple function of the age structure of the population. Population forecasts obtained by demographic techniques can thus be converted into workloads in a straightforward manner.

A more difficult, and more common, situation arises when the same individual has to perform a variety of tasks, the relative magnitudes of which vary with time. One way to proceed is to predict each workload series separately, in which case the main problem is deferred to the following stage when the workloads have to be converted into a manning requirement. However, is it likely that the various workload series will be correlated among themselves, and then it may be possible to reduce the dimensionality of the problem from the outset. To take an extreme case, if the levels of each kind of work move precisely in step it would suffice to take just one series—or an average of them all—for projection purposes. More generally, there may be sufficient correlation between the series to enable us to construct a small number of indices which, together, summarize most of the information.

Formally, the position is as follows. We have $k$ types of job being done and data are assumed to be available giving the magnitudes of each in the past. Let subscripts be used to distinguish types of work and let the series be denoted by

$$\begin{array}{lll} x_1(t), & x_1(t-1), & x_1(t-2), \ldots \\ x_2(t), & x_2(t-1), & x_2(t-2), \ldots \\ \vdots & \vdots & \vdots \\ x_k(t), & x_k(t-1) & x_k(t-2), \ldots \end{array}$$

The object is to forecast the manning level required to service the workloads arising from future values of these series. The first possibility noted above involves projecting each of these series and then converting the projections into manning requirements. The second approach is to replace the $k$ series by a smaller numer, $p$ say, which we write

$$\begin{array}{lll} y_1(t), & y_1(t-1), & y_1(t-2), \\ \vdots & \vdots & \vdots \\ y_p(t), & y_p(t-1), & y_p(t-2), \ldots \end{array}$$

where the $y$'s are functions of the $x$'s chosen in such a way as to preserve as much of the information in the original data set as possible. If such a transformation can be effected with $p$ much smaller than $k$, the forecasting exercise will be simpler because of the smaller number of series involved though this advantage must be offset against the effort of transforming the $x$'s into the $y$'s. The main advantage claimed for the method, however, is that it simplifies the problem of converting the workload projection to a manning requirement. One reason for this is that the transformation can be done in such a way that the $y$-series are uncorrelated.

There are two statistical techniques in common use whose purpose is to reduce the dimensionality of a problem in the manner described above (though they have to be applied with care to time series data). The first is a purely descriptive technique known as principal component analysis. Its object is to find linear transformations from the $x$'s to $y$'s of the form

$$y_i = \sum_{j=1}^{k} a_{ij} x_j \qquad (i = 1, 2, \ldots, k)$$

having the properties:

(a) that the $y$'s are uncorrelated;

(b) that as much as possible of the total variation in the $y$'s is accounted for by $y_1$, as much of what remains by $y_2$, and so on.

If as a result of the transformation, the first few $y$'s account for almost all of the variation, the remainder can be discarded and the objective set out above has been achieved. We would thus convert our $x$-series into a few $y$-series determined by the principal components and then carry out the projections. Details of the methodology of principal component analysis will be found, for example, in Kendall (1975), and Kendall and Stuart (1976). Computer packages for multivariate analysis usually contain routines for principal component analysis.

The second technique is called factor analysis and is described, for example, in Lawley and Maxwell (1971), Kendall (1975), and Harman (1967). The aims behind the technique are similar to those of principal component analysis in that the object is to reduce the dimensionality of the data. However, factor analysis starts from a statistical model of the situation and the object is then to fit the model to the data. Two applications to manpower forecasting have been made, of which that by Rowntree and Stewart in Smith (1976) provides a simple introduction to the basic ideas. This application examined the four kinds of work carried out by the Land Registry of the British Civil Service concerned with recording changes of ownership of land. Time series on each kind of work were available which led to the following correlation matrix:

| | $x_1$ | $x_2$ | $x_3$ | $x_4$ |
|---|---|---|---|---|
| $x_1$ | 1 | 0.911 | 0.917 | 0.823 |
| $x_2$ | | 1 | 0.980 | 0.923 |
| $x_3$ | | | 1 | 0.889 |
| $x_4$ | | | | 1 |

It is immediately clear from the matrix that there is a high degree of intercorrelation between the variables. It is thus plausible to suppose that all of them change in response to some unobserved variable like the 'level of economic activity'. If we denote this variable by $y$, one might suppose that the data were generated by a model of the form

$$x_i = a_i + b_i y + e_i \qquad (i = 1, 2, 3, 4) \tag{8.5}$$

where the $a$'s and $b$'s are unknown constants and the $e$'s are independent and identically distributed random variables with zero mean and unknown variance. If the $e$'s were all zero, all of the correlation coefficients in the above matrix would be 1. It is the fact that they are all large and near to 1 which suggests to us that a model like (8.5) might be appropriate. Further analysis shows that the model does fit the data and hence it is possible to replace the four series by a single series.

Equation (8.5) is described as a one-factor model because the four $x$-variables are expressed in terms of a single underlying (or latent) variable $y$. If there is more than one factor in the model it turns out that there are too many parameters for the model to be uniquely determined by the data. In fact, any orthogonal transformation of the factors will fit the data equally well. Factor analysis, therefore, tells us how well a model of given dimension fits the data but it does not uniquely determined the 'factors'. Factor analysts have proposed various 'rotations' (i.e. linear transformations of the $y$'s) for resolving the indeterminacy in ways which seem meaningful in the practical context.

A larger-scale application was made by Cameron and Nash (1974) where there were nine workload series. Their method was to first carry out a principal component analysis on the standardized workload series (i.e. transformed to make their variances 1). They found that the first three principal components accounted for 86 per cent of the total variation and, on this basis, decided that a three-factor model was appropriate according to which the workloads could be represented as

$$x_i = a_i + b_i y_1 + c_i y_2 + d_i y_3 + e_i \qquad (i = 1, 2, \ldots, 9). \tag{8.6}$$

They fitted this model and used a particular rotation (varimax) to arrive at the factors which were then projected by time series methods.

### Predicting diffusion

The foregoing methods assumed that the character of the work will not change significantly during the forecast period. This will clearly not be the case if a major technological change intervenes. Such changes and their effects can often be foreseen because of the time-lag between the introduction of an innovation and its adoption throughout an industry. An essential ingredient of the forecasting exercise must therefore be to predict the rate of diffusion of the new technology, since this determines the workload and hence the manning requirements. One example of such a study is Purkiss (1974) who was concerned with the rate at which a new steelmaking process would spread throughout the industry. Similar examples arise with the diffusion of such things as colour television, and new brands of consumer goods generally.

There is an extensive literature, both empirical and theoretical, on the diffusion of innovations. Some examples of the former will be found in Hagerstrand (1967) and Rogers (1962), while the theory is treated in Bartholomew (1973a) where further references will be found. Diffusion over time is often expressed by plotting the number of 'adoptors' as a function of time. In practice such curves are often approximated by one of two mathematical forms. To oversimplify somewhat, if information is mainly spread from one firm to another by personal contact or observation, then the growth curve is likely to be shaped like an ogive with growth starting slowly and then rising more rapidly, and finally tailing off in the later stages. The logistic growth curve often describes this shape reasonably well and has the added advantage of arising from a simple stochastic model of the diffusion process. The equation for a population of size $N$ is

$$n(T) = N(1 - e^{-aT})/(b + ce^{-aT}) \tag{8.7}$$

where $n(T)$ is the number of adopters at time $T$ and $a$, $b$, and $c$, are positive constants with $b > 1$. Such a curve rises from zero at $T = 0$ to an asymptote of $N/b$ as $T \to \infty$. The practical statistical problem is to estimate the parameters of this curve. This is a relatively simple matter if values of $n(T)$ are available for a range of $T$ spanning the whole curve. Then the curve can be fitted by a technique such as maximum likelihood—see Oliver (1966). In practice this situation will only arise where data on the diffusion of a similar innovation are already available and where one is prepared to assume that a new one will spread in the same manner. More usually, a limited amount of data will be available near the beginning of the curve and the object will then be to fit the curve and extrapolate it to those values of $T$ yet to occur. This was the course followed by Purkiss. If the time period over which observations on $n(T)$ are available is short, so that direct empirical evidence is available only on the concave part of the curve, it will be difficult to get a precise estimate of $b$ (which determines the saturation level). In that case a simplifying assumption is to take $b = 1$ which implies that all members of the population will eventually adopt the innovation. This may be a very reasonable assumption in, for example, a nationalized industry where all the 'firms' are under a common management. In such cases, however, the pattern of diffusion is more likely to follow the second form discussed below.

The second pattern is likely to be encountered when knowledge—and the adoption—of the innovation is derived from a common source such as a trade publication or advertising. Under certain simple assumptions the growth curve in this case will have an inverted exponential form with

$$n(T) = N(1 - e^{-aT})/b \tag{8.8}$$

This is, of course, a special case of (8.7) obtained by putting $c = 0$. When this growth pattern is followed the rate of growth declines steadily until the saturation level is reached. Similar remarks to those made above about fitting the curve apply here also, but if $b = 1$ (or is known) a simple graphical method is available. We may then write (8.8) in the form

$$T = -\frac{1}{a}\ln\left\{1 - \frac{n(T)}{N}\right\} \tag{8.9}$$

Thus if $T$ is plotted against $-\ln\{1 - n(T)/N\}$ the curve should be a straight line through the origin with slope $a^{-1}$. If the first part of the curve has been observed the remainder can then be predicted by extending the fitted straight line.

## 8.5 RELATING WORKLOADS TO MANNING LEVELS

### Productivity factors

A common and simple way of converting a workload into a manning requirement is to divide it by a productivity factor. The rationale behind this is that if it takes $N$ people to produce an amount $W$ of output, then it will take $aN$ to produce an amount $aW (a > 0)$. Thus let us define a productivity index, $P$, by

$$P = \frac{\text{amount of output}}{\text{number of people used to produce it}}$$

A forecasting equation can then be obtained by re-writing and re-interpreting this expression as follows:

$$\text{number required in future} = \text{predicted workload}/P \tag{8.10}$$

If the workload can be predicted by methods such as those discussed in the last section and if $P$ can be estimated from historical data or technical information, then the manning requirements follow at once from (8.10).

Although widely used, this method is open to serious objections as the assumptions on which it rests are highly questionable. Productivity, for example, usually depends on a multitude of factors and this makes it unreasonable to treat it as a constant. It might, for example, depend on the size of the workforce so that doubling the numbers might lead to more efficient (or less efficient) methods of working, leading to more (or less) than twice the output. Changes in technology, agreements with unions over manning levels, and systems of payment are other examples of factors which could lead to changes in $P$ during the forecast period.

A somewhat more sophisticated approach, which tries to take account of these considerations, is to make $P$ a function of time. This implies that values of $P$ in the forecast period will have to be predicted, like the workload series, on the basis of past data or technical knowledge.

Rather than pursue further refinements of this approach it seems preferable to adopt a broader and more flexible framework within which to study the relationship between manning levels, workloads and other factors.

## Regression methods

The aim here is to find a prediction equation linking the manning level to all the factors on which it is believed to depend. Let $N$ denote the number of people in the job category being considered and let $x_1, x_2, \ldots, x_k$ be the variables which are supposed to influence the value of $N$. These latter may include the volumes of different kinds of work, as in the Land Registry example, together with technological variables and indices or morale based on such things as absenteeism or accident rates. They may also include indicator variables showing the presence or absence of some attributes. The statistical problem is then to find a formula

$$N = f(x_1, x_2, \ldots, x_k) \qquad (8.11)$$

from which values of $N$ can be predicted. Notice that this can be viewed as a generalization of the productivity index approach which specified a relationship of the form $N \propto x$, where $x$ is the workload. In the work of Lapp and Thompson (1974) on predicting the demand for engineers in Ontario, $x$ was taken as the gross national product.

To implement a regression method in practice, historical data is required from which the form of (8.11) can be estimated. It is then necessary to predict the various series of $x$-variables so that predictions of $N$ can be obtained via (8.11).

There may be technical or other data to guide us in the choice of the function $f(.)$ (see, for example, the discussion of the Cobb–Douglas function at the end of this section) but failing this, a first step will often be to take a linear predictor of the form

$$N = b_0 + b_1x_1 + \ldots + b_kx_k \qquad (8.12)$$

Drui (1963) used this approach to find forecasting equations for various departments of a firm. In the finance department, for example, there were four independent variables as follows:

$x_1$ = total company employment;
$x_2$ = total direct employment (employed on manufacturing);
$x_3$ = sales;
$x_4$ = reports processed.

Since the finance department fulfilled a service function to the whole company, the variables selected were those which created work for that department. The fitted linear regression equation accounted for 86 per cent of the variation in manning levels over the period to which the data related.

Regression methods are very versatile and provide an attractive way of using historical data to establish manning requirements. However, there are many pitfalls for the unwary, some of which we shall now enumerate.

If some, or all, of the $x$-variables are highly correlated with one another (as they are likely to be if they represent levels of activity as in Drui's example) there may be difficulties in fitting the equation and in interpreting its coefficients. The former circumstances, known as the problem of *multicollinearity*, can be tackled in various ways. One, which is sometimes worth investigating, is to carry out a principal component analysis on the $x$-variables and to use some (or all) of the principal components in place of the original independent variables.

Even if the equation is successfully fitted, one must beware of placing too much faith in the interpretation of individual regression coefficients because their values, and even their signs, can be affected if some relevant independent variable is omitted through ignorance or lack of data. For example, if the estimate of $b_1$ turns out to be positive, it is natural to infer that an increase in $x_1$ will lead, on average, to an increase in $N$. However, if $x_1$ is negatively correlated with some other variable which has not been included in the equation, it could be the case that the inclusion of that variable would have changed the sign of $b_1$ and led to the opposite conclusion.

It often happens that there are many independent variables which might be expected to influence manning levels. If they are all included in the regression equation the resulting prediction formula may be unnecessarily cumbersome in the sense that almost as good a fit could have been obtained with fewer variables. Most computer software packages for regression analysis include facilities for making a selection of variables and these usually aim to include only those variables whose regression coefficients are statistically significant. However, the tests of significance which they employ depend upon distributional assumptions about the dependent variable which are highly suspect, if not meaningless, in the kind of applications we have in mind. At best, therefore, they are somewhat arbitrary rules for finding a good subset and their operation in practice sometimes produces results which users who are not statisticians find disconcerting. For example, variables which are 'known' to be important in determining manning levels may not appear in the equation. This may arise because their influence makes their presence felt through some other variable (or group of variables) with which they are highly correlated. In other words, any one variable of a group may serve as a proxy for the others, and which one happens to get into the equation will depend on which type of selection procedure is being used. The same phenomenon also has the effect of making the regression relationship appear to be highly unstable. Thus, for example, if the equation is fitted to data for two successive time periods, the group of variables included on the first occasion may be very different from those on the second. Both such equations may be equally good as predictors while appearing nonsensical if interpreted as expressions of a 'law' determining manning levels.

These considerations serve to emphasize that regression analysis is a tool for prediction rather than explanation but even then extrapolation beyond the range of the data is always hazardous and especially so in manpower forecasting. If the projected values of the independent variables lie outside the range spanned by those from which (8.12) is estimated there is no guarantee that the equation will describe the relationship accurately.

Even if the form of the relationship holds for all likely values of the $x$-variables it may change over time and so introduce errors into the forecasts. Methods of testing for such changes have been given by Brown, Durbin, and Evans (1975) and used by Cameron and Nash (1974).

Livingstone and Montgomery (1966) criticised Drui's use of regression analysis on the grounds that the use of least squares estimation on time series data was likely to lead to bias in the predictions. This effect arises if the error terms in the regression model are autocorrelated, which is likely to be the case with any time series. To combat this difficulty Livingstone and Montgomery advocated the use of lagged variables in the regression relationship. This has much to commend it on grounds of realism since there will often be a time-lag between the appearance of a requirement for manpower and the recruitment or transfer of people to meet that requirement. Thus one ought to replace (8.11) by a prediction equation of the more general form

$$N(t) = f(x_1(t-1), x_1(t-2), \ldots, x_2(t-1), x_2(t-2), \ldots, x_k(t-1), x_k(t-2), \ldots) \tag{8.13}$$

Such an approach was used by Halpern (1974) in a study which arose out of trying to forecast total Civil Service numbers in the United Kingdom. The $x$-variables in that case were series on public expenditure which could be forecast independently and hence used to forecast $N$ by means of a linear regression equation involving lagged variables. In that particular application it proved to be much more effective to work with the first differences of all variables rather than with the variables themselves. In other words, changes in such things as levels of public expenditure could be related to changes in manning levels in a more precise way than a relationship could be established between the variables themselves. This is often likely to be the case when the forecast changes in the $x$-variables are relatively small.

The choice of a linear regression equation may be plausible, as well as convenient, especially if it is known that the effects of the independent variables are additive as they would be if they represented different workloads. In other circumstances this assumption may be very wide of the mark and non-linear regressions must then be used. The choice of function can be approached empirically by fitting linear regressions using transformed variables. For example, logarithms would be appropriate if the effects of the

$x$'s were multiplicative rather than additive. This can be easily and speedily achieved using modern computer packages. Occasionally there may be technical or theoretical evidence about the form of the prediction equation. For example, at the macro-level, the Cobb–Douglas production function relates production ($P$) with capital employed ($C$) and labour ($L$) by an equation of the form

$$P \propto C^a L^{1-a} \qquad (0 < a < 1) \tag{8.14}$$

which may be expressed in a linear form by taking logarithms to give

$$\log L = \frac{1}{1-a}[\log P - a \log C - \text{constant}] \qquad (8.15$$

The parameter $a$ relates to the substitutability of capital and labour, and its value is determined from estimates of the regression coefficients obtained by fitting (8.15).

## 8.6 MISCELLANEOUS TECHNIQUES AND SUGGESTIONS

In many situations there will be information available to the planner which is not easily incorporated into any of the techniques mentioned in this chapter. Chiefly, this will arise because of the difficulty of quantifying with sufficient precision the many relevant factors in the manpower environment. Partly this will be a question of things like 'quality of life' which are difficult to quantify, and partly ignorance and uncertainty about the future values of critical parameters. Such information as managers have on such matters may not be easily extracted but should obviously be taken into account.

A general discussion of the problems which managers have in coping with uncertainty is given by Moore (1977) and the problem as it arises in manpower planning is treated in Bartholomew, Hopes and Smith (1976). An interesting, if specialized, attempt to incorporate uncertainty into a manpower demand forecast is given by Helps (1970) for the printing industry. He obtained subjective probability distributions for various parameters by asking managers to use their judgement to give 'upper', 'lower', and 'most likely' values. Helps then selected a beta-distribution to represent the uncertainty by making the end-points of its range coincide with the upper and lower limits and its mode with the most likely value. These probability distributions were then used to calculate probabilities of various future outcomes relevant to the planning process. This approach may be viewed as a first step on the road to a fully Bayesian approach to forecasting and decision making in which all uncertainties would be expressed in probabilistic terms. If this can be done in a convincing fashion there is much to be said for this approach, but much practical experience (as reported, for example, by Moore (1977) and Bartholomew, Hopes, and Smith (1976)) suggests that it will remain an unattainable ideal in many cases.

The problem of decision making under uncertainty is a wide one which cannot be taken up fully here. We content ourselves by reiterating two of the main points made by Bartholomew, Hopes and Smith (1976). One is the importance of undertaking sensitivity analyses in connection with any forecasting exercise. By this means it is possible to identify which assumptions in the forecasting model are critical and so to facilitate the creation of appropriate contingency plans. The other and related point is the importance of adopting a flexible planning posture. This requires a constant monitoring of the process so that undesirable trends can be identified and allowed for as soon as possible.

Demand forecasting is a hazardous business at best and it would certainly be unwise to place too much faith in any technique. Wherever possible, results should be cross-checked by using more than one method. For example, one might first forecast total numbers by a time series prediction method and then make independent forecasts for different parts of the organization on the basis of projected workloads. A check can then be made to see whether the two calculations are consistent.

It cannot be over-emphasized that in demand forecasting each problem must be considered on its merits and in its entirety. One or more of the techniques referred to in this chapter may be useful but the technique must always be adapted to the problem and not vice versa.

APPENDIX A

# The Program BASEQN

## A.1 INTRODUCTION

The program evaluates the basic transition model equation (4.4b). It is interactive in the sense that it prompts the user for both data and control commands as these are required: specimen printouts are shown in Section A.2. Since printouts from the program are used extensively in the examples and in the solutions to the exercises in Chapters 4 and 5, the reader is recommended to familiarize himself with their format; the outline description of the printouts given in Section 4.3 of Chapter 4 is sufficient for this purpose. However, although the data prompts are compatible with the notation used in the book and are mostly self explanatory, the responses to the recruitment term prompt ($R$), and the control prompt ($\langle * \rangle$) are rather more complex than their description in Section 4.3 suggests. This appendix is therefore intended for readers who wish to know more about the program or for those wishing to set up the program on their own systems (a BASIC listing is provided in Section A.4). The program will certainly aid the solution of the exercises and should prove a useful tool for simple projections. For readers without a BASIC compiler this appendix should provide a useful specification for producing a similar program in another language. As it stands, the program may well be inefficient in programming terms since it evolved over a number of years as a working program with each option added on as required. Because of its basic simplicity and cheapness, in terms of computer time, any more efficient rewriting of the listing is unlikely to repay the effort involved.

## A.2 EXAMPLES

As shown in Printouts A.1 and A.2

Printout A.1 An example run

```
    K =    ?3
    N =    ?180.145.35
    P =    ?.70,.10,0,     0,.85,.05.     0.0,.85
    R =    ?52,15,3,    1,0,0
  T,% =    ?5,YES
  <*>    ?  RUN

    T             1           2           3          TOTAL        R

    0        180(50%)   145(40%)    35(10%)     360(100%)
    1        178(48%)   156(42%)    40(11%)     374(104%)      70
    2        177(46%)   166(43%)    45(12%)     387(108%)      70
    3        176(44%)   173(44%)    49(12%)     398(111%)      70
    4        175(43%)   180(44%)    54(13%)     409(113%)      70
    5        174(42%)   185(44%)    58(14%)     418(116%)      70
  <*>    ?T=,10
   10        174(39%)   202(45%)    73(16%)     449(125%)      70
  <*>    ?T=,20
   20        173(37%)   213(45%)    87(18%)     473(131%)      70
  <*>    ?T=.99
   99        173(36%)   216(45%)    92(19%)     481(134%)      70
```

Printout A.2 Following on from Printout A.1 and illustrating the use of the control command

```
<*>     ?N=N(0)
<*>     ?R
  R =     ?.75,.20,.05,     -1,+,15
<*>     ?ROWP
I;P(I,J) J=1,K =  ?1,     .80,.05,0
<*>     ?PIJ
I;J;P(I,J)=     ?2,2,.86
<*>     ?PIJ=,2,3,.04
<*>     ?T%=,5,NO
<*>     ?PARAMS

N=

 180              145               35
P=

 .800000          5.00000E-02       0

 0                .860000           4.00000E-02

 0                0                 .850000
R=

 .750000          .200000           5.00000E-02
-1                +                 15
<*>     ?RUN

  T          1       2       3          TOTAL        R

  0        180     145      35     360(100%)
  1        197     148      39     384(107%)        71
  2        214     152      43     409(114%)        75
  3        231     157      47     435(121%)        79
  4        247     164      50     461(128%)        84
  5        264     171      53     488(136%)        88
<*>     ?+T=,3
  6        281     179      57     517(144%)        93
  7        298     187      60     546(152%)        98
  8        315     196      64     576(160%)       102
<*>     ?T=,10
 10        350     216      71     638(177%)       112
<*>     ?*R=,65,10,3,     1,0,0,    5,NO
 11        345     214      72     631(175%)        78
 12        341     211      73     625(174%)        78
 13        338     208      73     620(172%)        78
 14        335     206      74     615(171%)        78
 15        333     204      74     611(170%)        78
```

## A.3 EXPLANATION OF THE DATA PROMPTS AND THE CONTROL COMMANDS

The data prompts are given below in their normal order of appearance, and the sample responses which are shown underlined are those used in Printout A.1. The question mark following each prompt is a system signal signifying that data input is required (this symbol may vary on different computer systems). Data items in the responses are shown separated by commas (separators too may vary from one computer system to another), and spaces can also be inserted as required to improve the readability. For example, it is advisable to separate the rows in the data response to $P$.

**K = ? 3**

$K$ is the number of grades or classes in the model. The upper limit on this number depends only on the dimensions specified in the listing (see statement 100 on Printout A.3). The limit shown in Section A.4 is 10 although this can easily be changed. With a normal 72-character terminal the results table will be folded and the format disrupted for $K > 6$ if percentages are printed (see % prompt below), and for $K > 8$ if percentages are not printed.

**N = ? 180,145,35**

Since $N$ represents the initial grade sizes, $K$ numbers must be input in response to this prompt with the first corresponding to grade 1, the second to grade 2, and so on.

**P = ? .70,.10.0, 0,.85,.05, 0,0,.85**

$P$ is the transition matrix of flow rates, and $K^2$ numbers are required in row by row order. That is $p_{11}, p_{12}, \ldots, p_{1k}, p_{21}, p_{22}, \ldots, p_{2k}, \ldots, p_{k1}, \ldots, p_{kk}$.

**R = ? 52,15,3, 1,0,0**

The response to the $R$ prompt specifies the inflow. Although its format is not so straightforward as the other prompts, it does allow considerable flexibility in specifying the 'recruitment'. Its general form is $R_1, R_2, \ldots, R_k, A, B, C$. A total of $K + 3$ numbers are required, where the first $K$ numbers relate to each of the grades $1, 2, \ldots, K$ respectively, and the last 3 numbers specify any one of the 5 possible options. These options are defined and illustrated by the following examples.

(a) 52,15,3, 1,0,0

When $(A, B, C) = (1, 0, 0)$, $R_i$ is the (net) number entering grade $i$ each year. Thus in this example 52, 15, and 3 enter grades 1, 2, and 3 respectively every year.

(b) .6,.3,.1 100,+,10

The total (net) number entering each year starts at $A = (100)$ and increases arithmetically ($B = +$) by $C\,(=10)$ each year. These total numbers are spread over the grades by the distribution $R_i\,(=.6,.3,.1)$. Hence in this example the numbers entering each grade would be:

1st year: 100(.6,.3,.1) = (60,30,10)

2nd year: (100 + 10)(.6,.3,.1) = (66,33,11)

3rd year: (110 + 10)(.6,.3,.1) = (72,36,12)

If the number of entrants is to reduce then $C$ must be negative.

(c) .6,.3,.1, 100,*,1.2

This is similar to option (b) except that the total (net) number entering the system starts at $A\,(=100)$ and increases multiplicatively ($B = *$) by the factor $C\,(=1.2)$. This example would therefore imply the following recruits:

1st year: 100(.6,.3,.1) = (60,30,10)

2nd year: $100 \times 1.2(.6,.3,.1)$ = (72,36,12)

3rd year: $100 \times 1.2^2(.6,.3,.1)$ = (86.4,43.2,14.4)

If the number of entrants is to reduce then $C$ must be less than 1.

(d) .6,.3,.1, −1,+,50

When $A$ is $-1$ the program calculates the total (net) number of entrants required to make the total size of the system increase arithmetically ($B = +$) by $C\,(=50)$ per year. This total number of entrants is spread over the grades by the distribution $R_i\,(=.6,.3,.1)$. For contraction $C$ should be negative.

(e) .6,.3,.1, −1,*,1.1

This is similar to option (d) except that the total size of the system increases multiplicatively ($B = *$) by a factor of $C\,(=1.1)$ per year.

Note that options (d) and (e) give the required (pull) recruitment necessary to keep a system at pre-determined sizes. This required recruitment is shown in the results table in the column headed $R$.

T,%= ? 5, YES

Two items are required here. The first, which must be a number, is the time $T$ up to which the results of the projection will be printed out. This output will include the results for every year up to and including time $T$. (The user then has the option of continuing the projection—see control commands '$+T$' and '$T$ = '.) The second item specifies what if any percentages are to be shown in the results table. The only permissible responses are YES, NO, and YESO (any other response has the same effect as NO). With YES the grade sizes each year are expressed as percentages of the total size at that time and hence show the relative grade sizes or grade structure. The option YESO differs in that the size of each grade is expressed as a percentage of its own original size at time $T = 0$ and therefore shows the growth of each grade relative to its original size. The option NO produces no grade percentages. (The total size of the system is always shown as a percentage of the original total size at time $T = 0$: this indicates the growth of contraction of the system as a whole.)

*= ? RUN

This is the control command prompt which enables the user to control the program by responding with any of the list of commands detailed in Table A.1. These are illustrated in Printout A.2. This prompt occurs at the following points: at the end of the initial data input; after the results up to the specified time $T$ have been output; whenever the user presses the BREAK key on the terminal; and immediately following the input of certain commands. The main functions of these control commands are to:

(a) RUN the projection (with the current parameter values);
(b) correct any mistakes in the data input;
(c) re-set the parameters during a projection;
(d) change the parameters for another projection;
(e) finish or restart the program.

***Table A.1 The control commands and their effects***

In this table the / character is used to separate commands, and ⟨...⟩ represents data in whatever format is required.

| *Commands* | *Effects* |
|---|---|
| RUN | Causes the program to project up to and including the specified time $T$ using the current values of the parameters. The results are printed for every year up to and including $T$. |
| FINISH/END/Z | The user has finished the current session. (Each of these commands has the same effect.) |
| START/RESTART/BEGIN/A | Returns the program to the beginning of the initial data input phase. (Each of these commands has the same effect.) |

***Table A.1 Continued***

| *Commands* | *Effects* |
|---|---|
| N/P/R/T% | Each of these elicits the corresponding data prompts. These commands enable the user to correct or change the current values for any of these parameters. As soon as the appropriate data have been input in response to the prompt, the program responds with another command prompt ⟨*⟩. |
| PIJ/ROWP | Either of these can be used to save re-entering the whole of the matrix $P$ when only a few elements are to be changed. PIJ, which is used to change a single element $p_{ij}$, elicits a prompt for the three items $i$, $j$, and $p_{ij}$. ROWP, which is used to change a whole row, produces a prompt for the row number $i$, and its $k$ elements. |
| N=,⟨...⟩/P=,⟨...⟩<br>R=,⟨...⟩/T%=,⟨...⟩/<br>PIJ=,⟨...⟩/ROWP=,⟨...⟩ | These commands can also be used to change any of the parameters. They are more direct than the commands of the two previous sections since they short-cut the procedure of first responding with the parameter symbol and then waiting for the program before inputing the data. The main advantages of these prompts are in improving the readability of the printout and in reducing the space on the printout taken up by data amendments. This latter feature is particularly useful for parameter changes made during a projection since these appear in the middle of the results table. |
| T=,# | Produces further projections up to time# with the results printed *only* for time# (# should be a single number greater than the current $T$-value). |
| +T=,# | Produces further projections up to time# with results printed for *every year* up to time# (# should be a single number greater than the current $T$-value). |
| *R=,⟨...⟩ ⟨...⟩ | the first ⟨...⟩ represents data for the prompt $R$, and the second ⟨...⟩ data for $T$%. This is a special command enabling the user to change both $R$ and $T$% during a projection using only one line on the printout to avoid disrupting the results table. |
| N=N(0) | This resets $N$ to its previous initial values. Remember to use this option when re-running a projection, otherwise $N$ will start from its current value of $N(T)$. Note too that N(0) has a zero. |
| PARAMS | Prints the current values for all the parameters. |

## A.4 LISTING OF BASEQN

The following comments may be useful when setting up the program on your own system.

(a) The program uses the ON BREAK statement (see line 112 of Printout A.3) which immediately directs the program to a particular line (in this case line 355) if the break key is pressed at any time during program execution. Although useful this is not a standard BASIC facility, and line 112 can be deleted without affecting the rest of the program in any way.

(b) The program uses BASIC MAT operations. The format for these do not seem to be standard, particularly with respect to the specification of dimensions. Hence check these statements for your system and also the DIM statement.

(c) Beware of characters being peculiar to particular systems and terminals. For example, $ is sometimes £, and % may not be available.

(d) Check too whether your system rounds or truncates numbers before printing. The program assumes the former and does no rounding itself.

## Printout A.3 BASEQN listing in BASIC

```
100 DIM  N(1,10),P(10,10),R(1,10),X(1,10),Q(1,10)
110 PRINT"BASEQN  EVALUATES     N(T+1)=N(T)*P + R"
112 ON BREAK 355
115 A=0
124*
125* DATA INPUT SEGMENT
130 PRINT
134 PRINT"  K =";
135 INPUT K
136*
140 PRINT"  N =";
150 MAT INPUT N(1,K)
153 IF A=1 GOTO 360
154 D0=0.0000001
155 FOR J=1 TO K
156 M(J)=N(1,J)
157 D0=D0+N(1,J)
158 NEXT J
159*
160 PRINT"  P =";
170 MAT INPUT P(K,K)
175 IF A=1 GOTO 360
179*
180 PRINT"  R =";
182 MAT INPUT Q(1,K)
184 INPUT R,R$,E
186 IF R=-1 GOTO 700
195 IF C$="*R=" GOTO 210
196 IF A=1 GOTO 360
199*
200 PRINT"T,% =";
210 INPUT S,B$
212 IF C$="*R=" GOTO 214
213 GOTO 355
214 S=S-1
215 GOTO 270
217*
218* CALC AND OUTPUT SEGMENT
220 PRINT
223 T1=-1
224 T2=0
229* TABLE HEADING
230 PRINT"  T ";
235 FOR J=1 TO K
240 IF B$="YES" GOTO 255
242 IF B$="YESO" GOTO 257
245 PRINT USING 403,J;
250 GOTO260
255 PRINT USING 405,J;
256 GOTO 260
257 PRINT USING 408,J;
260 NEXT J
262 PRINT"          TOTAL";
265 PRINT"  R"
267 PRINT
270 FOR T=0 TO S
272 IF T1=-1 GOTO 305
```

```
289* TO SET MAT R
290 GOSUB 550
291*
303 MAT X=N*P
304 MAT N=X + R
305*
308 D=0.00000001
310 FOR J=1 TO K
313 D=D+N(1,J)
316NEXT J
318 T1=T1+1
319 IF T1=T2 GOTO 321
320 GOTO350
321 T2=T2+1
322 PRINT USING 402,T1;
323 PRINT"  ";
324 FOR J=1 TO K
325 IF B$="YES" GOTO 331
326 IF B$="YESO" GOTO 335
328 PRINT USING 403,N(1,J);
330 GOTO 340
331 PRINT USING 404,N(1,J),100*N(1,J)/D;
333 GOTO 340
335 PRINT USING 407,N(1,J),100*N(1,J)/M(J);
340 NEXT J
345 PRINT USING 406,D,100*D/D0;
347 IF T1=0 GOTO 349
348 PRINT USING 409,R1;
349 PRINT
350 NEXT T
352*
355 A=1
358*
359* CONTROL SEGMENT
360*
361 PRINT"<*>";
363 INPUT C$
364 IF C$="RUN"   GOTO 220
366 IF C$="N"  GOTO 136
367 IF C$="N=" GOTO 150
369 IF C$="P"   GOTO 159
370 IF C$="P=" GOTO 170
372 IF C$="R"    GOTO 179
373 IF C$="R=" GOTO 182
374 IF C$="*R=" GOTO 182
375 IF C$="T%" GOTO 199
376 IF C$="T%=" GOTO 210
377 IF C$="T="  gOTO 511
378 IF C$="+T="   GOTO 500
380 IF C$="START"   GOTO 115
381 IF C$="RESTART" GOTO 115
382 IF C$="BEGIN"   GOTO 115
383 IF C$="A"       GOTO 115
385 IF C$="END"    GOTO 999
386 IF C$="FINISH" GOTO 999
387 IF C$="Z"      GOTO 999
390 IF C$="PIJ"     GOTO 750
```

```
391 IF C$="PIJ="    GOTO 754
392 IF C$="ROWP"    GOTO 760
393 IF C$="ROWP="   GOTO 765
395 IF C$="N=N(0)" GOTO 520
397 IF C$="PARAMS" GOTO 446
399 PRINT"NO SUCH CONTROL";
400 GOTO 360
402: ##
403:  -###
404:-####(##%)
405:      -###
406:-######(###%)
407:-####(###%)
408:       -###
409: -####
445*
446* PARAM PRINT-CHECK SEGMENT
449 PRINT
450 PRINT"N="
455 MAT PRIN N
460 PRINT"P="
465 MAT PRINT P
470 PRINT"R="
475 MAT PRINT Q
477 PRINT R,R$,E
480 GOTO360
500 * FURTHER YEARS SAME TABLE
504 INPUT S
507 S=S-1
510 GOTO 270
511*PRINT ONLY SPECIFIED YEAR
513 INPUT T2
515 S=T2-T1-1
517 GOTO 270
520* N IS RESET TO ORIGINAL N(0)
523 FOR J=1 TO K
526 N(1,J)=M(J)
530 NEXT J
533 GOTO 360
549*
550*** SUB TO SET MAT R
552 IF R=-1 GOTO 556
554 IF R$="0" GOTO 640
555 GOTO 610
556* TOTAL SIZE SYSTEM FIXED
559 R1=0
562 FOR J=1 TO K
565 R1=R1+N(1,J)*W(J)
568 NEXT J
572 IF R$="+" GOTO 581
575 IF R$="*" GOTO 587
578 PRINT"ERROR IN R$ AT LINE 578"
581 R1=R1+E
584 GOTO 590
587 R1=R1+D*(E-1)
590 MAT R=(R1)*Q
595 GOTO 650
```

```
610** THIS SEG FOR R NE 1 AND R$ NE 0
611 R1=R
612 MAT R=(R1)*Q
613* CALC SCAL R READY FOR FOLLOWING YEAR
615 IF R$="+" GOTO 623
618 IF R$="*" GOTO 628
621 PRINT "ERROR IN R$ AT LINE 621"
623 R=R+E
625 GOTO 650
628 R=R*E
630 GOTO 650
640** THIS SEG FOR R$=0
642 MAT R=Q
643 R1=0
645 FOR J=1 TO K
646 R1=R1+R(1,J)
648 NEXT J
650 RETURN
651*
700*** CALC WAST RATES IF R=-1
703 FOR I=1 TO K
706 W(I)=1
709 FOR J=1 TO K
712 W(I)=W(I)-P(I,J)
715 NEXT J
718 NEXT I
720 GOTO 195
749*
750*** CHANGE INDIVIDUAL ELEMENTS OF P
752 PRINT"I;J;P(I,J)=";
754 INPUT I,J
755 INPUT P(I,J)
758 GOTO 360
759*
760*** CHANGE WHOLE ROW OF P
763 PRINT"I;P(I,J) J=1,K =";
765 INPUT I
767 FOR J=1 TO K
769 INPUT P(I,J)
772 NEXT J
775 GOTO 360
999 END
```

## APPENDIX B

# A Program for Calculating the Career Pattern from Promotion and Wastage Rates

### B.1 INTRODUCTION

The program calculates the career pattern corresponding to specified promotion and wastage transition rates. (The methodology is described in Sections 6.4 and 6.5.) It is assumed that duration within a grade is divided into three zones with boundaries *A* and *B*. (Since suffices and lower case letters are not generally available in programs all variables in this appendix are shown not as in Chapter 6, but as they appear on the terminal, e.g. *TP** instead of $T_p^*$, *A* instead of *a*, etc.) Below seniority *A* and above seniority *B* no one is promoted. The interval (*A*, *B*) is the promotion zone and we denote by *P*2 the per annum promotion (transition) rate for people in this interval. The wastage rates for the zones (0, *A*) and (*A*, *B*) are *W*1 and *W*2 respectively.

The career pattern statistics which the program calculates are:

- *PU*: the expected proportion of entrants to the grade who are promoted;
- *TP*: the average seniority of those promoted;
- *PU**: the chance of promotion for an individual if he is prepared to wait if necessary until he reaches the end of the promotion zone;
- *TP**: the expected time before promotion for this individual if he is promoted.

To calculate these quantities the program requires values for *A*, *B*, *P*2, *W*1, and *W*2.

The program can also be used if there is no promotion zone. For example, if people are equally likely to be promoted whatever time they have spent in the grade but there is a cut-off or retirement point *K*, say, then $A = 0$, $B = K$, and $W1 = 0$. Similarly, if there is no upper limit to promotion, *B* should be put equal to some suitably large number (this situation then corresponds to that of the Markov chain methods described in Section 6.6).

Another situation which may well occur in practice is that the boundaries of the promotion zone are known but the *zone*-specific rates are not. In this case if the *grade*-specific promotion and wastage rates (*P* and *W*) are known, the program can use an approximation to calculate *P*2 based on the values *A*, *B*, *K*, *P*, and *W*. The variables *PU*, etc., are then calculated using this value of *P*2 and $W1 = W2 =$ W.

The user can choose between detailed and abbreviated versions for both the data prompts and the printout. Abbreviated prompts are offered because they save time and because the detailed prompts can become tedious once the user is familar with the required data. The detailed output for the results is intended primarily to provide the user with a reminder of their interpretation.

The examples of Section B.2 illustrate the input and output formats, while Section B.3 contains further explanations and discussions of the data prompts and responses. Section B.4 gives summaries of the formulae used, and Section B.5 contains the listing of the program.

## B.2 EXAMPLES

See Printouts B.1, B.2, B.3, and B.4.

Printout B.1 Showing detailed prompts and results with input of zone specific rates (data from Table 6.4 in Example 6.3)

```
PROGRAM CALCULATES THE CAREER PATTERN FROM PROMOTION AND
WASTAGE TRANSITION RATES.  FOR METHODOLOGY SEE STMP CH 6.


DETAILED OR ABBREVIATED PROMPTS?   REPLY D/A  ?D

DETAILED OR ABBREVIATED RESULTS?   REPLY D/A  ?D



A, LOWER BOUNDARY OF PROMOTION ZONE =   ?6

B, UPPER BOUNDARY OF PROMOTION ZONE =   ?12

WHICH ASSUMPTION DO YOU WISH TO MAKE IN CALCULATING THE
PROMOTION RATE P*(=P2*) CONDITIONAL ON NOT LEAVING :
  1: PROMOTION BEFORE WASTAGE
  2: WASTAGE BEFORE PROMOTION
  3: PROMOTION AND WASTAGE SIMULTANEOUS
REPLY 1/2/3   ?3

1: DO YOU HAVE VALUES FOR THE ZONE SPECIFIC RATES, OR
2: DO YOU WISH TO USE THE APPROXIMATION FOR P2 WHICH
   REQUIRES GRADE SPECIFIC RATES AND THE CUT-OFF POINT ?
REPLY 1/2   ?1

P2, PROMOTION TRANSITION RATE WITHIN PROMOTION ZONE =   ?0.138

W1, WASTAGE TRANSITION RATE WITHIN FIRST (NON-PROMOTION)
     ZONE =   ?0.131

W2, WASTAGE TRANSITION RATE WITHIN PROMOTION ZONE =   ?0.089


THE PROMOTION RATE CONDITIONAL ON NOT LEAVING, P2*=0.145

THE PROPORTION OF ENTRANTS PROMOTED         (PU)=0.206

THE AVERAGE EXPERIENCE OF PROMOTEES         (TP)= 8.28

THE AVERAGE CHANCE OF PROMOTION FOR AN
  INDIVIDUAL WHO IS PREPARED IF NECESSARY
  TO NOT LEAVE FOR A TIME B= 12.0          (PU*)=0.609

THE EXPECTED TIME BEFORE PROMOTION FOR
  THIS INDIVIDUAL IF HE IS PROMOTED        (TP*)= 8.55


FINISH OR RESTART?   REPLY F/R  ?R
```

Printout B.2 As for Printout B.1 but with abbreviated prompts and results

```
DETAILED OR ABBREVIATED PROMPTS?   REPLY D/A  ?A

DETAILED OR ABBREVIATED RESULTS?   REPLY D/A  ?A

PROM ZONE : A,B =    ?6,12
WHICH ASSUMPTION FOR P2* (1/2/3)  ?3
1: INPUT ZONE RATES, OR 2: USE APPROX FOR P2 ?  (1/2)   ?1
P2,W1,W2=   ?0.138,0.131,0.089


P2*=0.145

PU=0.206   TP= 8.28     PU*=0.609   TP*= 8.55




FINISH OR RESTART?   REPLY F/R  ?R
```

Printout B.3 As for Printout B.1 but showing detailed prompts and results where grade-specific rates and the approximation for P2 are used (see Example 6.4)

```
FINISH OR RESTART?   REPLY F/R  ?R


DETAILED OR ABBREVIATED PROMPTS?   REPLY D/A  ?D

DETAILED OR ABBREVIATED RESULTS?   REPLY D/A  ?D



A, LOWER BOUNDARY OF PROMOTION ZONE =    ?6

B, UPPER BOUNDARY OF PROMOTION ZONE =    ?12

WHICH ASSUMPTION DO YOU WISH TO MAKE IN CALCULATING THE
PROMOTION RATE P*(=P2*) CONDITIONAL ON NOT LEAVING :
  1: PROMOTION BEFORE WASTAGE
  2: WASTAGE BEFORE PROMOTION
  3: PROMOTION AND WASTAGE SIMULTANEOUS
REPLY 1/2/3   ?3

1: DO YOU HAVE VALUES FOR THE ZONE SPECIFIC RATES, OR
2: DO YOU WISH TO USE THE APPROXIMATION FOR P2 WHICH
   REQUIRES GRADE SPECIFIC RATES AND THE CUT-OFF POINT ?
REPLY 1/2   ?2

K, THE CUTOFF OR RETIREMENT POINT =   ?15

P, PROMOTION TRANSITION RATE FOR GRADE =  ?0.046

W, WASTAGE TRANSITION RATE FOR GRADE =  ?0.124


THE CALCULATED VALUE FOR THE APPROXIMATE P2=0.196
THE WASTAGE RATES WILL BE TAKEN AS W1=W2=W=0.124

THE PROMOTION RATE CONDITIONAL ON NOT LEAVING, P2*=0.211

THE PROPORTION OF ENTRANTS PROMOTED         (PU)=0.250

THE AVERAGE EXPERIENCE OF PROMOTEES         (TP)= 7.97

THE AVERAGE CHANCE OF PROMOTION FOR AN
  INDIVIDUAL WHO IS PREPARED IF NECESSARY
  TO NOT LEAVE FOR A TIME B= 12.0         (PU*)=0.758

THE EXPECTED TIME BEFORE PROMOTION FOR
  THIS INDIVIDUAL IF HE IS PROMOTED       (TP*)= 8.33



FINISH OR RESTART?   REPLY F/R  ?R
```

Printout B.4 As for Printout B.3 but with abbreviated prompts and results.

```
DETAILED OR ABBREVIATED PROMPTS?   REPLY D/A   ?A

DETAILED OR ABBREVIATED RESULTS?   REPLY D/A   ?A

PROM ZONE : A,B =    ?6,12
WHICH ASSUMPTION FOR P2* (1/2/3)   ?3
1: INPUT ZONE RATES, OR 2: USE APPROX FOR P2 ?  (1/2)   ?2
K,P,W=  ?15,0.046,0.124


APPROX P2=0.196           W1=W2=W=0.124

P2*=0.211

PU=0.250   TP= 7.97     PU*=0.758   TP*= 8.33



FINISH OR RESTART?   REPLY F/R  ?F
```

## B.3 EXPLANATION OF THE DATA PROMPTS

The data prompts are shown below in upper case and are given in their order of appearance. Unless otherwise stated there are no traps in the program to reject inadmissible data: for example, all transition rates should be between 0 and 1 although the program does *not* check for this. Where appropriate the prompts show possible replies separated by /: for example, D/A. The example responses below are shown underlined and correspond to the examples of Section B.2.

DETAILED OR ABBREVIATED PROMPTS? REPLY D/A ?D
DETAILED OR ABBREVIATED RESULTS? REPLY D/A ?D

Use of the detailed prompts is recommended at least on the first run. The abbreviated prompts and printout are shown in Section B.2. Any responses other than *D* or *A* will produce the detailed prompts and printouts.

A, LOWER BOUNDARY OF PROMOTION ZONE= ?6

This number should be an integer greater than or equal to zero. If there is no promotion zone put $A = 0$.

B, UPPER BOUNDARY OF PROMOTION ZONE= ?12

This number should be greater than or equal to *A*. If there is no promotion zone put *B* equal to the maximum length of service. If there is no maximum length of service put *B* equal to some appropriately large number.

WHICH ASSUMPTION DO YOU WISH TO MAKE IN CALCULATING THE PROMOTION RATE P*(=PZ*) CONDITIONAL NOT ON LEAVING:
1: PROMOTION BEFORE WASTAGE
2: WASTAGE BEFORE PROMOTION
3: PROMOTION AND WASTAGE SIMULTANEOUS
REPLY 1/2/3 ?3

The choice will depend on what is known about the promotion process in the organization. Assumptions 1 and 2 above are the two extremes, so that the results given by assumption 3 should be between those of 1 and 2. Note that $P^*$ is only used in the calculation of $PU^*$ and $TP^*$, so the choice of this assumption will not affect the values of $PU$ and $TP$. Any numeric response other than 1, 2, or 3 gives assumption 3.

1: DO YOU HAVE VALUES FOR THE ZONE SPECIFIC RATES, OR
2: DO YOU WISH TO USE THE APPROXIMATION FOR P2 WHICH REQUIRES GRADE SPECIFIC RATES AND THE CUT-OFF POINT

REPLY 1/2 ?1

If the rates specific to the zones are available (i.e. P2, W1, and W2 see below) use Option 1. Option 2 is only recommended if the zones are known but only grade-specific rates are available. The approximation is based on the assumption that the seniority distribution within the grade has a simple steady state form. (The prompts which result if Option 2 is chosen are shown at the end of this section).

P2, PROMOTION TRANSITION RATE WITHIN PROMOTION ZONE= ?0.138

W1, WASTAGE TRANSITION RATE WITHIN FIRST (NON-PROMOTION) ZONE= ?0.131

W2, WASTAGE TRANSITION RATE WITHIN PROMOTION ZONE= ?0.089

$P2$ and $W2$ are the rates specific to the zone $(A, B)$, and $W1$ is specific to $(0,A)$.

[*The results are now printed as in Printout B*.1.]

If the option for the approximation for $P2$ is used the prompts are as below from the following prompt onwards (see Printout B.3 in Section B.2).

1: DO YOU HAVE VALUES FOR THE ZONE SPECIFIC RATES, OR
2: DO YOU WISH TO USE THE APPROXIMATION FOR P2 WHICH REQUIRES GRADE SPECIFIC RATES AND THE CUT-OFF POINT

REPLY 1/2 ?2

K, THE CUT-OFF OR RETIREMENT POINT= ?15

This is the average maximum time which a person can spend in the grade. If there is no maximum put $K$ equal to some appropriately large number. $K$ should always be greater than or equal to $B$. $K$ helps specify the seniority distribution used in the approximation. Since this distribution is very skewed, the value of $K$ can sometimes be critical. It may therefore be advisable, particularly if the actual value of $K$ is uncertain, to check the sensitivity of the results by repeating the calculations using various values of $K$.

P, PROMOTION TRANSITION RATE FOR GRADE= ?0.046

W, WASTAGE TRANSITION RATE FOR GRADE= ?0.124

These are the grade-specific rates which apply to the whole grade.

[*The results are now printed as in Printout B.3.*]

## B.4 SUMMARY OF FORMULAE

$$PU = S1^A P2(1 - S2^{(B-A)})/(1 - S2)$$
$$TP = A + (1 - S2)^{-1} - \tfrac{1}{2} - (B - A)S2^{(B-A)}/(1 - S2)^{(B-A)})$$
$$PU^* = 1 - S2^{*(B-A)}$$
$$TP^* = A + (1 - S2^*)^{-1} - \tfrac{1}{2} - (B - A)S2^{*(B-A)}/(1 - S2^{*(B-A)})$$

where $S2 = (1 - W2 - P2)$, $S2^* = (1 - P2^*)$, and the other variables are as previously defined. The symbol * is used here as part of the variable label and should not be confused with its use as a multiplication sign. The conditional chance of promotion

$P2^* = P2$ for assumption (a)

$= P2/(1 - W2)$ for assumption (b)

$= 1 - S2^{P2/(P2+W2)}$ for assumption (c).

The approximation for $P2$ is given by

$$P2 = P(1 - S^k)/(S^A - S^B).$$

## B.5 LISTING

The comments made in Section A.4 on possible problems in setting up the program also apply here.

Printout B.5 BASIC listing of program

```
100 PRINT "PROGRAM CALCULATES THE CAREER PATTERN FROM PROMOTION AND "
104 PRINT "WASTAGE TRANSITION RATES.  FOR METHODOLOGY SEE STMP CH 6."
108 PRINT
112 PRINT
116 PRINT"DETAILED OR ABBREVIATED PROMPTS?   REPLY D/A";
120 INPUT C$
124 PRINT
128 PRINT"DETAILED OR ABBREVIATED RESULTS?   REPLY D/A";
132 INPUT B$
136 IF C$="A" GOTO 656
140 *************************************************************
144 *INPUT SEGMENT USING DETAILED DATA PROMPTS
148 *************************************************************
152 PRINT
156 PRINT
160 PRINT
164 PRINT
168 PRINT"A, LOWER BOUNDARY OF PROMOTION ZONE =";
172 INPUT Z1
176 PRINT
180 PRINT"B, UPPER BOUNDARY OF PROMOTION ZONE =";
184 INPUT Z2
188 PRINT
192 PRINT "WHICH ASSUMPTION DO YOU WISH TO MAKE IN CALCULATING THE"
196 PRINT"PROMOTION RATE P*(=P2*) CONDITIONAL ON NOT LEAVING :"
200 PRINT "  1: PROMOTION BEFORE WASTAGE"
204 PRINT"  2: WASTAGE BEFORE PROMOTION"
208 PRINT"  3: PROMOTION AND WASTAGE SIMULTANEOUS"
212 PRINT "REPLY 1/2/3";
216 INPUT H1
220 PRINT
224 PRINT"1: DO YOU HAVE VALUES FOR THE ZONE SPECIFIC RATES, OR"
228 PRINT"2: DO YOU WISH TO USE THE APPROXIMATION FOR P2 WHICH"
232 PRINT"   REQUIRES GRADE SPECIFIC RATES AND THE CUT-OFF POINT ?"
236 PRINT"REPLY 1/2";
240 INPUT H2
244 IF H2=1 GOTO 264
248 IF H2=2 GOTO 320
252 GO TO 220
256 *
260 *  1: INPUT FOR ZONE RATES
264 PRINT
268 PRINT"P2, PROMOTION TRANSITION RATE WITHIN PROMOTION ZONE =";
272 INPUT P2
276 PRINT
280 PRINT"W1, WASTAGE TRANSITION RATE WITHIN FIRST (NON-PROMOTION)"
284 PRINT"     ZONE =";
288 INPUT W1
292 PRINT
296 PRINT"W2, WASTAGE TRANSITION RATE WITHIN PROMOTION ZONE =";
300 INPUT W2
304 PRINT
308 GOTO 400
312 *
316 *  2: INPUT FOR APPROX P2
320 PRINT
324 PRINT"K, THE CUTOFF OR RETIREMENT POINT =";
```

```
328 INPUT Z3
332 IF Z3<Z2 GOTO 320
336 PRINT
340 PRINT"P, PROMOTION TRANSITION RATE FOR GRADE =";
344 INPUT P
348 PRINT
352 PRINT"W, WASTAGE TRANSITION RATE FOR GRADE =";
356 INPUT W
360 *  CALC OF APPROX P2
364 W1=W
368 W2=W
372 S=1-P-W
376 P2=P*(1-S**Z3)/(S**Z1-S**Z2)
380 IF P2>(1-W2) GOTO 712
384 IF P2<0.0 GOTO 712
388 ***************************************************************
392 * CALCULATIONS SEGMENT FOR PU, TP, PU*, TP*.
396 ***************************************************************
400 S1=1-W1
404 S2=1-W2-P2
408 Z=Z2-Z1
412 R1=S2**Z
416 IF H1=3 GOTO 428
420 IF H1=1 GOTO 436
424 IF H1=2 GOTO 444
428 Q=1-S2**(P2/(P2+W2))
432 GOTO448
436 Q=P2
440 GOTO 448
444 Q=P2/(1-W2)
448 R2=(1-Q)**Z
452 A =(S1**Z1)*P2*(1-R1)/(1-S2)
456 B =Z1 + 1/(1-S2) - 0.5 - Z*R1/(1-R1)
460 C =1 - R2
464 D =Z1 + 1/Q - 0.5 - Z*R2/(1-R2)
468 ***************************************************************
472 *  PRINTOUT SEGMENT
476 ***************************************************************
480 IF B$="A" GOTO 564
484 PRINT
488 IF H2=1 GOTO 508
492 PRINT
496 PRINT
500 PRINT USING 788,P2
504 PRINT USING 792,W
508 PRINT
512 PRINT USING 784,Q
516 PRINT
520 PRINT USING 796,A
524 PRINT
528 PRINT USING 800,B
532 PRINT
536 PRINT USING 804
540 PRINT USING 808
544 PRINT USING 812,Z2,C
548 PRINT
552 PRINT USING 816
```

```
556 PRINT USING 820,D
560 GOTO 596
564 PRINT
568 IF H2=1 GOTO 580
572 PRINT
576 PRINT USING 772,P2,W
580 PRINT
584 PRINT USING 776,Q
588 PRINT
592 PRINT USING 780,A,B,C,D
596 PRINT
600 PRINT
604 PRINT
608 PRINT
612 PRINT
616 PRINT"FINISH OR RESTART?   REPLY F/R";
620 INPUT A$
624 IF A$="R" GOTO 108
628 IF A$="F" GOTO 824
632 GOTO 616
636 PRINT
640 GOTO 564
644 *****************************************************************
648 * INPUT SEGMENT USING ABBREVIATED DATA PROMPTS
652 *****************************************************************
656 PRINT
660 PRINT "PROM ZONE : A,B =";
664 INPUT Z1, Z2
668 PRINT "WHICH ASSUMPTION FOR P2* (1/2/3)";
672 INPUT H1
676 PRINT "1: INPUT ZONE RATES, OR 2: USE APPROX FOR P2 ?  (1/2)";
680 INPUT H2
684 IF H2 = 2  GOTO 700
688 PRINT "P2,W1,W2=";
692 INPUT P2,W1,W2
696 GOTO 400
700 PRINT "K,P,W=";
704 INPUT Z3,P,W
708 GOTO 364
712 *****************************************************************
716 * ERROR MESSAGES AND FORMAT STATEMENTS
720 *****************************************************************
724 PRINT
728 PRINT USING 756,P2
732 PRINT USING 760
736 PRINT USING 764
740 PRINT USING 768
744 PRINT
748 PRINT
752 GOTO388
756 :THE APPROXIMATE P2=#.###  SINCE THIS SHOULD LIE BETWEEN 0 AND 1-W2
760 :SUGGESTS THAT THE ZONES ARE NOT COMPATIBLE WITH THE GRADE SPECIFIC
764 :RATES. REMEMBER THE APPROXIMATION ASSUMES THE SENIORITY
768 :DISTRIBUTION WITHIN GRADE IS STEADY STATE.
772 :APPROX P2=#.###           W1=W2=W=#.###
776 :P2*=#.###
780 :PU=#.###    TP=##.##      PU*=#.###    TP*=##.##
```

```
784 :THE PROMOTION RATE CONDITIONAL ON NOT LEAVING, P2*=#.###
788 :THE CALCULATED VALUE FOR THE APPROXIMATE P2=#.###
792 :THE WASTAGE RATES WILL BE TAKEN AS W1=W2=W=#.###
796 :THE PROPORTION OF ENTRANTS PROMOTED          (PU)=#.###
800 :THE AVERAGE EXPERIENCE OF PROMOTEES          (TP)=##.##
804 :THE AVERAGE CHANCE OF PROMOTION FOR AN
808 :   INDIVIDUAL WHO IS PREPARED IF NECESSARY
812 :   TO NOT LEAVE FOR A TIME B=###.#           (PU*)=#.###
816 :THE EXPECTED TIME BEFORROMOTION FOR
820 :   THIS INDIVIDUAL IF HE IS PROMOTED         (TP*)=##.##
824 END
```

# Bibliography and Author Index

The following list contains all the references in the book and the pages on which they are cited are given in square brackets. It also contains other references concerned with the statistical aspects of manpower planning.


Aitchison, J. (1955). Contribution to the discussion on Lane and Andrew (1955). [54]

Aitchison, J., and J. A. C. Brown (1966). *The Lognormal Distribution*, Cambridge University Press, London. [59, 62]

Almond, G. *See* Young and Almond (1961).

Al-Nuaimi, A. *See* Martel and Al-Nuaimi (1973).

Andrew, J. E. *See* Lane and Andrew (1955).

Andrews, P. *See* Wheeler and Andrews (1972).

Annoni, A. J. *See* Milkovich, Annoni, and Mahoney (1972).

Armacost, R. L. *See* Oliver and coworkers (1972).

Ashdown, P. L. (1974). 'Manpower planning in the forestry commission'. *Personnel Review*, **3**, 26–33.

Balinsky, W., and A. Reisman (1973). 'A taxonomy of manpower–educational planning models'. *Socio-Economic Planning Sciences*, **7**, 13–18.

Barone, S. *See* Kwak, Garrett, and Barone (1977).

Bartholomew, D. J. (1959). 'Note on the measurement and prediction of labour turnover'. *J. R. Statist. Soc.* **A122**, 232–239. [48, 60, 75]

Bartholomew, D. J. (1963a). 'A multi-stage renewal process'. *J. R. Statist. Soc.* **B25**, 150–168. [9, 74, 153]

Bartholomew, D. J. (1963b). 'An approximate solution of the integral equation of renewal theory'. *J. R. Statist. Soc.* **B25**, 432–441. [138]

Bartholomew, D. J. (1963c). 'Two-stage replacement strategies'. *Operat. Res. Quart.*, **14**, 71–87. [153]

Bartholomew, D. J. (1971). 'The statistical approach to manpower planning'. *Statistician*, **20**, 3–26.

Bartholomew, D. J. (1972). Contribution to the discussion of Cox (1972). [39]

Bartholomew, D. J. (1973a). *Stochastic Models for Social Processes* (2nd ed.), Wiley, London. (1st ed. 1967). [9, 49, 50, 51, 74, 75, 87, 92, 110, 111, 137, 138, 153, 180, 184, 209, 217, 222, 241]

Bartholomew, D. J. (1973b). 'A model of completed length of service'. *Omega*, **1**, 235–240.

Bartholomew, D. J. (1975a). 'Errors of prediction in Markov chain models'. *J. R. Statist. Soc.*, **B37**, 444–456. [110]

Bartholomew, D. J. (1975b). 'A stochastic control problem in the social sciences' (with discussion). *Bull. Int. Statist. Inst.*, **46**, 670–680. [111, 209, 222]

Bartholomew, D. J. (1976a). 'Statistical problems of prediction and control in manpower planning'. *Math. Scientist.*, **1**, 133–144. [209]

Bartholomew, D. J. (1976b). 'Renewal theory models in manpower planning'. *Symposium Proceedings Series No. 8, The Institute of Mathematics and Its Applications*, 57–73. [153]

Bartholomew, D. J. (1977a). 'Manpower planning literature: statistical techniques of mapower analysis'. *Department of Employment Gazette*, **1977**, 1093–1095. [11]

Bartholomew, D. J. (1977b). 'The analysis of data arising from stochastic processes', in O'Muircheartaigh and Payne (1977), 145–174. [106, 111]

Bartholomew, D. J. (1977c). 'Maintaining a grade or age structure in a stochastic environment'. *Adv. Appl. Prob.*, **9**, 1–17. [111, 209, 222]

Bartholomew, D. J. (ed.) (1977d). *Manpower Planning*, Penguin Modern Management Readings, Penguin Books, Harmondsworth, Middlesex. [11, 37, 153, 221]

Bartholomew, D. J., and A. D. Butler (1971). 'The distribution of the number of leavers for an organization of fixed size', in Smith (1971b), 417–426.

Bartholomew, D. J., R. F. A. Hopes, and A. R. Smith (1976). 'Manpower planning in the face of uncertainty'. *Personnel Review*, **5**, 5–17. [55, 110, 234, 246]

Bartholomew, D. J., and B. R. Morris (eds) (1971). *Aspects of Manpower Planning*, English Universities Press, London and American Elsevier, New York.

Bartholomew, D. J., and A. R. Smith (eds) (1971). *Manpower and Management Science*, English Universities Press, London, and D. C. Heath and Co., Lexington, Mass. [221]

Bell, D. J. (1974). *Planning Corporate Manpower*, Longman, London.

Bell, D. J. *See also* Jones and coworkers (1967).

Blight, B. J. N. (1968). 'A note on a modified exponentially weighted predictor'. *J. R. Statist. Soc.* **B30**, 318–320 [236]

Blom, A. J., and A. J. Knights (1976). 'Long-term manpower forecasting in the Zambian mining industry'. *Personnel Review*, **5**, 18–28. [11]

Bowey, A. M. (1969). 'Labour stability curves and labour stability'. *Brit. J. Indust. Rel.*, **7**, 69–84. [75]

Bowey, A. M., (1974). *A Guide to Manpower Planning*, Macmillan, London. [68, 73]

Box, G. E. P., and Jenkins, B. M. (1970). *Time Series Analysis Forecasting and Control*, Holden-Day, San Francisco. [237]

Brown, J. A. C. *See* Aitchison and Brown (1955).

Brown, R. G. (1963). *Smoothing, Forecasting and Prediction of Discrete Time Series*. Prentice-Hall, New Jersey. [236]

Brown, R. L., J. Durbin, and J. M. Evans (1975). 'Techniques for testing the constancy of regression relationships over time'. *J. R. Statist. Soc.*, **B37**, 149–192. [245]

Bryant, D. T. (1972). 'Recent developments in manpower research'. *Personnel Review*, **1**, 14–31.

Bryant, D. T. (1973). 'Manpower planning and techniques'. *Business Horizons*, **16**, 69–78.

Burack, E. H., and J. W. Walker (1972). *Manpower Planning and Programming*, Allyn and Bacon, Boston.

Burr, I. W. (1942). 'Cumulative frequency functions'. *Ann. Math. Statist.*, **13**, 215–232. [75]

Butler, A. D. (1971). 'The distribution of numbers promoted and leaving in a graded organization'. *Statistician*, **20**, 69–84.

Butler, A. D. *See also* Bartholomew and Butler (1971).

Cameron, M. H., and J. E. Nash (1974). 'On forecasting the manpower requirements of an organization with homogeneous workloads'. *J. R. Statist. Soc.*, **A137**, 200–218. [237, 240, 245]

Center, A. *See* Jones and coworkers (1967).

Charnes, A., W. W. Cooper, and R. J. Niehaus (1972). *Studies in Manpower Planning*, Office of Civilian Manpower Management, Department of the Navy, Washington D.C. [10, 222]

Chatfield, C., and D. Prothero (1973). 'Box–Jenkins seasonal forecasting: problems in a case study'. *J. R. Statist. Soc.*, **A136**, 295–336. [237]

Chiang, C. L. (1968). *Introduction to Stochastic Processes in Biostatistics*, John Wiley, New York. [37, 174, 183]

Clark, H. L. (1974). 'Problems and progress in Civil Service manpower planning in the United States', in Clough, Lewis, and Oliver (eds) (1974), 227–239.

Clough, D. J., C. G. Lewis, and A. L. Oliver (eds) (1974). *Manpower Planning Models*, English Universities Press, London.

Clowes, G. A. (1972). 'A dynamic model for the analysis of labour turnover'. *J. R. Statist. Soc.*, **A135**, 242–256. [50]

Coleman, D. *See* Jones and coworkers (1967).

Coleman, J. (1973). *The Mathematics of Collective Action*, Heinemann Educational Books Ltd., London. [38]

Cooper, W. W. *See* Charnes, Cooper, and Niehaus (1972).

Cox, D. R. (1961). 'Prediction by exponentially weighted moving averages and related methods'. *J. R. Statist. Soc.*, **B23**, 414–422. [236]

Cox, D. R. (1970). *Analysis of Binary Data*, Methuen, London. [38]

Cox, D. R. (1972). 'Regression models and life tables'. *J. R. Statist. Soc.*, **B34**, 187–220. [38]

Cronin, D. (1977). *Theory and Applications of the Log-logistic Growth Function*, M.A. thesis, Polytechnic of Central London. [75]

Cullingford, G., and D. Scott (1973). 'Optimality and manpower planning'. *Personnel Review*, **2**, 38–48.

Cullingford, G. *See also* Scott and Cullingford (1974).

Daellenbach, H. G. (1976). 'Note on a stochastic manpower smoothing and production model'. *Operat. Res. Quart.*, **27**, 573–579.

Davies, G. S. (1973). 'Structural control in a graded manpower system'. *Man. Sci.*, **20**, 76–84. [111]

Davies, G. S. (1975). 'Maintainability of structures in Markov chain models under recruitment control'. *J. Appl. Prob.*, **12**, 376–382.

Dawson, D. A., and F. T. Denton (1974). 'Some models for simulating Canadian manpower flows and related systems'. *Socio-Economic Planning Sciences*, **8**, 233–248. [112]

Denton, F. T. *See* Dawson and Denton (1974).

Dill, W. R., D. P. Gaver, and W. L. Weber (1966). 'Models and modelling for manpower planning'. *Man. Sci.*, **13**, B142–B166.

Drandell, M. (1975). 'Composite forecasting methodology for manpower planning utilizing objective and subjective criteria'. *Academy of Management Journal*, **18**, 510–519.

Drui, A. B. (1963). 'The use of regression equations to predict manpower requirements'. *Man. Sci.* **8**, 669–677. [243]

Durbin, J. *See* Brown, Durbin, and Evans (1975).

Evans, J. M. *See* Brown, Durbin, and Evans (1975).

Feichtinger, G. (1976). 'On the generalization of stable age distributions to Gani-type person-flow models'. *Adv. App. Prob.*, **8**, 433–445.

Feichtinger, G., and A. Mehlmann (1976). 'The recruitment trajectory corresponding to particular stock sequences in Markovian person-flow models'. *Math. of Operat. Res.*, **1**, 175–184.

Forbes, A. F. (1971a). 'Markov chain models in manpower systems', in Bartholomew and Smith (1971), 93–113. [97, 107, 110, 112]

Forbes, A. F. (1971b). 'Non-parametric methods of estimating the survivor function'. *Statistician*, **20**, 27–52. [28, 29, 37]

Forbes, A. F. (1973). 'The relationship between promotion rates and promotion prospects'. Institute of Manpower Studies, University of Sussex, Brighton. [174, 178, 183, 184]

Forbes, A. F. (1976). 'An advisory service on computer based manpower models'. Institute of Manpower Studies, University of Sussex, Brighton. [183]

Forbes, A. F. (1977). *A Note on a Discrete Approximation for the Zone Specific Promotion Rate*, Institute of Manpower Studies, Brighton. [179]

Forbes, A. F., R. W. Morgan, and A. J. Rowntree (1975). 'Manpower planning models in use in the Civil Service Department Statistics Division'. *Personnel Review*, **4**, 23–35. [222]

Fougstedt, G. (1961). 'Problems of forecasting the future supply of persons with university training and the demand for their services'. *Bull. Int. Statist. Inst.*, **39**, 4376.

Gani, J. (1963). 'Formulae for projecting enrolments and degrees awarded in universities'. *J. R. Statist. Soc.*, **A126**, 400–409. [9, 92, 112, 184]

Garrett, W. A. *See* Kwak, Garrett, and Barone (1977).

Gaver, D. P. *See* Dill, Gaver, and Weber (1966).

Glen, J. J. (1977). 'Length of service distributions in Markov manpower models'. *Operat. Res. Quart.*, **28**, 975–982. [112]

Gray, D. A. *See* Kahalas and Gray (1976).

Gray, D. H. (1966). *Manpower Planning*, Institute of Personnel Management, London.

Grinold, R. C. (1976a). 'Input policies of a longitudinal manpower flow model'. *Man. Sci.*, **22**, 570–575.

Grinold, R. C. (1976b). 'Manpower planning with uncertain requirements'. *Operations Research*, **24**, 387–399.

Grinold, R. C., and K. T. Marshall (1977). *Manpower Planning Models*. North-Holland, New York and Amsterdam. [11, 110, 111]

Grinold, R. C. and K. T. Marshall (1978). 'Manpower Planning under uncertain conditions'. *TIMS Studies in the Management Sciences*, **8**, 209–217.

Grinold, R. C., and R. E. Stanford (1974). 'Optimal control of a graded manpower system'. *Man. Sci.*, **20**, 1201–1216. [110, 111]

Hagerstrand, T. (1967). *Innovation Diffusion as a Spatial Process*, The University of Chicago Press, Chicago and London. [241]

Halpern, J. (1974). 'A forecasting technique with an application to the Civil Service', in Clough, Lewis, and Oliver (1974).

Harman, H. H. (1967). *Modern Factor Analysis*, University of Chicago Press. [239]

Harrison, P. J. (1965). 'Short-term sales forecasting'. *App. Statist.*, **14**, 102–139. [236]

Harrison, P. J. (1967). 'Exponential smoothing and short-term sales forecasting'. *Man. Sci.*, **13**, 821–842. [236]

Hayne, W. J. and K. T. Marshall (1977). 'Two-characteristic Markov-type manpower flow models'. *Naval Research Logistics Quarterly*, **24**, 235–256.

Hedberg, M. (1961). 'The turnover of labour in industry, an actuarial study'. *Acta. Sociologica*, **5**, 129–143. [13]

Helps, I. G. (1970). 'A method of estimating craft manpower requirements at industry level'. *Operat. Res. Quart.*, **21** 341–352. [246]

Herbst, P. G. (1963). 'Organizational commitment: a decision model'. *Acta Sociologica*, **7**, 34–45. [50, 53]

Hershey, J. C. *See* Wandel and Hershey (1976).

Hill, J. M. M. *See* Rice and coworkers (1950).

Holt, C. C. (1957). *Forecasting Seasonality and Trends by Exponentially Weighted Moving Averages*, O.N.R. Research Memorandum No. 52, Carnegie Institute of Technology. [236]

Hopes, R. F. A. (1973). 'Some statistical aspects of manpower planning in the Civil Service'. *Omega*, **1**, 65–180. [9, 153, 183]

Hopes, R. F. A. *See also* Bartholomew, Hopes, and Smith (1976).

Hopkins, D. S. P. (1974). 'Faculty early retirement programs'. *Operat. Res.*, **22**, 455–467. [112]

Hopkins, D. S. P. *See also* Oliver and coworkers (1972).

Hyman, R. (1970). 'Economic motivation and labour stability'. *Brit. J. Indust. Rel.*, **8**, 159–178. [75]

Jenkins, G. M. *See* Box and Jenkins (1970).

Jessop, W. N. (1966). *Manpower Planning*, English Universities Press, London and American Elsevier, New York.

Johnson, W. D. and G. G. Koch (1978). 'Linear models analysis of competing risks for grouped survival times'. *Int. Statist. Review*, **46**, 21–51.

Jones, G. (1978). 'A method of re-interpreting promotion rates in terms of promotion prospectuses'. *Applied Statistics*, **27**, 58–68. [183]

Jones, G., D. J. Bell, D. Coleman, and A. Center (1967). *Perspectives in Manpower Planning on Edinburgh Group Report*, Institute of Personnel Management, London.

Jones, R. C., S. R. Morrison, and R. P. Whiteman (1973). 'Helping to plan a bank's manpower resources'. *Operat. Res. Quart.*, **24**, 365–374.

Judge, G. C. *See* Lee and coworkers (1970).

Kahalas, H. and D. A. Gray (1976). 'A quantitative model for manpower decision making'. *Omega*, **4**, 685–697.

Kay, R. (1977). 'Proportional hazard regression models and the analysis of censored survival data'. *App. Statist.*, **26**, 227–237.

Keenay, G. A., R. W. Morgan, and K. H. Ray (1977a). 'An analytical model for company manpower planning'. *Operat. Res. Quart.*, **28**, 983–996. [169]

Keenay, G. A., R. W. Morgan, and K. H. Ray (1977b). 'The Camel model: a model for career planning in a hierarchy'. *Personnel Review*, **6**, 43–50. [169, 183]

Keenay, G. A. *See also* Morgan and coworkers (1973).

Kemeny, J. G., and L. Snell (1960). *Finite Markov Chains*, Van Nostrand, New York. Second printing (1976), Springer-Verlag, Berlin. [95, 180]

Kendall, M. G. (1975). *Multivariate Analysis*, Griffin, High Wycombe, Bucks.

Kendall, M. G. (1976). *Time Series* (2nd ed.), Griffin, High Wycombe, Bucks. [235, 239]

Kendall, M. G., and A. Stuart (1976). *The Advanced Theory of Statistics*, Vol. 3 (3rd ed.), Griffin, High Wycombe, Bucks. [239]

Knights, A. J. *See* Blom and Knights (1976).

Koch, G. G. *See* Johnson and Koch (1978).

Kwak, N. K., W. A. Garrett and S. Barone (1977). 'A stochastic model of demand forecasting for technical manpower planning'. *Man. Sci.*, **23**, 1089–1098.

Lane, K. F., and J. E. Andrew (1955). 'A method of labour turnover analysis'. *J. R. Statist. Soc.*, **A118**, 296–323. [9, 28, 53, 54, 68, 69, 70, 71, 72]

Lapp, P. A., and I. W. Thompson (1974). 'Supply and demand for engineering manpower related to the University system in Ontario', in Clough, Lewis, and Oliver (1974), 293–310. [243]

Laslett, R. E. (1972). *A survey of mathematical methods of estimating the supply and demand for manpower*. Engineering Training Board. Occasional Paper No. 1.

Lawley, D. N., and A. E. Maxwell (1971). *Factor Analysis as a Statistical Method*, Butterworths, London. [239]

Lawrence, J. R. (1973). 'Manpower and personnel models in Britain'. *Personnel Review*, **2**, 4–26.

Lee, T. C., G. C. Judge, and A. Zellner (1970). *Estimating the Parameters of the Markov Probability Model from Aggregate Time Data*, North Holland Publishing Co. [106]

Lewin, C. G. (1971), 'A manpower planning study'. *Operat. Res. Quart.*, **22**, 99–116.

Lewis, C. G. (ed.) (1969). *Manpower Planning: a Bibliography*, English Universities Press, London.

Lewis, C. G. *See also* Clough, Lewis, and Oliver (1974).

Lilien, G. L., and A. G. Rao (1975). 'Model for manpower management'. *Man. Sci.*, **21**, 1447–1457.

Lindsey, J. K. (1973). *Inferences from Sociological Survey Data—A Unified Approach*, Elsevier Scientific Publishing Company, Amsterdam, London, New York. [38]

Livingstone, J. L., and D. B. Montgomery (1966). 'The use of regression equations to predict manpower requirements: critical comments'. *Man. Sci.*, **12**, 616–618. [245]

McCarthy, C., and T. M. Ryan (1977). 'Estimates of voter transition probabilities from the British General Election of 1974'. *J. R. Statist. Soc.*, **A140**, 78–85. [106]

MacCrimmon, K. R. *See* Vroom and MacCrimmon (1968).

McLean, S. I. (1975). 'A comparison of the lognormal and transition models of wastage'. *Statistician*, **25**, 281–294.

McLean, S. I. (1976). 'The two-stage model of personnel behaviour'. *J. R. Statist. Soc.*, **A139**, 205–217.

McLean, S. I. (1978). 'Continuous-time stochastic models of a multi-grade population'. *J. Appl. Prob.*, **15**, 26–37.

Mahoney, T. A., and G. T. Milkovich (1971). 'The internal labour market as a stochastic process', in Bartholomew and Smith (1971), 75–91. [112]

Mahoney, T. A., G. T. Milkovich, and N. Weiner (1977). 'A stock and flow model for improved human resource measurement'. *Personnel*, May–June, 57–66. [6]

Mahoney, T. A. *See also* Milkovich, Annoni, and Mahoney (1972).

Mapes, R. (1968). 'Promotion in static hierarchies'. *J. Man. Studies*, **5**, 365–379.

Marland, M. W. (1977). 'Officer manpower planning in the Royal Air Force'. *Statistical News*, May, 37.22–37.23.

Marshall, K. T., and R. M. Oliver (1970). 'A constant work model for student attendance and enrolment'. *Operat. Res.*, **18**, 193–206.

Marshall, K. T. *See also* Grinold and Marshall (1977), (1978) and Hayne and Marshall (1977).

Marshall, M. L. (1971). 'Some statistical methods for forecasting wastage'. *Statistician*, **20**, 53–68.

Marshall, M. L. (1974). 'Fitting the two-term exponential and two-parameter lognormal distributions to grouped and censored data'. *App. Statist.* **23**, 313–322. [57, 63, 64]

Marshall, M. L. (1975). 'Equilibrium age distributions for graded systems'. *J. R. Statist. Soc.*, **A138**, 62–69.

Martel, A., and A. Al-Nuaimi (1973). 'Tactical manpower planning via programming under uncertainty'. *Operat. Res. Quart.*, **24**, 571–585.

Maxwell, A. E. (1977). *Multivariate Analysis in Behavioural Research*, Chapman and Hall, London.

Maxwell, A. E. *See also* Lawley and Maxwell (1971).

Mehlmann, A. (1977). 'A note on the limiting behaviour of discrete-time Markovian manpower models with inhomogeneous independent Poisson input'. *J. App. Prob.*, **14**, 611–613.

Mehlmann, A. *See also* Feichtinger and Mehlmann (1976).

Milkovich, G. T., A. J. Annoni and T. H. Mahoney (1972). 'The use of the Delphi procedures in manpower forecasting'. *Man. Sci.*, **19**, 381–388.

Milkovitch, G. T. *See also* Mahoney and Milkovitch (1971); Mahoney and coworkers (1977); and Valliant and Milkovitch (1977).

Miller, S. *See* van der Merwe and Miller (1971).

Montgomery, D. B. *See* Livingstone and Montgomery (1966).

Moore, P. G. (1977). 'The managers' struggles with uncertainty' (with discussion). *J. R. Statist. Soc.*, **A140**, 129–165. [246]

Morgan, R. W. (1971). 'The use of a steady state model to obtain the recruitment, retirement and promotion policies of an expanding organization', in Bartholomew and Smith (1971), 283–291. [9, 222]

Morgan, R. W., G. A. Keenay, and K. H. Ray (1974). 'A steady state model for career planning', in Clough, Lewis, and Oliver (1974). [9, 166, 169]

Morgan, R. W. *See also* Forbes and coworkers (1975); Keenay and coworkers (1977a) and (1977b).

Morris, B. R. *See* Bartholomew and Morris (1971).

Morrison, S. R. *See* Jones, R. C., and coworkers (1973).

Moya-Angeler, J. (1976). 'A model with shortage of places for educational and manpower systems'. *Omega*, **4**, 719–730.

Muth, J. F. (1960). 'Optimal properties of exponentially weighted forecasts'. *J. Amer. Statist. Ass.*, **55**, 299–306.

Nash, J. E. *See* Cameron and Nash (1974).

Niehaus, R. J. *See* Charnes and coworkers (1972).

Nielson, G. L., and A. R. Young (1973). 'Manpower Planning: a Markov chain application'. *Public Personnel Management*, March–April, 133–143. [112]

Oliver, A. L. *See* Clough, Lewis, and Oliver (1974).

Oliver, F. R. (1966). 'Aspects of maximum likelihood estimation of the logistic growth function'. *J. Amer. Statist. Ass.*, **61**, 697–705. [241]

Oliver, R. M., D. S. P. Hopkins, and R. L. Armacost (1972). 'An equilibrium flow model of a university campus'. *Operat. Res.* **20**, 249–264. [221]

Oliver, R. M. *See also* Marshall and Oliver (1970).

O'Muircheartaigh, C. A., and C. D. Payne (1977). *The Analysis of Survey Data*: Vol. 1, *Exploring Data Structures*; Vol. 2, *Model Fitting*. Wiley, London.

Parkhouse, J. (1977). 'Simple model for medical manpower studies'. *Brit. Med. J.*, 20 August, 530. [6]

Payne, C. D. (1977). 'The log-linear model for contingency tables', in O'Muircheartaigh and Payne (1977), 105–144. [38]

Payne, C. D. *See also* O'Muircheartaigh and Payne (1977).

Pollard, J. H. (1966). 'On the use of the direct matrix product in analysing certain stochastic population models'. *Biometrika*, **53**, 397-415. [110]

Price, P. C. (1971). 'Mathematical models of staff structure evaluation', in Bartholomew and Morris (1971), 98–108. [11]

Prothero, D. *See* Chatfield and Prothero (1973).

Purkiss, C. J. (1974). *Manpower Planning: a Contribution of Concepts and Practice.* Ph.D. Thesis, University of Lancaster. [240]

Rao, A. G. *See* Lilien and Rao (1975).

Ray, K. (1977). 'Managerial manpower planning—a systematic approach'. *Long Range Planning*, **10**, 21–30.

Ray, K. H. *See also* Morgan and coworkers (1973); Keenay and coworkers (1977a) and (1977b).

Reisman, A. *See* Balinsky and Reisman (1973).

Rice, A. K., J. M. M. Hill, and E. L. Trist (1950). 'The representation of labour turnover as a social process'. *Human Relations*, **3**, 349–381. [9, 38, 47, 49]

Rice, J., and M. Rosenblatt (1976). 'Estimation of the log-survivor function and hazard function'. *Sankhya*, **A38**, 60–78.

Robinson, D. (1974). 'Two-stage replacement strategies and their application to manpower planning'. *Man. Sci.*, **21**, 199–208.

Rogers, E. (1962). *Diffusion of Innovations*, Free Press, Glencoe, Illinois. [241]

Rosenblatt, M. *See* Rice and Rosenblatt (1976).

Rowland, K. M., and M. G. Sovereign (1969). 'Markov chain analysis of internal manpower supply'. *Industrial Relations*, **9**, 88–99. [112]

Rowntree, A. J. (1970). 'Stationary population models', in Smith (1970), 15–22.

Rowntree, A. J. (1971). 'A new entry to the Civil Service', in Bartholomew and Smith (1971), 267–281.

Rowntree, A. J., and P. A. Stewart (1976). 'Estimating manpower needs—II: statistical methods', in Smith (1976), 36–53.

Rowntree, A. J. *See also* Forbes and coworkers (1975).

Sales, P. (1971). 'The validity of the Markov chain model for a class of the Civil Service'. *Statistican*, **20**, 85–110. [107, 108, 110, 112]

Sawtell, R. A., and P. Sweeting (1975). 'A practical guide to company manpower planning'. *Personnel Review*, **4**, 33–40.

Scott, D., and G. Cullingford (1974). 'Transition to a desired manpower structure'. *Omega*, **2**, 793–803.

Scott, D. *See also* Cullingford and Scott (1973).

Seal, H. L. (1945). 'The mathematics of a population composed of *k* stationary strata each recruited from the stratum below and supported at the lowest level by a uniform number of annual entrants'. *Biometrika*, **33**, 226–230. [8]

Silcock, H. (1954). 'The phenomenon of labour turnover'. *J. R. Statist. Soc.*, **A117**, 429–440. [9, 13, 47, 48, 49, 50, 62, 70]

Smith, A. R. (ed.) (1970). *Some Statistical Techniques of Manpower Planning*, CAS Occasional Paper No. 15, HMSO, London. [221]

Smith, A. R. (1971a). 'Developments in manpower planning'. *Personnel Review*, **1**, 44–54.

Smith, A. R. (ed.) (1971b). *Models for Manpower Systems*, English Universities Press. London.

Smith, A. R. (ed.) (1976). *Manpower planning in the Civil Service*, Civil Service Studies No. 3, HMSO, London. [9, 151, 234]

Smith, A. R. *See also* Bartholomew and Smith (1971); Bartholomew, Hopes, and Smith (1976).

Snell, L. *See* Kemeny and Snell (1960).

Sovereign, M. G. *See* Rowland and Sovereign (1969).

Stainer, G. (1971). *Manpower Planning*, Heinemann, London.

Stanford, R. E. *See* Grinold and Stanford (1974).

Stewart, P. A. *See* Rowntree and Stewart (1976).

Stewman, S. (1975). 'An application of the job vacancy chain model to a Civil Service internal labour market'. *J. Math. Soc.*, **4**, 37–59. [112, 142, 143, 144, 153]

Stuart, A. *See* Kendal and Stuart (1976).

Sweeting, P. *See* Sawtell and Sweeting (1975).

Teather, D. (1971). *The Estimation of Transition Probabilities of a Simple Markov Chain with Applications in Manpower Planning*, M.Sc. Thesis, University of London. [106].

Thompson, I. W. *See* Lapp and Thompson (1974).

Thonstad, T. (1969). *Education and Manpower; Theoretical Models and Empirical Applications*, Oliver and Boyd, Edinburgh and London. [112]

Trist, E. L. *See* Rice and coworkers (1950).
Vajda, S. (1947). 'The stratified semi-stationary population'. *Biometrika*, **34**, 243–254. [8]
Vajda, S. (1948). 'Introduction to a mathematical theory of a graded stationary population'. *Bull. de l'Ass. Actuair. Suisses*, **48**, 251–273. [8]
Vajda, S. (1975). 'Mathematical aspects of manpower planning'. *Operat. Res. Quart.*, **26**, 527–542. [111]
Vajda, S. (1978). *Mathematics of Manpower Planning*, Wiley, Chichester.
Valliant, R., and G. T. Milkovitch (1977). 'Comparison of semi-Markov and Markov models in a personnel forecasting application'. *Decision Sciences*, **8**, 465–477.
Van der Merwe, R., and S. Miller (1971). 'The measurement of labour turnover'. *Human Relations*, **24**, 233–254. [72]
Vassiliou, P. C. G. (1976). 'A Markov chain model for wastage in manpower systems'. *Operat. Res. Quart.*, **27**, 57–70. [112]
Vassiliou, P. C. G. *See also* Young and Vassiliou (1974).
Vimont, C. (1969). 'La prevision de la demande de main-d'œuvre'. *Bull. Int. Stat. Inst.*, **43**, 341–358.
Vroom, V. H., and K. R. MacCrimmon (1968). 'Towards a stochastic model of managerial careers'. *Admin. Sci. Quart.*, **13**, 26–46.
Walker, J. W. (1969). 'Forecasting manpower needs'. *Harvard Bus. Rev.*, **47**, 152–164.
Walker, J. W. *See also* Burack and Walker (1972).
Wandel, S. E., and J. C. Hershey (1976). 'Evaluation of nurse staffing policies using a manpower planning and scheduling model'. *Operational Research '75*, North Holland Publishing Co.
Ward, D. H. (1963). 'Comparison of different systems of exponentially weighted prediction'. *Statistician*, **13**, 173–185. [236]
Weber, W. L. (1971). 'Manpower planning in hierarchical organisations: a computer simulation approach'. *Man. Sci.*, **18**, 119–144.
Weber, W. L. *See also* Dill and coworkers (1966).
Weiner, N. *See* Mahoney and coworkers (1977).
Wellman, G. (1972). 'Practical obstacles to effective manpower planning'. *Personnel Review*, **1**, 32–47.
Wheeler, B., and P. Andrews (1972). 'Cost of an age structure'. *Personnel Management*, **4**, 32–39.
White, H. C. (1969). 'Control and evaluation of aggregate personnel: flows of men and jobs'. *Admin. Sci. Quart.*, **14**, 4–11.
White, H. C. (1970). *Chains of Opportunity*, Harvard University Press, Cambridge, Mass. [142, 143, 153]
Whiteman, R. P. *See* Jones, R. C., and coworkers (1973).
Wilson, N. A. B. (1969). *Manpower Research*, English Universities Press, London.
Winters, P. R. (1960). 'Forecasting sales by exponentially weighted moving averages'. *Man. Sci.* **6**, 324–342. [236]
Wise, D. A. (1975). 'Personal attributes, job performance and probability of promotion'. *Econometrica*, **43**, 913–932. [183]
Wishart, D. (1976). 'Manpower supply models IV: The Mansim model', in Smith (1976), 140–165. [11]
Young, A. (1965). *The Remuneration of University Teachers 1964–65*, Association of University Teachers, London.
Young, A. (1965). 'Models for planning recruitment and promotion of staff . *Brit. J. Indust. Rel.*, **3**, 301–310.
Young, A. (1969). *The Remuneration of University Teachers 1967–68*, Association of University Teachers, London.

Young, A. (1971). 'Demographic and ecological models for manpower planning', in Bartholomew and Morris (1971), 75–97. [13, 92, 110]

Young, A., and G. Almond (1961). 'Predicting distributions of staff'. *Comp. J.*, **3**, 246–250. [9, 112]

Young, A., and P. C. G. Vassiliou (1974). 'A non-linear model on the promotion of staff'. *J. R. Statist. Soc.*, **A137**, 584–595. [110, 236]

Young, A. R. *See* Nielson and Young (1973).

Zellner, A. *See* Lee and coworkers (1970).

# Subject Index

*Applied Probability and Statistics (continued)*

DUNN and CLARK · Applied Statistics: Analysis of Variance and Regression

ELANDT-JOHNSON · Probability Models and Statistical Methods in Genetics

FLEISS · Statistical Methods for Rates and Proportions

GALAMBOS · The Asymptotic Theory of Extreme Order Statistics

GIBBONS, OLKIN, and SOBEL · Selecting and Ordering Populations: A New Statistical Methodology

GNANADESIKAN · Methods for Statistical Data Analysis of Multivariate Observations

GOLDBERGER · Econometric Theory

GOLDSTEIN and DILLON · Discrete Discriminant Analysis

GROSS and CLARK · Survival Distributions: Reliability Applications in the Biomedical Sciences

GROSS and HARRIS · Fundamentals of Queueing Theory

GUTTMAN, WILKS, and HUNTER · Introductory Engineering Statistics, *Second Edition*

HAHN and SHAPIRO · Statistical Models in Engineering

HALD · Statistical Tables and Formulas

HALD · Statistical Theory with Engineering Applications

HARTIGAN · Clustering Algorithms

HILDEBRAND, LAING, and ROSENTHAL · Prediction Analysis of Cross Classifications

HOEL · Elementary Statistics, *Fourth Edition*

HOLLANDER and WOLFE · Nonparametric Statistical Methods

HUANG · Regression and Econometric Methods

JAGERS · Branching Processes with Biological Applications

JESSEN · Statistical Survey Techniques

JOHNSON and KOTZ · Distributions in Statistics
Discrete Distributions
Continuous Univariate Distributions-1
Continuous Univariate Distributions-2
Continuous Multivariate Distributions

JOHNSON and KOTZ · Urn Models and Their Application: An Approach to Modern Discrete Probability Theory

JOHNSON and LEONE · Statistics and Experimental Design in Engineering and the Physical Sciences, Volumes I and II, *Second Edition*

KEENEY and RAIFFA · Decisions with Multiple Objectives

LANCASTER · An Introduction to Medical Statistics

LEAMER · Specification Searches: Ad Hoc Inference with Nonexperimental Data

McNEIL · Interactive Data Analysis

MANN, SCHAFER, and SINGPURWALLA · Methods for Statistical Analysis of Reliability and Life Data

MEYER · Data Analysis for Scientists and Engineers

OTNES and ENOCHSON · Applied Time Series Analysis: Volume I, Basic Techniques

OTNES and ENOCHSON · Digital Time Series Analysis

PRENTER · Splines and Variational Methods

RAO and MITRA · Generalized Inverse of Matrices and Its Applications

SARD and WEINTRAUB · A Book of Splines